# Transportation Technologies for a Sustainable Future

Renewable energy options for road, rail, marine and air transportation

**Online at: https://doi.org/10.1088/978-0-7503-5306-9**

# IOP Series in Renewable and Sustainable Power

The IOP Series in Renewable and Sustainable Power aims to bring together topics relating to renewable energy, from generation to transmission, storage, integration, and use patterns, with a particular focus on systems-level and interdisciplinary discussions. It is intended to provide a state-of-the-art resource for all researchers involved in the power conversation.

**Series Editor**
**Professor David Elliott**
*Open University, UK*

**About the Editor**
David Elliott is emeritus Professor of Technology Policy at the Open University, where he developed courses and research on technological innovation, focusing on renewable energy policy. Since retirement, he has continued to write extensively on that topic, including a series of books for IOP Publishing and a weekly blog post for *Physics World* (physicsworld.com/author/david-elliott)

**About the Series**
Renewable and sustainable energy systems offer the potential for long-term solutions to the world's growing energy needs, operating at a broad array of scales and technology levels. The IOP Series in Renewable and Sustainable Power aims to bring together topics relating to renewable energy, from generation to transmission, storage, integration, and use patterns, with a particular focus on systems-level and interdisciplinary discussions. It is intended to provide a state-of-the-art resource for all researchers involved in the power conversation.

We welcome proposals in all areas of renewable energy including (but not limited to): wind power, wave power, tidal power, hydroelectric power, PV/solar power, geothermal power, bioenergy, heating, grid balancing and integration, energy storage, energy efficiency, carbon capture, fuel cells, power to gas, electric/green transport, and energy saving and efficiency.

Authors are encouraged to take advantage of electronic publication through the use of colour, animations, video, data files, and interactive elements, all of which provide opportunities to enhance the reader experience.

**Do you have an idea for a book you'd like to explore?**
We are currently commissioning for the series; if you are interested in writing or editing a book please contact Caroline Mitchell at caroline.mitchell@ioppublishing.org.

# Transportation Technologies for a Sustainable Future

Renewable energy options for road, rail, marine and air transportation

**Richard A Dunlap**
*Department of Physics and Atmospheric Science, Dalhousie University, Halifax, Nova Scotia, Canada*

**IOP** Publishing, Bristol, UK

ISBN 978-0-7503-5306-9 (ebook)
ISBN 978-0-7503-5304-5 (print)
ISBN 978-0-7503-5307-6 (myPrint)
ISBN 978-0-7503-5305-2 (mobi)

DOI 10.1088/978-0-7503-5306-9

Version: 20230201

IOP ebooks

British Library Cataloguing-in-Publication Data: A catalogue record for this book is available from the British Library.

Published by IOP Publishing, wholly owned by The Institute of Physics, London

IOP Publishing, No.2 The Distillery, Glassfields, Avon Street, Bristol, BS2 0GR, UK

US Office: IOP Publishing, Inc., 190 North Independence Mall West, Suite 601, Philadelphia, PA 19106, USA

# Contents

# Preface

From the time when our ancient ancestors first made use of fire for heating and cooking, we have depended on energy extracted from nature to provide for our needs. Until the end of the 19th century, this energy was provided mainly by the combustion of wood. Around 1900, coal supplanted wood as our major source of energy, and this was replaced by petroleum products around 1950. In the latter part of the 20th century, it became increasingly obvious that our sources of energy could not remain the same indefinitely. By the end of the 20th century, it also had become obvious that anthropogenic emissions, largely carbon dioxide from fossil fuel use, were responsible for global climate change. By this time, it had become apparent, both from a resource utilization standpoint and from an environmental standpoint, that it was necessary for human society to transition to sustainable energy sources.

Considerable work has been undertaken over the past few decades on the development of sustainable low-carbon energy sources. Much of this has dealt with electricity produced by harvesting energy from the environment. Hydroelectricity, solar photovoltaics, and wind energy are the principal examples of such developments. At present, about one quarter of the electricity generated globally comes from these low-carbon renewable sources, while another 10% comes from low-carbon nuclear power stations. However, about two-thirds of the world's electricity is still produced by the combustion of fossil fuels. Going forward, the replacement of fossil-fuel-fired electricity generating facilities with environmentally sustainable electricity sources will be a major task.

Transportation is a major component of our energy use and accounts for more than a third of the global total. At present, about 95% of transportation energy comes from fossil fuel sources, and any comprehensive approach intended to reduce fossil fuel use must deal with transportation energy needs. To date, comparatively little progress has been made towards converting transportation technologies to low-carbon energy sources. This is, to a large extent, due to the particular requirements for transportation energy. In nearly all cases, the source of energy for transportation must be portable and of sufficient specific energy density. Fossil fuels are ideal in this respect and, thus far, have remained relatively inexpensive. The economics of transportation energy technologies are quite different than those of electricity generation. In the case of electricity generation, infrastructure is developed by the utility provider and the consumer has little direct input into the technology that is utilized to produce the electricity that is delivered to their home. In the case of transportation, at least for private vehicles, industry provides most of the fueling infrastructure, such as bioethanol stations or public electric vehicle charging stations, while the consumer makes decisions concerning the choice of vehicle. The incentive to develop a fueling infrastructure depends on the availability of customers with appropriate vehicles, while the incentive for consumers to purchase specific vehicles depends on the availability of fuel. This situation has been at least one factor in the lack of growth of technologies such as hydrogen-fueled vehicles. There are numerous options for low-carbon energy that may be considered in the context of various modes of transportation, including road transportation, rail transportation, marine transportation, and air transportation. The path forward requires a careful consideration of the advantages and disadvantages of each of these options.

This book begins with a review of our energy use and its implications for future resource availability and the global environment. It also reviews possible future approaches to our overall energy system. The various approaches to low-carbon transportation energy are considered for the different major modes of transportation. The advantages and disadvantages of each approach are analyzed, and the book provides an objective overview of the options that are available or could be available in the near future to provide guidance for the reader in making their own transportation choices. Finally, the way in which transportation energy requirements can be integrated into our overall approach to sustainable and environmentally friendly energy is considered.

This book is intended for a general reader with some background in high-school-level science. Some background material is covered in more technical detail in the appendices.

# Author biography

## Richard A Dunlap

Richard A Dunlap received a BS from Worcester Polytechnic Institute (Physics 1974), an A.M from Dartmouth College (Physics 1976), and a PhD from Clark University (Physics 1981). Since 1981, he has been a professor in the Department of Physics and Atmospheric Science at Dalhousie University and currently holds a position as Research Professor. He was Faculty of Science Killam Research Professor from 2001 to 2006 and Director of the Dalhousie University Institute for Research in Materials from 2009 to 2015. Professor Dunlap's research interests include nuclear spectroscopies, magnetic materials, quasicrystals, critical phenomena, and advanced battery materials. He has published more than 300 refereed research papers and his previously published books include *Experimental Physics: Modern Methods* (Oxford 1988), *The Golden Ratio and Fibonacci Numbers* (World Scientific 1997), *An Introduction to the Physics of Nuclei and Particles* (Brooks/Cole 2004), *Sustainable Energy* (Cengage, 1st edn 2015, 2nd edn 2019), *Novel Microstructures for Solids* (IOP/Morgan & Claypool 2018), *Particle Physics* (IOP/Morgan & Claypool 2018), *The Mössbauer Effect* (IOP/Morgan & Claypool 2019), *Lasers and Their Application to the Observation of Bose–Einstein Condensates* (IOP/Morgan & Claypool 2019), *Electrons in Solids—Contemporary Topics* (IOP/Morgan & Claypool 2019), and *Energy from Nuclear Fusion* (IOP Publishing 2021).

# Chapter 1

# Our current transportation technologies and the need for change

## 1.1 Introduction

Human society uses large amounts of energy. It is now an essential component of our daily lives. The earliest humans used only the energy that was provided by their own bodies. This came from the food that they consumed and typically amounted to about 2000 calories per day, which is equivalent to an average continuous power expenditure of about 100 W (for a discussion of the relationship between energy and power, see appendix B). Throughout history, human energy consumption has increased, first with the use of fire for heating and cooking and later with the domestication of animals for agricultural needs and transportation. This was followed by the widespread use of fossil fuels. At present, the average person in an industrialized country has a continuous consumption of about 10 kW, i.e. 100 times that of early humans. About half of this energy consumption might be considered to be personal energy, while about half may be considered to be societal energy. Personal energy is the energy that we use in our homes for heating, cooking, and appliances, as well as our personal transportation. Societal energy is energy that is used in industry for manufacturing, businesses, and commercial transportation; it is averaged over society. One of the most important uses of energy is for transportation. This is also the use that poses some of the greatest challenges for the future development of new energy technologies and is the subject of this book.

It is important to consider in some detail what energy is used for. It turns out that most of the energy that we use is wasted. Energy exists in a number of different forms, such as thermal energy associated with objects that are hot, mechanical energy associated with things that are moving, and chemical energy that is associated with the bonds between atoms. Energy is wasted because the energy that we get from nature is generally not in the form that we want to use, and the conversion processes that convert one form of energy into another are not 100% efficient.

doi:10.1088/978-0-7503-5306-9ch1  

Overall, energy is conserved (in the case of nuclear energy, it is the total mass/energy that is conserved) but conversion processes typically do not convert all of one form of energy into the form that we want. In addition, in many cases we have to input additional energy into the process in order to effectively make use of the energy that nature provides. We can see these features in a couple of common examples of energy use.

First, we can consider the use of a rechargeable LED flashlight. Its energy may begin as chemical energy contained in a coal deposit. The coal is mined and transported to a power station. The coal is then burned, and the chemical energy is converted into heat. The heat is used to produce steam which turns a turbine. The turbine then turns a generator which produces electricity. The electricity is distributed to the grid, and we use that electricity to charge the battery in the flashlight. This process converts electrical energy into stored chemical energy. When we turn on the flashlight the chemical energy is converted back into electrical energy and the LED converts this electrical energy into light. This process ultimately converts chemical energy into light, but you can see that there are many steps in between. Some important points to note in this example are that energy needs to be input into the process, primarily for the mining and transport of the coal.

A second example is driving an automobile. Transportation of any kind inevitably represents mechanical energy, because an object is moved from one point to another. In the case of a gasoline-powered vehicle, the energy begins as chemical energy stored in an oil deposit. The oil is extracted and transported to a refinery where gasoline and other hydrocarbons are separated. The gasoline is transported to a service station and delivered to the customer. The automobile engine burns the gasoline, converting chemical energy to heat and the heat drives the pistons in the engine to produce mechanical energy, which moves the vehicle. Again, we see the need to input processing energy and the various energy conversion processes involved.

The above examples emphasize an important point, namely, the difference between primary energy sources, i.e. the energy that we extract or harvest from nature (such as the chemical energy associated with coal or crude oil), and end-user energy, such as the light produced by the flashlight or the mechanical energy that transports a vehicle from one point to another.

In both of these examples, there are a number of energy conversions between the primary energy that is extracted from nature and the end-user energy that is required in each particular case. While the overall efficiency of each application depends on all the conversions that are required, the most notable loss of energy is in the conversion of heat into mechanical energy that is achieved by burning a fossil fuel. It is, therefore, important to look in more detail at the conversion of heat into mechanical energy. Any device, such as the steam turbine or the internal combustion gasoline engine, which converts heat to mechanical energy is referred to as a 'heat engine.' Heat engines can never be 100% efficient, as they always leave some of the heat unconverted. A diagram of the energy flow in a heat engine is shown in figure 1.1. A steam turbine might be 35% efficient and a gasoline engine is typically only about 20% efficient. An analysis of the energy conversions involved in a

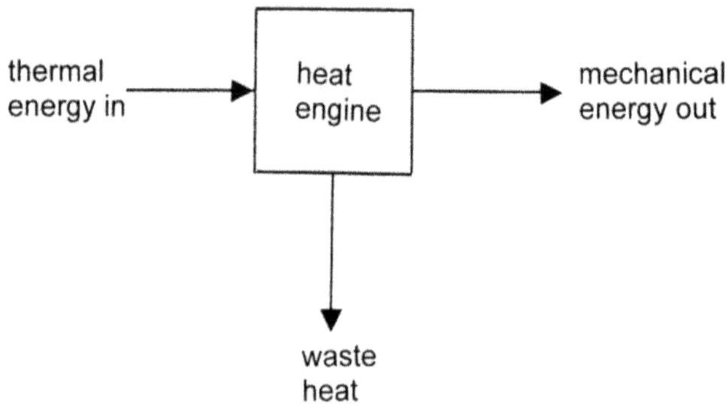


**Figure 1.1.** Energy flow in a typical heat engine which converts thermal energy to mechanical energy, showing the loss of energy due to waste heat.

particular transportation technology is a crucial component of understanding its viability and will form an important part of the discussions throughout this book.

## 1.2 A breakdown of transportation energy use

Energy use, i.e. the amount of energy used and the sources of that energy, varies in different countries according to the specific needs of that country. Some factors that affect energy use include the per capita gross domestic product (GDP), types of industry, population density, climate, and natural resources. In general, countries with a high per capita GDP are highly industrialized and consume more energy per capita than countries with a low per capita GDP. Countries with heavy industries, such as automobile manufacturing, use more energy than countries with light industries, such as small electronics production or banking. Countries with low population density typically use more energy than countries with high population density because people tend to travel more, and goods must be transported over greater distances. Countries with extreme climates, particularly those with cold climates, use more energy because of the greater need for heating or air conditioning. Finally, the energy sources that are most commonly used are often governed by the resources that are available nationally. The major countries that consume the most energy per capita are Norway and Canada. Both have high per capita GDPs, cold climates, and relatively low population densities. It is also interesting that both countries have extensive hydroelectric resources and generate the majority of their electricity by this method, which accounts for 98% of the electricity generated for Norway and 60% for Canada. While the overall primary energy sources differ considerably between countries on the basis of the above considerations, the energy breakdown for the United States serves as a good example of the situation that is common in many industrialized countries. Table 1.1 shows the distribution of primary energy sources in the United States as of 2019. The table shows that 80% of all primary energy used in the United States comes from fossil fuels. This is very close to the worldwide average for the percentage of fossil fuel use, 84% in 2019 (BP 2020).

**Table 1.1.** Breakdown of primary energy sources in the United States in 2019. Data are adapted from https://flowcharts.llnl.gov/content/assets/images/energy/us/Energy_US_2019.png.

| Energy source | Percentage of total use |
|---|---|
| Petroleum | 36.6 |
| Natural gas | 32.0 |
| Coal | 11.4 |
| Nuclear fission | 8.4 |
| Biofuels | 5.0 |
| Wind | 2.7 |
| Hydroelectric | 2.5 |
| Solar | 1.0 |
| Geothermal | 0.2 |

**Table 1.2.** Breakdown of energy use by sector in the United States in 2019. Data are adapted from the U.S. Energy Information Administration (EIA 2019).

| Sector | Percentage of total use |
|---|---|
| Electricity generation | 34.3 |
| Transportation | 23.9 |
| Industrial | 22.2 |
| Residential | 12.2 |
| Commercial | 7.2 |

A breakdown of end-user energy use in the United States is shown in table 1.2. Transportation includes personal vehicles as well as commercial vehicles, railways, marine transportation, and aircraft. A further breakdown of transportation energy is discussed below. Commercial energy includes retail stores and office buildings, while residential energy primarily includes heat for single-family homes and apartment buildings. Industrial energy is that which is used in manufacturing.

Transportation is an essential component of our modern society. While we are probably most familiar with the personal transportation that we use in the form of passenger vehicles, buses, and commercial aircraft, there is also an enormous transportation infrastructure that moves goods to support our society. Table 1.3 gives a breakdown in terms of the energy consumed by the different modes of transportation that are used worldwide. The next three chapters of the present book deal primarily with road transportation, as this comprises the largest fraction of transportation energy use. Other modes of transportation, i.e. rail, marine, and air, are discussed in chapter 5.

We can also look at the sources of energy used for transportation. Table 1.4 shows a breakdown of the energy sources used for transportation in the United States.

**Table 1.3.** Percentages of world energy used for different types of transportation. Data are adapted from www.eia.gov/outlooks/ieo/pdf/transportation.pdf.

| Type of transportation | Percentage of total |
|---|---|
| Passenger vehicles | 62.2 |
| Road freight | 22.4 |
| Air | 9.5 |
| Marine | 4.1 |
| Rail | 1.7 |

**Table 1.4.** Breakdown of the energy used for transportation in the United States in 2019. Data are adapted from https://flowcharts.llnl.gov/content/assets/images/energy/us/Energy_US_2019.png.

| Energy | % Total |
|---|---|
| Fossil fuels | 94.7 |
| Biofuels | 4.8 |
| Electricity | 0.5 |

At present, about 95% of transportation energy comes from fossil fuels (petroleum or natural gas), about 5% comes from biofuels (for example, ethanol additives to gasoline), and less than 1% from electricity.

There are two factors that require serious consideration when dealing with transportation energy: efficiency and portability. With regard to efficiency, it is interesting to go back and look at the difference between primary energy and end-user energy as it specifically relates to transportation. It is obvious that transportation energy must ultimately take the form of mechanical energy in order to move objects from one location to another. While there are cases in which we can harvest mechanical energy from the environment, for example, using wind turbines, in virtually all cases, conversion of the harvested energy into other forms is necessary in order for it to be utilized for transportation. Since, at present, nearly all transportation energy is obtained from either fossil fuels or biofuels, this primary energy must be converted into mechanical energy by means of a heat engine. This is obviously the case for vehicles powered by internal combustion engines, but it is also the case for many electric vehicles which utilize electricity that has been produced by a coal-fired power plant to charge batteries. Since the efficiency of a heat engine is limited by the basic laws of thermodynamics, the conversion of primary energy to transportation energy is much less than 100% efficient. In fact, at present, about two thirds of the primary energy that is used for transportation is lost as wasted heat given off to the environment. This has two adverse consequences: first, much more primary energy must be extracted from the environment than is ultimately needed for transportation, and second, the excess heat that is deposited in the environment may have adverse environmental effects.

The second transportation factor that requires consideration is portability. This is most important for road transportation, such as passenger vehicles, trucks, and buses. It is necessary for vehicles to utilize an energy source that has sufficiently high energy density that a reasonable amount (i.e. by mass and volume) of onboard fuel can provide an acceptable vehicle range. Portability is less of a concern for marine transportation, where the mass and/or volume of fuel is somewhat less critical, and in rail transportation, where an external energy source, i.e. a third rail or catenary, may be feasible.

In the next two sections, the need to change the transportation energy infrastructure is discussed in terms of resource availability and environmental consequences.

## 1.3 Future energy needs

In order to understand how energy choices will affect our future environment, it is important to consider several points. These include:

- the amount of energy we will need in the future;
- the sources of energy that will be used in the future;
- strategies to reduce atmospheric greenhouse gases.

In this section, we look at predictions for the amount of energy that will be needed in the future. The second point is considered in sections 1.4–1.6, and the final point is considered in section 1.7.

It is, perhaps, most informative to begin by looking at the per capita primary energy consumption of some countries, as shown in figure 1.2. It is clear from the figure that around 1970 there was a significant increase in the per capita energy consumption in North America. This was due to an increase in mobility and the use of electronics during that period. There has also been a slight downward trend in North American per capita energy use since around 2000. This is due to an increased awareness of future energy concerns and an increase in conservation efforts and device efficiency. Figure 1.3 shows the total energy use in different regions. It can be seen that the total energy use in North America has remained more or less constant for the past 20 years. This is because the slight decrease in per capita energy consumption has been compensated for by a slight increase in population. The situation in many other highly industrialized countries is similar, as can be seen in the data for Europe shown in figure 1.3.

Energy consumption in many less industrialized countries follows somewhat different trends. Figure 1.2 shows that since about 2000, the per capita energy consumption in China has approximately tripled. A somewhat smaller although still significant increase has occurred in India. This trend results from an increase in industrialization and an increase in personal energy consumption in these countries. This trend in per capita energy use, combined with significant population growth, has resulted in the total energy consumption trend as shown in figure 1.3 for Asia.

Thus, it is clear from the above examples that the total world energy consumption in the future will be a function of population growth and the per capita energy

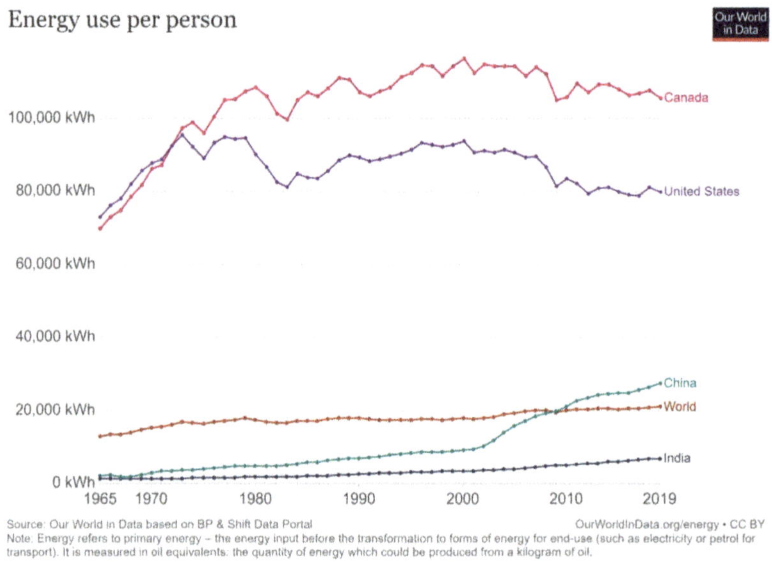


**Figure 1.2.** Per capita annual primary energy consumption in Canada, the United States, and China along with the world average from 1965 to 2019. Note 100 000 kWh = 360 GJ. Reproduced from OurWorldInData (2022). CC BY 4.0.

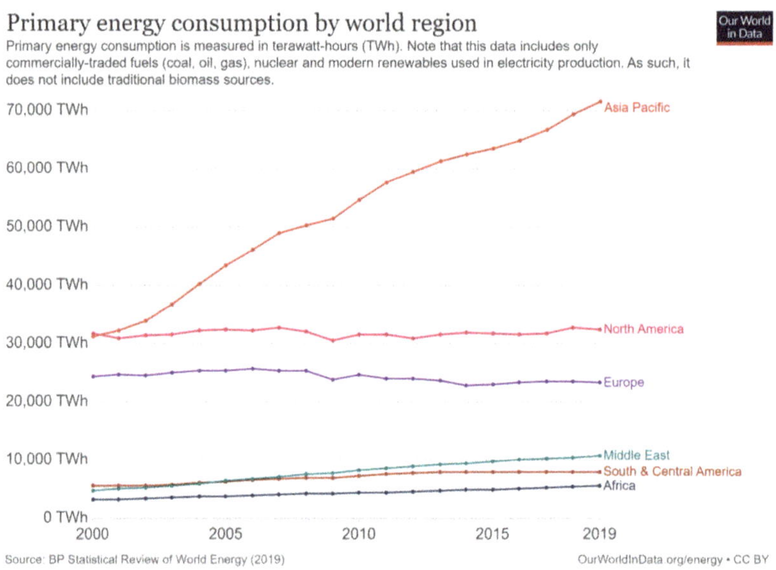


**Figure 1.3.** Total annual primary energy consumption in different regions of the world. Note 100 000 kWh = 360 GJ. Reproduced from OurWorldInData (2022). CC BY 4.0.

consumption in various countries. While we might generally expect relatively constant overall energy use in highly industrialized countries, as has been the case in recent years, we might also expect increasing total energy use in developing countries that have not yet reached the same degree of industrialization as North America and Europe. The combined uncertainty of the results obtained from models of population growth and per capita energy needs in different parts of the world leads to a considerable range of possible energy requirements in the future. Figure 1.4 shows the mean of a number of predictions for future energy use through the end of the current century. The mean predicted primary energy consumption worldwide in 2100 is about 1650 EJ per year, compared to the current consumption of about 580 EJ per year and shows an increase by a factor of 2.7 over the next 80 years. In determining the need to develop new energy infrastructure in the future, it is important to account for this anticipated growth. The figure also shows that there is a substantial range of predicted values for world energy consumption in 2100 ranging from a low of 514 EJ per year (slightly less than the current value) to over 2200 EJ per year, based on different population and economic growth models.

The mean value for world energy use in 2100 from figure 1.4 can be viewed in the context of per capita energy use. Using the United Nations (2019) estimate of 11 billion for the world population in 2100, the data from the figure gives an average per capita primary energy consumption of about 150 GJ per year. This may be compared to the current world average per capita annual energy consumption of 81 GJ and is close to current values for most European countries and Japan.

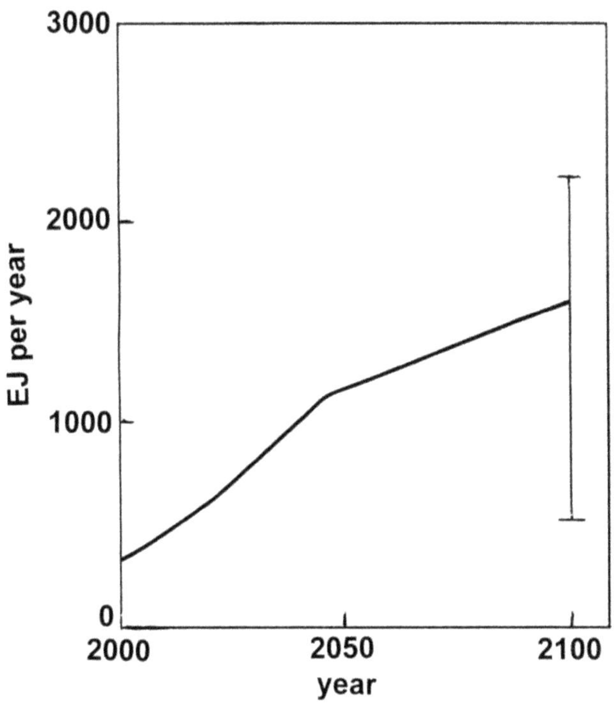


**Figure 1.4.** Predicted total world primary energy use until the end of the 21st century. The line gives the mean of the models considered and the vertical bar at year 2100 shows the range of predicted values in that year. Based on models described in IPCC (2000).

This increase in per capita energy consumption is basically indicative of the increase in industrialization and expected living standards in developing countries. Details of energy needs based on the ratio of primary energy to end-user energy are discussed in chapter 6.

## 1.4 Pollution and global climate change

Our choices for future energy sources largely depend on two factors:

- a consideration of the environmental consequences of energy use;
- the availability of resources.

The first point is considered in this section and the second point is considered in section 1.5.

The utilization of energy by society has a number of adverse environmental effects. The most significant among these is the emission of various pollutants that occurs during the combustion of fossil fuels. A discussion of the major chemical and particulate pollutants that result from fossil fuel use follows. Specific pollutants that act as greenhouse gases are then discussed, along with their environmental consequences.

### 1.4.1 Chemical and particulate pollution

*Carbon monoxide*—Carbon monoxide results from the incomplete combustion of carbon-containing fuels. This is described by the reaction

$$2C + O_2 \rightarrow CO,$$

which typically occurs when there is insufficient oxygen present to produce complete oxidization of the carbon. The production of carbon monoxide is most common when the combustion of carbon-containing compounds occurs in an enclosed space, such as in an internal combustion automobile engine. Carbon monoxide is an important pollutant because of its adverse health effects when it is inhaled. Carbon monoxide reacts readily with hemoglobin in the blood and displaces oxygen. In sufficient quantities, it can be fatal.

*Nitrogen compounds*—Nitrogen–oxygen compounds may be formed whenever any material undergoes combustion in air. If the temperature exceeds about 1100 °C, nitrogen in the atmosphere is oxidized by the process

$$N_2 + O_2 \rightarrow 2NO,$$

thereby forming nitric oxide. The conditions that lead to the formation of nitric oxide are common in internal combustion engines. The nitric oxide may be further oxidized to form nitrogen dioxide by reaction with ozone in the atmosphere:

$$NO + O_3 \rightarrow NO_2 + O_2.$$

Nitrogen dioxide is a toxic brown gas that is responsible for the color of the smog that collects in many cities.

*Hydrocarbons*—Hydrocarbons form under the same conditions as carbon monoxide and represent the unburned component of hydrocarbon-containing fuels. As is the case with carbon monoxide, hydrocarbon pollution is most commonly produced by internal combustion automobile engines. Hydrocarbon pollution is a contributory factor to lung disease.

*Sulfur dioxide*—Sulfur dioxide is produced by the oxidation of sulfur impurities that are common in many fossil fuels. This situation is most common for coal-fired power stations. Sulfur dioxide reacts with oxygen in the atmosphere according to the process

$$2SO_2 + O_2 \rightarrow 2SO_3,$$

which is followed by a reaction with water vapor, namely

$$SO_3 + H_2O \rightarrow H_2SO_4.$$

The final product of these reactions is sulfuric acid, which is highly corrosive. The acid rain that results from the formation of sulfuric acid in the atmosphere is highly damaging to stone and concrete structures, as well as painted surfaces.

*Particulate matter*—The combustion of solid fuel, such as wood and coal, results in the production of particulate matter pollution. As the chemical composition of the particulate matter depends on the composition of the fuel, its environmental effects are somewhat uncertain. However, this form of pollution typically acts as a respiratory irritant.

*Carbon dioxide*—Carbon dioxide is the principal emission that results from the combustion of any fuel that contains carbon. For pure carbon, the reaction is

$$C + O_2 \rightarrow CO_2.$$

Virtually all potential fuels, with the exception of hydrogen and ammonia (which are discussed later in this book), contain carbon and produce carbon dioxide. Carbon dioxide pollution is discussed in detail in the section below on the greenhouse effect.

It is clear from the above discussions that some pollutants are closely associated with our current transportation technologies. These include carbon monoxide, nitrogen compounds, and hydrocarbons. Other pollutants, such as sulfur compounds and particulates, are more closely related to the use of coal. Table 1.5 gives

**Table 1.5.** Current percentages of various pollutants in the United States that result from vehicle use. Data adapted from Dunlap (2019).

| Pollution | Percentage produced by vehicles |
|---|---|
| Carbon monoxide | 60 |
| Nitrogen compounds | 35 |
| Hydrocarbons | 35 |
| Sulfur dioxide | ~0 |
| Particulate matter | ~0 |

**Table 1.6.** Improvement in passenger vehicle emissions in the United States as a result of government emission control standards between 1970 and 2004. Data are in grams per kilometer. Adapted from Dunlap (2019).

| Year | Carbon monoxide | Nitrogen oxides | Hydrocarbons |
|---|---|---|---|
| 1970 | 21.1 | 3.1 | 2.5 |
| 2004 | 1.1 | 0.04 | 0.056 |

the percentages of the various pollutants that have been described above that result from the energy used for transportation. It is clear that transportation has an important impact on hazardous pollutants in the environment. Emission control standards that have been imposed on vehicle manufacturers over the years have done much to mitigate the emission of pollutants, especially from light-duty vehicles. Examples of improvements made in vehicle emissions over the past half century or so are given in table 1.6. In addition to environmental improvements that have resulted from increased emission control standards for vehicles, much has been done in recent years to reduce pollution from power stations, particularly those which are coal-fired. Scrubbers are effective at reducing sulfur dioxide emissions by reacting power station exhaust gases with calcium-containing compounds such as $CaO$ or $CaCO_3$. Mechanical filters or electrostatic precipitators are effective at reducing the particulate matter in power station exhaust gas.

Despite advances in emission controls in many countries, pollution remains an important concern and can have serious adverse effects on human health, which are typically concentrated in urban and industrial areas. Such effects are often temporary, occurring primarily at particular times of the day or week and are often exacerbated during particular seasons or by certain weather conditions.

### 1.4.2 The greenhouse effect

Clearly, the most serious environmental effect of our energy use is global climate change. This is the result of the greenhouse effect and is caused by the presence of certain pollutants in our atmosphere. The greenhouse effect arises because of the wavelength dependence of the interaction of electromagnetic radiation with certain molecules. Radiation from the Sun has wavelengths in the range of about 200 nm to 2000 nm. This radiation includes the ultraviolet region (<380 nm), the visible region (380 nm–700 nm), and the infrared region (>700 nm) and has a maximum intensity at around 480 nm (in the yellow region of the visible spectrum). The sunlight that is incident on the Earth's outer atmosphere interacts with the Earth in different ways. It may either be reflected back into space by the atmosphere, absorbed by the atmosphere, transmitted through the atmosphere and reflected by the Earth's surface, or transmitted through the atmosphere and absorbed by the Earth's surface. The radiation that is absorbed by the surface warms the Earth. This warm surface reradiates energy at a much longer wavelength than that of the incident solar radiation and has a peak intensity at about 9.7 μm. Without the presence of the

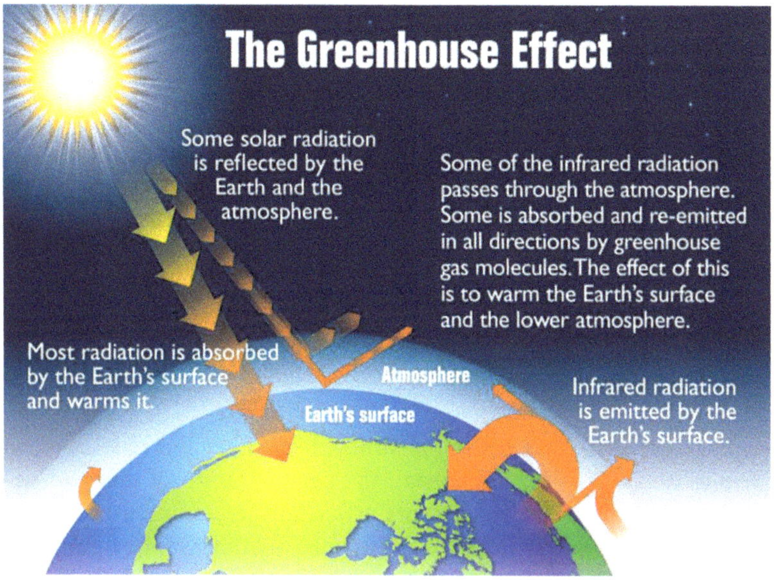


**Figure 1.5.** Illustration of the greenhouse effect. Reproduced from EPA (2022). Image stated to be in the public domain.

atmosphere, this reradiated energy would be lost to space. However, with the atmosphere, this reradiated energy can be transmitted through the atmosphere, or it can be reflected back to the Earth, further increasing the surface temperature. The greenhouse effect occurs because certain molecular species in the atmosphere are relatively transparent to short-wavelength radiation (i.e. solar radiation) but more reflective to long-wavelength radiation (i.e. energy reradiated from the surface). Thus, the presence of these molecular species (referred to as greenhouse gases) traps solar energy near the Earth's surface and raises its temperature. This is the same effect that occurs when an automobile is left in the Sun. The vehicle's windows act similarly to the Earth's atmosphere because they transmit short-wavelength radiation and reflect long-wavelength radiation, thereby trapping solar energy inside the vehicle and raising the temperature. The details of the greenhouse effect in the Earth's atmosphere are shown graphically in figure 1.5.

The most important greenhouse gases in the Earth's atmosphere are given in table 1.7. The table shows the relative infrared absorptions per molecule for the different greenhouse gas species, the current concentrations of these gases in the atmosphere, and their relative contribution to the greenhouse effect. Clearly, more than two thirds of the greenhouse effect is the result of carbon dioxide in the atmosphere, primarily because of its large concentration. However, other greenhouse gases are significant, mainly because of their greater infrared absorption per molecule. The presence of greenhouse gases in the atmosphere (specifically carbon dioxide) is essential for the existence of life on the planet. Without the greenhouse effect that the atmosphere provides, the equilibrium temperature of the Earth's surface would be about −19 °C, which is inconsistent with the evolution of life.

**Table 1.7.** The relative importance of greenhouse gases. Data adapted from Dunlap (2019) and www.eea.europa.eu/data-and-maps/indicators/atmospheric-greenhouse-gas-concentrations-10/assessment.

| Greenhouse gas | Relative infrared absorption per molecule | Current atmospheric concentration by volume in 2015 (ppm) | Relative atmospheric greenhouse effect (%) |
|---|---|---|---|
| Carbon dioxide ($CO_2$) | 1 | 402.9 | 68 |
| Methane ($CH_4$) | 25 | 1.835 | 19 |
| Nitrous oxide ($N_2O$) | 298 | 0.3277 | 7 |
| Chlorofluorocarbon ($CCl_2F_2$) | 10 900 | 0.000 530 | 6 |

The presence of greenhouse gases in the atmosphere raises the Earth's mean surface temperature to about +15 °C. Increasing the concentration of greenhouse gases in the atmosphere obviously increases the Earth's surface temperature, and this will affect the environmental conditions on Earth in a way that will be detrimental to our society.

Human activity is the major contributor to increasing greenhouse gases at present and energy production is the most significant source of the greenhouse gases that are produced by human society. Figure 1.6 shows the relative contributions of different human activities in the United States to the production of greenhouse gases. Transportation is the largest single contributor to greenhouse gases, followed closely by electricity generation. It is clear that carbon dioxide is the most significant (although not the only important) greenhouse gas. Transportation, along with electricity generation, account for over half of the carbon dioxide that is emitted worldwide. Human activities are also a major contributor to atmospheric methane, the second most important greenhouse gas. These activities include agriculture, biomass burning, organic landfill waste, and methane that escapes during fossil fuel production.

The correlation between anthropogenic greenhouse gas emissions and global climate has been studied in significant detail for the past several decades. Figure 1.7 shows the results of an extensive study of ice cores from Antarctica. The data show periodic variations in temperature that have occurred over the past 400 000 years that are very well correlated with atmospheric carbon dioxide and methane levels. The reasons for these fluctuations in atmospheric chemistry and climate are not precisely known, but it is obvious that these are not the result of industrialized human activity. There is, however, some evidence that relates these climatic variations to changes in the cosmic ray flux that results from the rotation of the Sun around the galactic core. What is most significant in this analysis is a comparison of this historical data based on ice core samples with more recent ice core and direct measurements of the concentrations of these greenhouse gases. Figure 1.7 shows that over the past 400 000 years or so, atmospheric carbon dioxide levels have varied between about 180 ppm and 300 ppm, while atmospheric methane levels have varied between about 0.35 ppm and 0.7 ppm. Recent measurements of

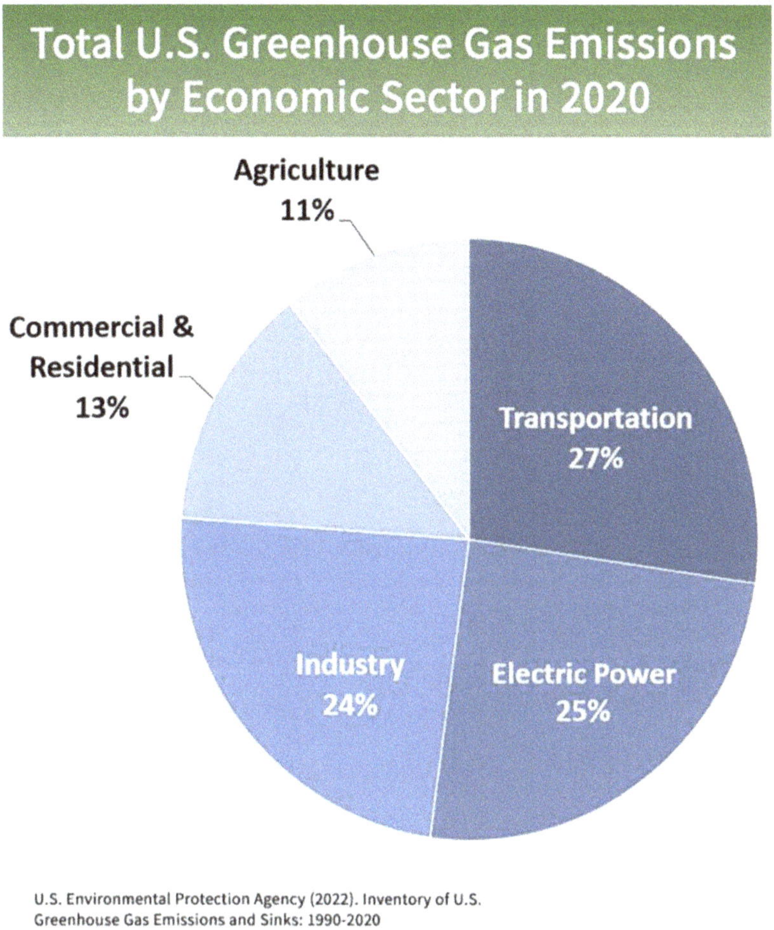


**Figure 1.6.** Distribution of the sources of greenhouse gases in the United States in 2020. Reproduced from EPA (2022). Image stated to be in the public domain.

atmospheric carbon dioxide and methane concentrations are shown in figures 1.8 and 1.9, respectively. In addition, figure 1.8 shows the recent correlation between global temperature and atmospheric carbon dioxide concentration. Both these figures show that recent greenhouse gas levels have been increasing consistently in recent years and that current concentrations, namely, over 400 ppm for carbon dioxide and about 1.85 ppm for methane, are well above the historical values for the past 400 000 years.

Three conclusions are obvious from the analysis of the above data:

- levels of atmospheric greenhouse gases are higher at present than they have been at any time during the past 400 000 years;
- levels of atmospheric greenhouse gases are currently increasing;
- there is a clear correlation between atmospheric greenhouse gas concentrations and global temperatures.

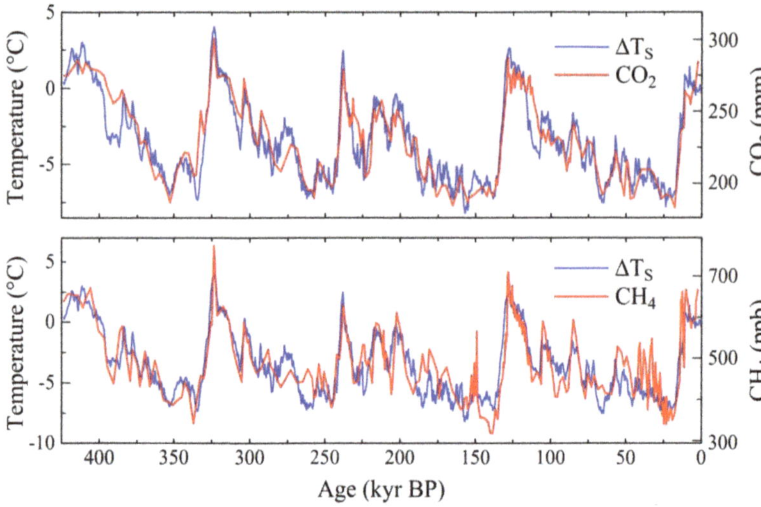


**Figure 1.7.** A comparison of historical atmospheric carbon dioxide ($CO_2$) and methane ($CH_4$) levels and temperature anomalies, as determined from the analysis of ice core samples from Vostok, Antarctic. Reproduced from NASA (2005). Image stated to be in the public domain.

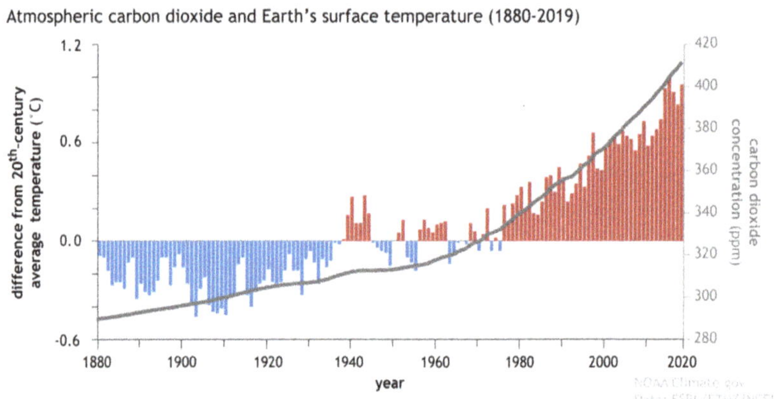


**Figure 1.8.** Correlation between atmospheric carbon dioxide levels (gray line) and temperature anomalies (blue and red bars) for the past 140 years. Reproduced from NOAA Climate.gov (2022). Image stated to be in the public domain.

The reasons for the features noted above are not directly obvious from an analysis of the data. However, the overwhelming scientific consensus is that the recent rises in carbon dioxide and other greenhouse gas concentrations in the atmosphere are the results of human activity.

### 1.4.3 The consequences of global climate change

It is clear from the above discussion that the major manifestation of increased greenhouse concentration in the atmosphere is an increase in global temperatures.

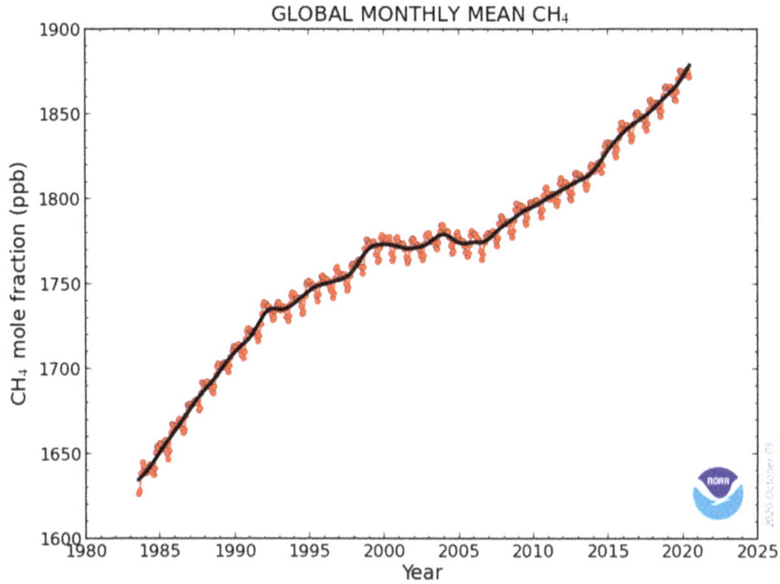


**Figure 1.9.** Measured atmospheric methane levels since about 1983. Reproduced from NOAA (2022). Image stated to be in the public domain.

The effects of these climatic changes are very far-reaching and influence many aspects of the environment. Some of the most significant environmental changes that are related to greenhouse gas emissions are discussed briefly below.

*1.4.3.1 Decreased number and extent of glaciers*

Increasing global temperatures have resulted in glacial melting. This is most obvious from a visual observation of glacial extents. A significant example of glacial melting has occurred in Iceland, where Okjökull (meaning 'Ok glacier' in Icelandic) has decreased in area from about 38 $km^2$ to less than 1 $km^2$ between the early 20th century and the early 21st century. In 2019, Okjökull was officially declassified as a glacier.

*1.4.3.2 Ice sheet loss*

Along the same lines as glacial melting, global warming has also resulted in the loss of ice mass in the Arctic and Antarctic. This is most noticeable from the Greenlandic and Antarctic ice sheets, where current annual loss rates are 286 Gt and 127 Gt, respectively.

*1.4.3.3 Increased ocean heat content*

As global temperatures increase, the heat absorbed by the upper layers of the ocean increases, leading to an increase in the total heat content of the oceans. The corresponding increase in ocean temperature can have significant effects on ocean ecology.

#### *1.4.3.4 Arctic sea ice decline*

Increasing ocean temperatures have resulted in the melting of sea ice. This is most apparent in the Arctic, where, according to current trends, it is expected that much of the Arctic will be ice-free during summer months by the end of the century.

#### *1.4.3.5 Increased sea level*

The melting of sea ice and ice sheets that results from global warming will give rise to increased sea levels. The thermal expansion of water will also contribute to increased ocean water volume. Since the late 19th century, sea levels have been observed to rise by about 20 cm. The continuation of this trend due to increased greenhouse gas concentration in the atmosphere can have highly adverse effects on both the natural habitat as well as human society in coastal regions. Because of its large heat capacity, the ocean will respond more slowly to global temperature changes than the atmosphere and, as a result, sea levels will continue to rise for many decades, even if greenhouse gas levels are reduced.

#### *1.4.3.6 Ocean acidification*

When the concentration of carbon dioxide in the atmosphere is increased, its partial pressure in the atmosphere is also increased. In order to maintain equilibrium between the carbon dioxide in the atmosphere and that which is dissolved in the oceans, the concentration of carbon dioxide in seawater will increase. About 45% of the carbon dioxide that is released into the atmosphere remains in the atmosphere, about 35% is dissolved in the oceans, while the remaining 20% or so is sequestered by terrestrial plants. Some of the carbon dioxide that is dissolved in the ocean forms carbonic acid ($H_2CO_3$). Some of this carbonic acid will dissociate by the process

$$H_2CO_3 \rightarrow 2H^+ + CO_3^{2-},$$

and the resulting $H^+$ ions will increase the acidity of seawater. Since seawater is normally alkaline, increasing the concentration of $H^+$ ions will bring it closer to neutral. It has been estimated that since industrialization, the surface ocean pH has changed from about 8.25 to 8.14. Decreasing ocean pH can have adverse effects on calcifying marine organisms, such as corals and those that have calcium carbonate ($CaCO_3$) exoskeletons (e.g. crustaceans and molluscs). This, in turn, can have adverse consequences for organisms that are higher up in the food chain.

#### *1.4.3.7 Ocean deoxygenation*

Phytoplankton produce oxygen in the surface layers of the ocean by photosynthesis. This oxygen is normally carried to lower layers of the ocean by the mixing of the surface layer with the denser layers below it. The increased ocean stratification that results from increased temperature and decreased salinity of surface layers due to global warming and the input of fresh water from melting ice sheets, respectively, reduces mixing and decreases oxygen concentration in the lower layers of the ocean. This leads to a decrease in the ability of marine organisms to survive in the lower regions of the ocean.

*1.4.3.8 Altered ocean circulation patterns*

Changes in ocean temperature and salinity gradients, as noted above, will lead to changes in ocean circulation patterns. Perhaps the most notable of these will be changes to the Gulf Stream and its northern extension, the North Atlantic Drift, which brings warmer water from the Gulf of Mexico to north Atlantic regions, including eastern Canada and northwestern Europe. Models suggest that global climate change could lead to a weakening of the Gulf Stream and a corresponding cooling of some regions, such as Scandinavia.

*1.4.3.9 Changes in precipitation patterns*

Changes in weather patterns will result in increased precipitation in some regions and decreased precipitation in others. These precipitation features will lead to increased drought and floods. Decreased precipitation in many agricultural regions, such as the Midwest and the Southwest in the United States, will lead to decreased crop production. Wildfires will occur with increased frequency and severity in regions with reduced precipitation.

*1.4.3.10 More frequent and severe hurricanes*

Global climate change is expected to result in more frequent extreme weather events. Perhaps the most notable of these will be the increased frequency, severity, and duration of hurricanes. These effects have been observed since the early 1980s.

*1.4.3.11 Altered geographic ranges of some species*

Ecological changes, such as temperature changes and changes in precipitation patterns affect the habitats of wildlife and can lead to decreased populations or even extinction of some species, as well as changes in their geographical range.

*1.4.3.12 Thawing of permafrost*

Permafrost is ground that is continuously frozen in regions where the ambient temperature remains sufficiently low. It occurs primarily in the Northern Hemisphere in the Arctic regions of Alaska, Canada, Greenland, and Siberia. Substantial quantities of greenhouse gases including carbon dioxide, methane, and nitrous oxide are trapped in permafrost. When permafrost melts as a result of global warming, these gases will be released to the atmosphere, leading to further global warming.

## 1.5 Future fossil fuel use

As discussed above, the vast majority of our current energy is supplied by fossil fuels. This is particularly true for the energy used for transportation. Fossil fuel resources are limited and at some point will no longer be able to fulfill a significant portion of our energy needs. We may also decide to transition to other energy sources before fossil fuels are depleted. This transition may be linked to environmental concerns but may also be linked to the rising fossil fuel prices that will result from diminishing supplies. Determining future fossil fuel use is an important component of predicting the environmental impact of our energy use. There are two approaches to understanding how fossil fuels may be used in the future. We can look at available fossil fuel resources, as this will limit what is available for use, or we can look at the future demand for fossil fuels. Here, we consider two fairly straightforward approaches to predicting future fossil fuel production based on resource availability. These are: (a) an analysis of reserve-to-production ratios and (b) Hubbert theory. We then consider different scenarios for future fossil fuel demand.

### 1.5.1 Reserve-to-production ratios

The simplest method of estimating the longevity of our fossil fuel resources is to calculate the ratio of how much of the resource is remaining and how much we use per year at present. This is known as the reserves-to-production ratio, as summarized in table 1.8 for the different fossil fuels; it gives the number of years of use that remain at the current rate of consumption. This approach may not be accurate, as it assumes that fuel will be utilized at a constant rate and that all known reserves are equally viable. However, this method can provide a rough estimate of the longevities of the different fossil fuels. A more mathematical approach is described next.

### 1.5.2 Hubbert theory

In the 1950s, American geophysicist Marion King Hubbert developed a mathematical approach to describing the utilization of natural resources, specifically oil in the United States. This model can, in general, be applied to any resource and has been successful in describing the utilization of coal, petroleum, and natural gas. There are three basic assumptions behind the Hubbert model:

1. The utilization of a resource begins slowly when the resource is determined to be useful. This is because the infrastructure required to use the resource must be developed.
2. The utilization of the resource increases as the infrastructure is developed and reaches a maximum when half of the available resource has been consumed.
3. The utilization of the resource decreases after half of the resource has been consumed and falls to zero when the resource has been exhausted or becomes no longer viable. This is because, as the resource becomes depleted, the efficiency of its extraction decreases and the cost of its use increases.

**Table 1.8.** Reserve-to-production ($r/p$) ratios for fossil fuels for 2019. Data adapted from BP (2020) and IEA (2020). Note: bbl = barrels = 158.97 L, $t = 10^3$ kg = 2025 lbs.

| Resource | Known reserves ($r$) (2019) | Annual production ($p$) (2019) | $r/p$ (years) | End year |
|---|---|---|---|---|
| Oil | $1.73 \times 10^{12}$ bbl | $3.47 \times 10^{10}$ bbl | 50 | 2069 |
| Natural gas | $1.99 \times 10^{14}$ m$^3$ | $3.93 \times 10^{12}$ m$^3$ | 51 | 2070 |
| Coal | $1.07 \times 10^{12}$ t | $7.92 \times 10^{9}$ t | 135 | 2154 |

Further mathematical details of the Hubbert model may be found in Dunlap (2020a). Typically, the annual production rate of a resource is taken as an indicator of its utilization, and in the Hubbert model the production rate of a resource is expected to follow the logistic curve shown in figure 1.10.

The validity of the Hubbert model has been tested on a variety of subsets of fossil fuel data. Figure 1.11 shows the data for the annual production rate of Pennsylvania anthracite coal. Anthracite coal is the most desirable form of coal because it has the highest carbon content (and hence highest energy content) as well as the lowest concentration of impurities (which reduce the energy content and add to the emission of chemical pollutants). Pennsylvania had the most significant anthracite coal resources in the United States, and these were utilized extensively beginning around the middle of the 19th century, as illustrated in figure 1.11. The utilization data shows good agreement with the shape of the curve predicted by the Hubbert model, except for an additional small peak during the 1940s that corresponds to increased production as a result of increased demand related to World War II. Production diminished to near zero by the beginning of the 21st century as a result of the depletion of economically viable resources. At present, Pennsylvania anthracite production amounts to about 1.6 Mt per year out of a total U.S. anthracite production of about 4.5 Mt per year (note Mt = $10^6$ metric tonnes = $10^9$ kg). These numbers may be contrasted with the Pennsylvania anthracite production of about 90 Mt per year around the time of World War I. The Hubbert model is also applicable to a number of other natural resources such as whale oil and whale bone (see Höök 2010), the production of which decreased and ended when it became unviable for economic, environmental, and social reasons.

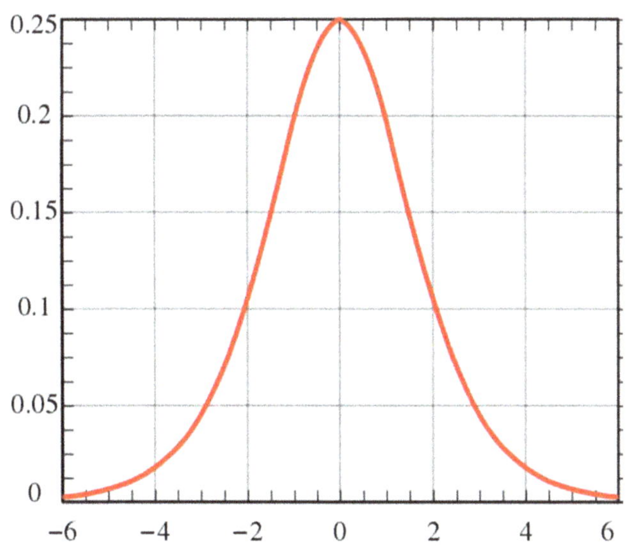


**Figure 1.10.** Logistic curve described by the Hubbert model for the utilization of a resource. This image (Rubber Duck 2015) has been obtained by the author from the Wikimedia website, where it is stated to have been released into the public domain. It is included within this chapter on that basis.

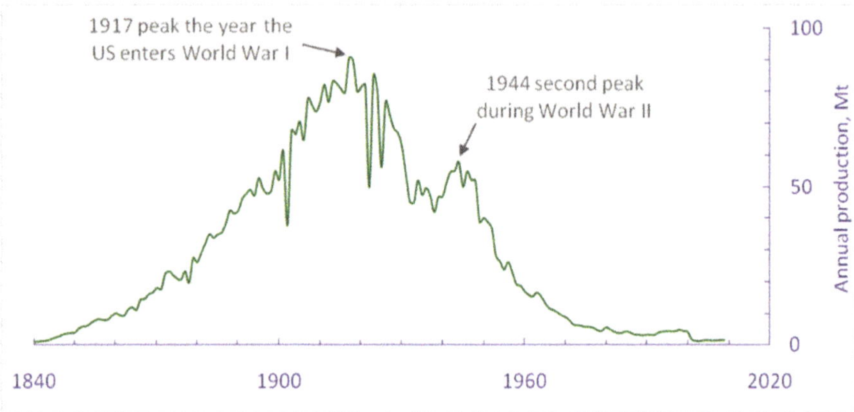


**Figure 1.11.** Production of Pennsylvania anthracite. Reproduced with permission from Rutledge (2011). Available at: http://rutledge.caltech.edu/.

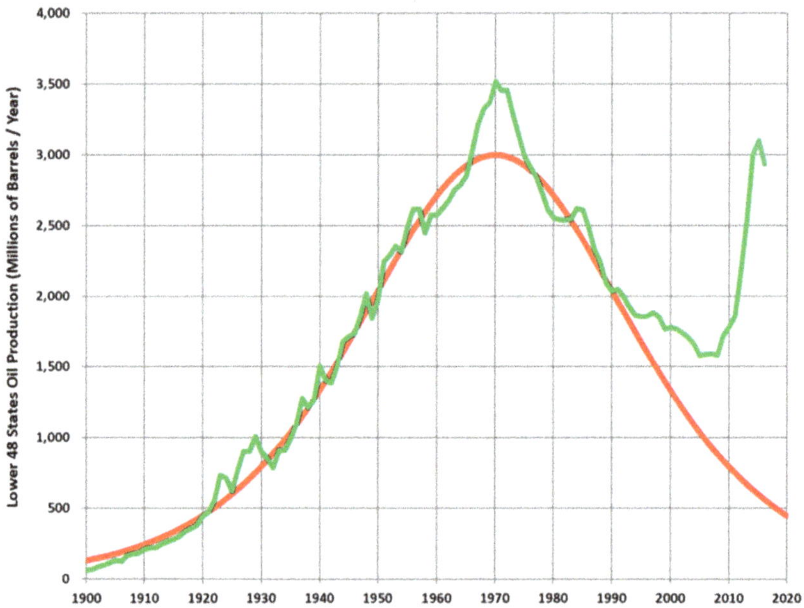


**Figure 1.12.** United States crude oil production for the contiguous 48 states (green curve) and Hubbert's prediction from 1956 (red curve). Oil produced in Alaska is not included, as this was not included in Hubbert's original analysis. Note: 1 barrel = 158.97 L. This image (Plazak 2015) has been obtained by the author from the Wikimedia website where it was made available under a CC BY-SA 4.0 licence, https://creativecommons.org/licenses/by-sa/4.0/deed.en. It is included within this chapter on that basis. It is attributed to Plazak.

The Hubert model must be applied to resource utilization data with some caution. An example of a case in which the application of the model is not straightforward is crude oil production in the United States. These data are shown in figure 1.12.

While there was some increased production around 1970 as a result of increased demand for vehicle fuel, the data shows reasonably good agreement with the Hubbert model shown by the red curve in the figure up until the early 2000s. After that, the production curve is substantially greater than the Hubbert prediction. Such deviations can occur for several reasons. First, significant previously unknown resources may be discovered and thereby increase the total amount of the resource available. Second, as is the case in figure 1.12, new technologies may be developed that allow resources that were previously considered economically or technically unusable to become viable. In figure 1.12, the red curve, which follows the original prediction from the 1950s by Hubbert, considers the utilization of what is referred to as 'conventional' oil. Around 2000, technologies that allowed for the extraction of what is referred to as 'tight' oil became common. Tight oil includes oil that is more difficult to extract from underground resources as well as oil that is trapped in shale, i.e. shale oil. Fracking and oil shale processing technologies have made these sources of crude oil economically viable. The availability of these resources, as well as the desire of the United States to reduce dependency on imported foreign oil, led to substantially increased production from tight oil sources. It is interesting to note that if only conventional oil production is plotted, the production curve continues to be reasonably well described by the Hubbert curve. While the addition of fracking and other non-conventional technologies has very noticeably increased oil production, the total amount of oil available (given by the area under the curve) has been increased by only a fairly modest amount. This is because, as the graph shows, the new peak is quite narrow (in years) and the overall longevity of the resource has not been increased very significantly. In order to properly analyze data such as that shown in figure 1.11 it is necessary to incorporate two Hubbert curves, one for conventional oil and one for tight oil. Additional Hubbert curves can be incorporated to account for additional resource discoveries or other new technologies, and such analysis is referred to as a multi-Hubbert analysis.

The application of suitable multi-Hubbert analyses to world fossil fuel production provides some highly significant information related to the anticipated longevity of these resources. Table 1.9 gives the half-peak years for oil, natural gas, and coal resources worldwide based on a 2012 analysis (Maggio and Cacciola 2012). The half-peak year is the year in which production rate will drop to one half of the peak production rate, which is also the time at which about 85% of the total resource will

**Table 1.9.** Half-peak years for world oil, natural gas, and coal production based on a multi-Hubbert analysis. Data are adapted from Maggio and Cacciola (2012).

| Resource | Half-peak year |
|---|---|
| Oil | 2052 |
| Natural gas | 2078 |
| Coal | 2115 |

have been used. The half-peak years shown in the table are in reasonable agreement with the results of the simple reserve-to-production ratios. Both approaches suggest that on a timescale of 40–50 years, oil and natural gas resources will be severely depleted, and coal resources will be severely depleted on a timescale of about 100 years.

### 1.5.3 Analysis of future fossil fuel demand

The analysis based on the Hubbert model as described above provides some interesting insight into future energy use. However, it is far from perfect. Production is more or less driven by demand, and it can be seen that in periods of increased demand there is a corresponding increase in production which results in deviations from the Hubbert model. It is also clear that the results of the analysis depend critically on the validity of the assumptions and the input parameters. Since Hubbert's first analysis in the 1950s, the estimated time of peak production has moved forward more or less continuously. This is due to the discovery (or utilization) of new resources (e.g. offshore oil or Alaskan oil) or the implementation of new technologies that make more of the resource viable (e.g. fracking, oil extraction from tar sands). This is clearly the reason for the non-logistic-curve behavior shown in figure 1.12. Although many limited sets of data, such as that shown in figure 1.11, are reasonably well described by the Hubbert model, it has also been questioned whether this approach is generally a valid method of analyzing energy resource utilization (Sorrell *et al* 2009).

Moving forward, the Hubbert model gives us some expectations concerning the longevity of available resources, within the limits described above. However, an assessment of the effects of our energy use on the environment is related to our actual energy use, rather than the energy resources that are available. It has, therefore, become beneficial to analyze future energy demand as a measure of environmental impact. The demand (for fossil fuel resources) may be less in the future than the availability of such resources would imply. This can occur for several (somewhat related) reasons including:

- reduction in future fossil fuel use as a means of mitigating climate change;
- reduction in fossil fuel use as a result of an increase in the capacity of low-carbon energy sources;
- reduction in fossil fuel use because of increasing costs relative to more sustainable options.

The last point can result from increasing fossil fuel costs due to decreasing supply; alternatively, it can result from decreasing sustainable energy costs due to improved technologies. Thus, determining future fossil fuel energy demand depends on a number of factors, most of which are not easy to predict. These not only include a number of scientific and technical issues related to renewable energy development but also political, economic, and social factors. It is possible, however, to estimate future demand based on different energy development scenarios. As an example, figure 1.13 shows some general features for world oil demand based on

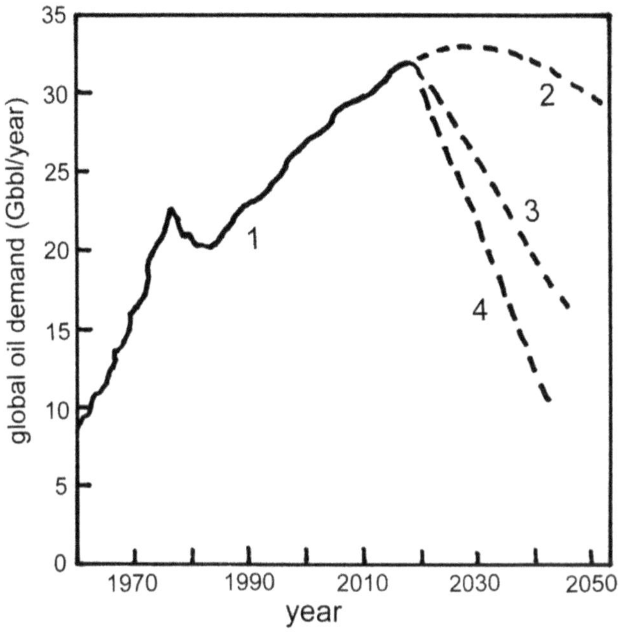


**Figure 1.13.** Global oil demand as a function of time: (1) historical production driven by demand, (2) future demand limited by resources, (3) sustainable future demand, and (4) net zero future demand.

considerations discussed in BP (2021) and IEA (2022). Historical data is shown up to about 2021 (curve 1), and various scenarios for future oil demand are illustrated. In the case of demand limited by resources (curve 2), current predictions along the lines of the Hubbert analysis of past oil production are shown; these illustrate peak production occurring within a decade or so. Sustainable future demand (curve 3) refers to the case in which the global temperature increases by the end of the century are limited to a maximum of 2.0 °C, which may be further reduced to a maximum of 1.5 °C if effective atmospheric carbon capture technology is implemented (as discussed in section 1.7). The net zero case (curve 4) represent the most aggressive reduction in fossil fuel use and results in a maximum global temperature increase of 1.5 °C, independent of carbon capture.

The discussion above clearly indicates that the use of fossil fuels in the future is dictated by demand, as long as this remains below the limits of availability. This demand is, at least to some extent, governed by the choices that society makes in its approach to energy sources, as well as economic and technical considerations.

In recent years it has become apparent that there are other factors that can have a direct influence on our energy use. The COVID-19 pandemic which spread world-wide in 2020 influenced overall energy use. Travel restrictions, as well as the economic downturn, that resulted from the pandemic reduced overall energy use and fossil fuel use in particular. This trend is apparent in figure 1.13, where a slight decrease is seen around 2020. It has been speculated that fossil fuel use will return to its pre-COVID-19 trends, but this has not occurred as of mid-2022.

Beginning in early 2022, the war in Ukraine has had severe effects on energy use in many parts of the world as a result of the disruption of fossil fuel distribution. The future short-term and long-term effects of this war and the COVID-19 pandemic on our overall energy use and the transition to carbon-free energy sources required to address global climate change still remain to be seen.

## 1.6 Sustainable energy sources for the future

The discussion above has emphasized two important reasons for the development of new energy sources: the depletion of world fossil fuel resources and the environmental consequences of continued carbon emissions. It is clear that our current energy mix will not be sustainable very far into the future and that transitioning to low-carbon energy sources as soon as possible is essential to minimize the effects of global climate change.

Sustainable low-carbon energy sources typically either produce electricity or a fuel in which chemical energy is stored. These two forms of energy are readily applicable to transportation as, for example, electricity for battery electric vehicles or fuel for internal combustion engine vehicles, respectively. In addition, renewable energy may make a minor contribution in the form of mechanical energy harvested from the wind that is used to power marine vehicles. A brief description of the carbon-free energy technologies that produce electricity or chemical fuels is given below. Nuclear energy, although not renewable, is low-carbon and is included in the list.

At present about 65% of electricity worldwide is produced from fossil fuels, about 25% is produced from renewable energy sources and about 10% is produced from nuclear energy. In this section, major low-carbon electricity-producing technologies are discussed in order of decreasing total world installed capacity, as summarized in table 1.10.

### 1.6.1 Hydroelectric energy

Hydroelectricity is the most extensively utilized renewable energy resource. It is also the oldest renewable source of electricity and was first developed as a commercial supplier of electricity to the public in the 1880s. Figure 1.14 shows the Milltown Hydroelectric Station, which is the oldest operational hydroelectric generating

**Table 1.10.** Installed worldwide generating capacities of the major carbon-free electricity sources at the end of 2019. Data adapted from Dunlap (2019), the International Atomic Energy Agency (IAEA 2020), and the International Renewable Energy Agency (IRENA 2020).

| Energy source | World capacity (GW) |
|---|---|
| Hydroelectric | 1190 |
| Wind | 623 |
| Solar | 586 |
| Nuclear fission | 392 |
| Biofuels | 124 |

**Figure 1.14.** Milltown Hydroelectric Station located on the Canada–United States border between New Brunswick and Maine. Image: Richard A Dunlap.

facility in North America. It was constructed in 1881 and has a generating capacity of 4 MW.

Hydroelectric facilities can be roughly divided into two categories: high head facilities and run-of-the-river facilities. High head facilities are large-capacity installations in which a substantial dam is used to create a reservoir. Water from the reservoir flows to turbines and generators located at the bottom of the dam, thereby generating electricity. Run-of-the-river facilities have a minimal dam or no dam and largely use the kinetic energy of the flowing water of the river to turn turbines to generate electricity.

While hydroelectric power does not in itself generate greenhouse gases, it can contribute to global climate change indirectly. First, high head facilities in particular are constructed using enormous quantities of concrete. The production of concrete involves the heating of limestone ($CaCO_3$) to produce CaO by the reaction

$$CaCO_3 \rightarrow CaO + CO_2,$$

thereby emitting $CO_2$ to the atmosphere. The creation of a reservoir utilizes land that in many cases was previously occupied by forests or other vegetation. Reservoirs for high head facilities can cover a very large area. For example, the reservoir that supplies the Three Gorges Dam on the Yangtze River in China covers an area of 1084 km$^2$. The creation of a reservoir eliminates vegetation that would otherwise sequester carbon dioxide from the atmosphere. In addition, organic matter that is covered by water decays and thereby emits carbon dioxide and methane. This problem is most significant in tropical regions. Run-of-the-river facilities do not share these environmental concerns, but typically have more variable capacity factors, as discussed below.

In recent years, there has been considerable growth in the utilization of hydroelectric energy in some developing countries. Figure 1.15 shows that this growth is notable in China and Brazil. At present, hydroelectric energy represents the largest component of renewable electricity generated worldwide (about 17% of total electricity generation).

### 1.6.2 Wind energy

As shown in table 1.10, wind currently has the second largest installed capacity of all low-carbon electricity sources. Figure 1.16 also shows that installed wind capacity

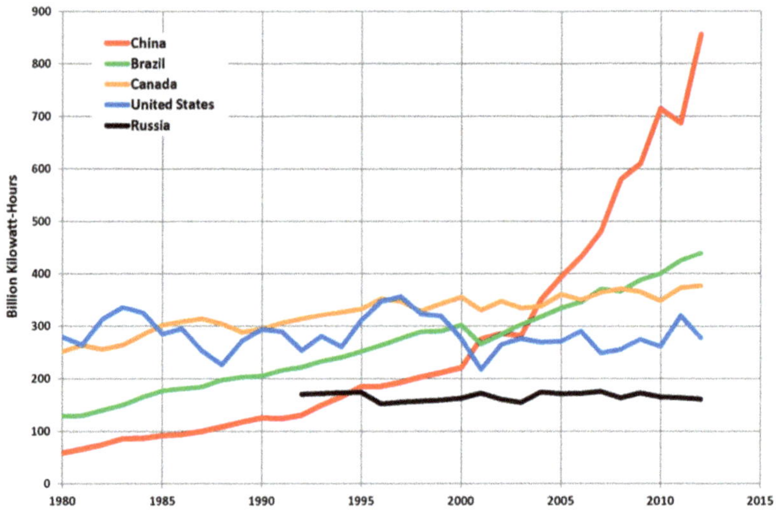


**Figure 1.15.** Growth of hydroelectric power in the top five countries. This image (Plazak 2013) has been obtained by the author from the Wikimedia website where it was made available under a CC BY-SA 3.0 licence, https://creativecommons.org/licenses/by-sa/3.0/deed.en. It is included within this chapter on that basis. It is attributed to Plazak.

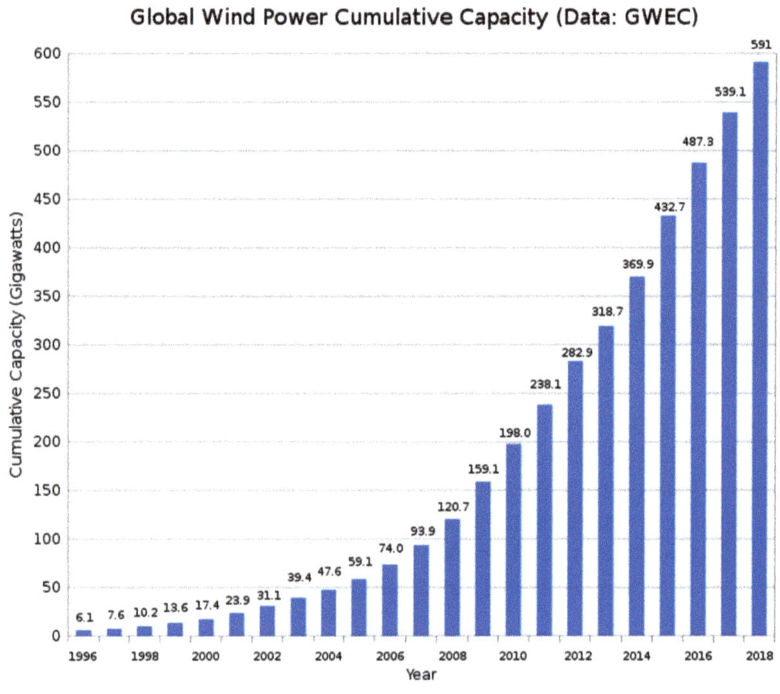


**Figure 1.16.** Increase in electricity produced by wind since the mid-1990s. This image (Delphi234 2014) has been obtained by the author from the Wikimedia website, where it is stated to have been released into the public domain. It is included within this chapter on that basis.

has increased substantially over the past two decades in almost all regions of the world. At present, about 15% of worldwide wind capacity is offshore, most of this located in Europe, where high population densities in many countries make offshore locations attractive.

Wind energy has a number of very attractive characteristics. Some of the most significant are as follows:

*Cost-effective*—Wind is a cost-effective method of generating electricity (see below).

*Well-established technology*—Wind turbine technologies are well established, although research to increase generator efficiency and improve turbine design continues.

*Wide distribution of resources*—While not all locations are ideal for the placement of wind turbines, most countries have wind resources, either onshore, offshore, or both, that are suitable for wind energy development. Offshore locations are attractive because their wind velocities are typically greater than those of onshore locations, and the use of land resources in areas of high population density can be avoided. However, as discussed later in the chapter, economic factors favor onshore wind energy development.

*Potential dual land use*—Wind farms typically consist of large numbers of individual turbines, and these must be spaced adequately in order to effectively make use of wind energy. The land between wind turbines can be utilized for other purposes, such as agriculture.

*Low risk*—While all methods of harvesting and utilizing energy involve some degree of risk, wind energy is one of the safest energy technologies. Since the 1970s, there have been about 200 human fatalities related to the wind power industry, mostly involving occupational workers. The risk of wind energy to the general public is quite small and typically involves accidents related to broken turbine blades or ice throws.

Although the potentially adverse effects that result from the utilization of wind energy are typically less severe than for many other energy sources, the more significant concerns are described below.

*Noise*—The potential for possible health risks associated with the noise produced by wind turbines has been one of the major concerns for the widespread implementation of this technology. In most jurisdictions, regulations are in place that prohibit the construction of wind turbines within certain distances of residential buildings. At 1 km, the sound level is about 30 dB, comparable to the ambient sound level in a quiet rural area.

*Shadow flicker*—On a sunny day, the shadows of the blades of a wind turbine move, leading to a flickering as viewed by an observer on the ground. This flickering typically occurs once every second or so (depending on the rotational frequency of the turbine) and causes annoyance to residents living in its path. This, combined with noise levels as well as the potential for blade accidents or ice throws, requires the placement of turbines at suitable distances from residential areas.

*Effects on wildlife*—While bird (and bat) fatalities are used as an argument against the development of wind energy, recent studies by the U.S. Fish and Wildlife

**Table 1.11.** Estimated bird mortality from some different anthropogenic causes in the United States. Data adapted from the U.S. Fish and Wildlife Services (2020).

| Hazard | Bird deaths per year |
|---|---|
| Domestic cats | 2 400 000 000 |
| Windows | 599 000 000 |
| Vehicles | 214 500 000 |
| Electric power lines | 25 500 000 |
| Wind turbines | 234 012 |

Services (2020), summarized in table 1.11, give the estimated bird mortalities from different anthropogenic causes. It is clear that the number of bird fatalities from other causes is much more significant than from wind turbines. Another way to look at this issue is to compare bird mortality for different energy technologies per unit energy generated. Specifically, fossil-fuel-generated electricity causes bird death as a result of mining operations, air pollution, and global warming and results in about 17 times the bird mortality per unit of electricity generated than that of wind-generated electricity (Sovacool 2009).

*Deforestation*—The removal of trees for the purpose of constructing a wind farm has the potential effect of decreasing carbon sequestration from the atmosphere. Thus, the choice of locations for wind farms should consider this possible negative environmental consequence.

### 1.6.3 Solar energy

Solar energy will undoubtedly be a major component of power in a sustainable future. Solar energy has been effectively used on a residential and community scale to provide direct thermal energy. It is, however, most likely to contribute in a substantial way to future energy as a source of electricity. There are two approaches to producing electricity from solar energy: thermal generation (sometimes called concentrated solar power), which uses solar radiation to heat a working fluid and generates electricity using a heat engine and photovoltaics, which generate electricity directly using semiconducting devices. At present, more than 99% of the installed solar grid capacity utilizes photovoltaic technology, and this is discussed further in this section.

The extensive development of solar photovoltaics began in the early 1990s, as illustrated in figure 1.17. This figure shows that over the past thirty years or so, the growth of photovoltaic capacity worldwide has been exponential (as evidenced by the approximately straight line on the semi-log plot). While most photovoltaic cells have been made of silicon, the technology has shifted from monocrystalline silicon to multicrystalline silicon over this time frame, as seen in figure 1.18. A major factor that has contributed to the growth of photovoltaics has been the cost of photovoltaic cells. Figure 1.19 shows the decrease in the cost of cells per unit power output. At present, solar photovoltaics is one of the most cost-effective methods of electricity generation for the grid.

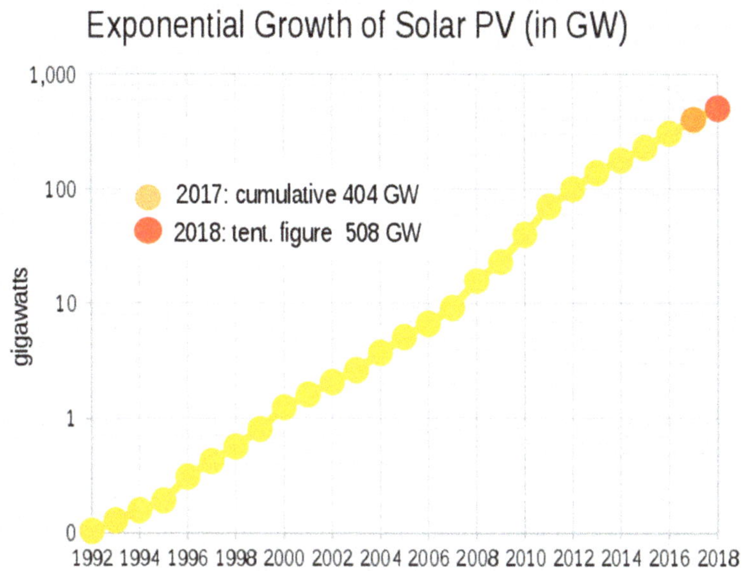


**Figure 1.17.** Total world installed photovoltaic capacity as a function of year from 1992 to 2018. This image (Rfassbind 2014a) has been obtained by the author from the Wikimedia website, where it is stated to have been released into the public domain. It is included within this chapter on that basis.

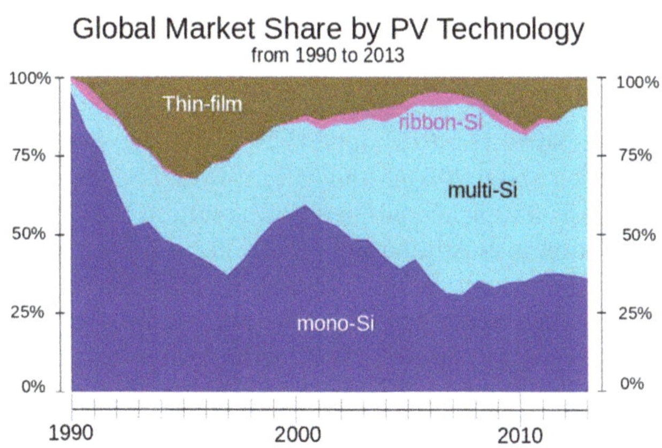


**Figure 1.18.** Different photovoltaic technologies used as a function of time from 1990 to 2013. Crystalline silicon includes monocrystalline silicon (mono-Si) and multicrystalline silicon (multi-Si), while thin-film technologies include amorphous silicon, CdTe, and copper–indium–gallium–selenide (CIGS). This image (Rfassbind 2014b) has been obtained by the author from the Wikimedia website, where it is stated to have been released into the public domain. It is included within this chapter on that basis.

Solar energy has a very low density at the surface of the Earth, and, as a result, a substantial land area is required to produce quantities of electrical power that are comparable to those generated by fossil fuel or nuclear power stations. In the United

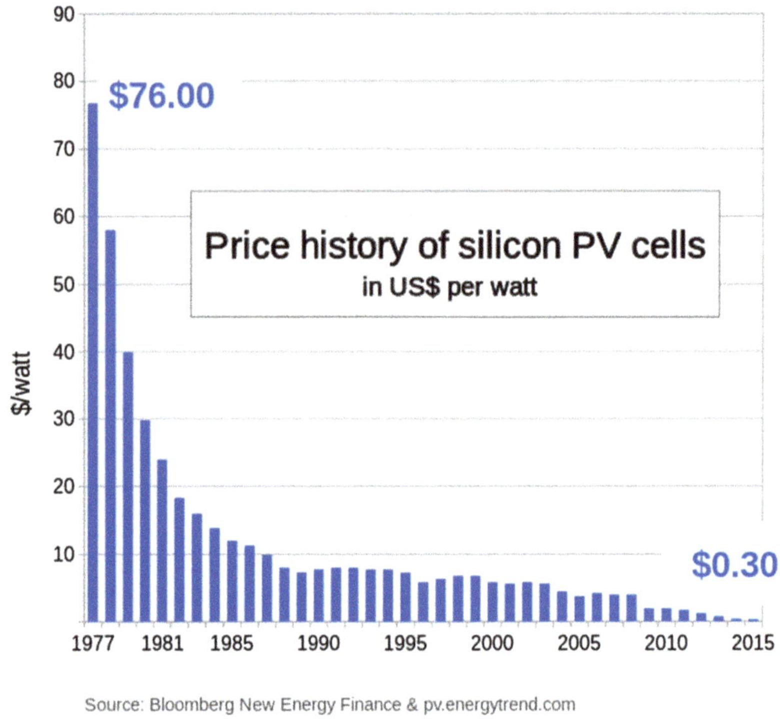


**Figure 1.19.** Price history chart of crystalline silicon solar cells in USD per watt since 1977. This image (Rfassbind 2015) has been obtained by the author from the Wikimedia website, where it is stated to have been released into the public domain. It is included within this chapter on that basis.

States, for example, the average annual 24 h radiation ranges from about 150 W $m^{-2}$ in the Northeast and Northwest to about 245 W $m^{-2}$ in the Southwest. This means that an area of between about 20 $km^2$ and 35 $km^2$ would be required to produce the same average output as a typical coal-fired power station (~1 GW). A typical solar photovoltaic installation is illustrated in figure 1.20.

At first glance the utilization of solar energy through the application of photovoltaic technology would seem to have few potentially adverse environmental consequences. While this is generally true, it is important to consider the negative effects of extensive global solar power generation. The large number of photovoltaic panels and the corresponding large land area that are needed to provide significant electrical power have certain drawbacks. While this is also the case for wind power, the situation for solar power is somewhat different, as the panels block the Sun's light from directly reaching the land beneath them. This makes land use for large-scale agriculture incompatible with solar photovoltaics. However, the spaces between the rows of solar panels can be used to grow grass, and pilot projects have involved the use of this land for grazing by livestock such as sheep. Although the significant manufacturing and infrastructure construction means that the risks associated with solar photovoltaics, particularly for occupational workers, are greater than might be expected, they are somewhat less than for wind power.

**Figure 1.20.** A portion of the Dau Tieng Solar Power Project, Tay Ninh Province, Vietnam. This facility has a capacity of 600 MW and covers an area of 5 $km^2$. Note that the electricity generated is related to the capacity factor, as discussed below. This image (TammyLe 2019) has been obtained by the author from the Wikimedia website where it was made available under a CC BY-SA 4.0 licence, https://creativecommons.org/licenses/by-sa/4.0/deed.en. It is included within this chapter on that basis. It is attributed to TammyLe.

Some of the reasons for this relate to their less demanding transportation requirements (as no huge components, such as wind turbine rotors, need to be transported as single units) and less hazardous maintenance conditions.

### 1.6.4 Nuclear fission

Nuclear fission refers to nuclear reactions in which a heavy nucleus is broken up into two (usually lighter) nuclei. Since the sum of the nuclear binding energy is greater for the two lighter nuclei than for the original heavier nucleus, excess energy is given off as kinetic energy of the nuclei. This kinetic energy is ultimately turned into heat when the reaction by-products interact with the atoms in a working fluid such as water. This heat may be used to produce electricity by means of a heat engine in the same way that heat from the combustion of coal is used to produce steam to run turbines and generate electricity.

Fission reactors designed for the purpose of producing electricity were first constructed in the 1950s and became common in many countries in the 1970s. Figure 1.21 shows the Seabrook Nuclear Generating Station in New Hampshire, a typical example of a pressurized water reactor. The construction of this station began in the mid-1970s during the period of the largest growth of nuclear fission energy, but it did not become operational until 1990. The variation in interest in nuclear power is illustrated in figure 1.22, where the number of new reactor construction starts are plotted as a function of year from 1954 to 2015. After around 2000, there has been a worldwide increase in the number of new nuclear reactors, but this activity must be balanced against aging reactors that have been decommissioned. In fact, as shown in figure 1.23, the total number of nuclear power

**Figure 1.21.** Seabrook Nuclear Generating Station in Seabrook, New Hampshire. This facility is a 1246 MW pressurized water reactor. Construction began in 1976 and the station became operational in 1990. Image: Richard A Dunlap.

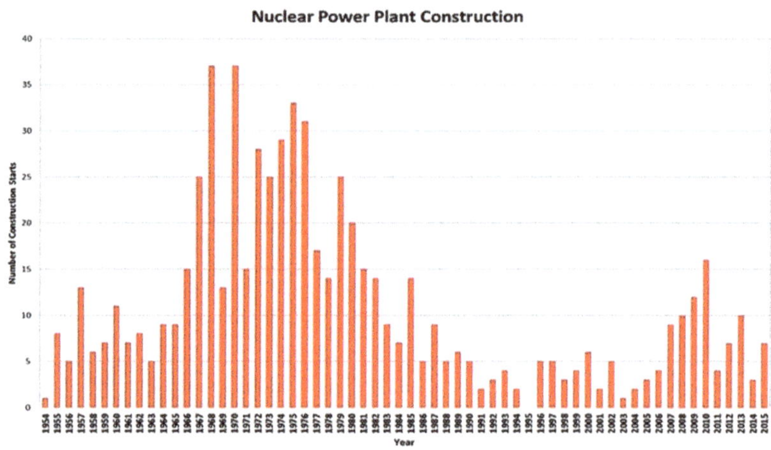


**Figure 1.22.** The number of new nuclear power station construction starts per year from 1954 to 2015. This image (Ypna 2013) has been obtained by the author from the Wikimedia website where it was made available under a CC BY-SA 3.0 licence, https://creativecommons.org/licenses/by-sa/3.0/deed.en. It is included within this chapter on that basis. It is attributed to Ypna.

reactors and the total nuclear generating capacity have changed very little since the late 1980s.

When assessing the future contribution of nuclear fission to our energy supply, it is necessary to consider resource availability, safety, security, and waste disposal, as discussed below.

*Fission energy resources*—The vast majority of operational fission power reactors worldwide are thermal neutron reactors. These reactors utilize energy derived from the fission of $^{235}U$, which accounts for only 0.72% of naturally occurring uranium. At present, nuclear power provides about 11% of all electricity worldwide and this represents about 4% of all energy. A simple analysis based on the

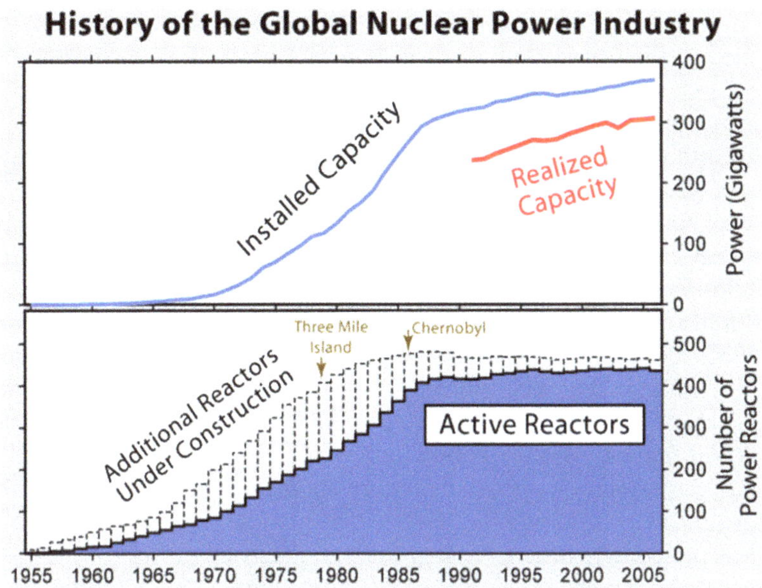


**Figure 1.23.** Installed global nuclear reactor capacity, along with actual generated output (top) and number of operational nuclear reactors and reactors under construction (bottom), from 1955 to 2007. This image (Rhode 2007) has been obtained by the author from the Wikimedia website where it was made available under a CC BY-SA 3.0 licence, https://creativecommons.org/licenses/by-sa/3.0/deed.en. It is included within this chapter on that basis. It is attributed to Robert A Rohde.

reserve-to-production ratio provides a lifetime for the remaining uranium resources of about 53 years (Dunlap 2021). Thus, under these conditions, we cannot expect nuclear fission power to provide a long-term solution for carbon-free electricity. A fast breeder reactor can use $^{238}U$ (which accounts for the remaining 99.28% of naturally occurring uranium) or naturally occurring thorium-232 ($^{232}Th$) as a fuel. At present, there are two operational commercial fast breeder reactors, compared to about 450 thermal neutron reactors worldwide. Certainly, expanding the use of fast breeder reactors would greatly increase the longevity of fission power by enabling the use of more plentiful fuels.

*Safety*—There have been numerous incidents and accidents at commercial power reactor facilities as a result of equipment failure or human error. The three most serious and well-known accidents are those that occurred at Three Mile Island in 1979, Chernobyl in 1986, and Fukushima in 2011. These accidents are certainly cause for concern, and there are no guarantees that similar or even more serious events will not happen in the future. However, current fission reactor designs are much improved over the older designs that were involved in these accidents and have implemented technology to drastically reduce the possibility of such future events.

*Security*—Nuclear fission reactors utilize fuel containing fissile material, particularly $^{235}U$. Such materials can be utilized for the construction of nuclear weapons. It is therefore important for international agencies, such as the International Atomic Energy Agency (IAEA) to monitor uranium production programs worldwide in order to distinguish between activities related to peaceful energy production and those related to the preparation of fissile material for nuclear weapons.

*Waste disposal*—Perhaps the greatest concern for the continuation of nuclear fission power is the question of waste disposal. The by-products of the fission process are radioactive and can present a significant health risk for many tens or hundreds of years. Thus, the spent reactor fuel must be stored safely until the activity level has decreased to a point where it is no longer a danger. A portion of the spent fuel can be reprocessed and reused as reactor fuel, although this may cause additional security concerns related to the possible illicit acquisition of weapons-grade material. At present most spent fuel is stored in sealed containers in underground locations. It is of interest to note the quantity of waste that is produced by nuclear reactors. Since around 1970, the total amount of high-level radioactive waste that has been generated by all of the fission reactors worldwide is about 400 000 tonnes. About 100 000 tonnes of this has been reprocessed, leaving about 300 000 tonnes for disposal. This corresponds to the volume of a cube about 40 m on a side.

As noted previously, nuclear fission reactor construction has decreased considerably since its high point in the early 1970s (see figure 1.22). Although the nuclear accidents at Three Mile Island (in 1979) and Chernobyl (in 1986) were a factor in swaying public opinion against nuclear power, a fairly continuous decline from the mid-1970s to the early 2000s can be seen in the figure. Recent analyses, however, (see, for example, Ritchie (2020) and the references therein) have concluded that nuclear fission energy is among the safest low-carbon options for future electricity generation (as well as one of the least carbon producing). Since around 2000, there has been some growth in new reactor starts, including reactors with improved safety features, as well as fast breeder reactors.

### 1.6.5 Biofuels

Biofuels can contribute to future transportation energy needs in two ways: first, as a fuel for power stations that contribute to grid power, which can then be used for battery electric vehicles (as discussed in chapter 2) or vehicles that utilize electrofuels (as discussed in chapter 3) and second, for direct use in vehicles using internal combustion engines or fuel cells.

The use of incinerators to burn municipal solid waste as an alternative to other means of waste disposal, such as landfill, has been in use for well over a hundred years. In the later part of the 20th century the use of heat from such combustion became of interest for the production of electricity, and today there are over 500 such facilities worldwide. This process is referred to as waste-to-energy (WTE), although the term WTE is also used to designate the process of converting solid waste to liquid or gaseous fuels (see chapter 4). Figure 1.24 shows a WTE plant in northeast England that burns 390 000 tonnes of municipal waste per year. The overall environmental benefits of WTE are uncertain (see Dunlap (2019)), although the production of carbon dioxide from the combustion of waste is generally considered to be a desirable alternative to the production of methane during its decomposition in a landfill site.

**Figure 1.24.** Teesside Energy from Waste Plant is a WTE power station in Haverton Hill, UK. It began operation in 1998 and has a capacity of 29.2 MW. This image (Fintan264 2009) has been obtained by the author from the Wikimedia website where it was made available under a CC BY-SA 3.0 licence, https://creativecommons.org/licenses/by-sa/3.0/deed.en. It is included within this chapter on that basis. It is attributed to Fintan264.

In recent years, wood has been used as a fuel for electricity generation. In this case, wood is either burned directly to produce heat for a steam boiler or it is heated to produce a gaseous fuel by gasification, which is then burned (see section 4.6). Wood fuel typically takes the form of wood chips or waste from construction or land clearing. Wood-fired power stations may be purpose-built facilities or may be former coal-fired power stations that have been converted to burn wood. The Bay Front Generating Station in Ashland, Wisconsin is an example of the latter category of facility and is illustrated in figure 1.25. Wood-fired power stations tend to be small (typically tens of MW) compared to coal-fired stations, which usually have capacities in the range of hundreds of MW to 1 GW.

The alternative approach to using biomass for transportation energy is the production of a liquid or gaseous fuel that can be used directly in an internal combustion engine or to power a fuel cell. This approach includes the use of purpose-grown crops, such as corn, to produce ethanol or the use of municipal or agricultural waste to produce biogas. Details of the possible use of biomass to produce transportation fuel are presented in chapter 4 for road vehicles and in chapter 5 for other modes of transportation.

### 1.6.6 Carbon footprints of energy technologies

As noted in the sections above for various sustainable electricity options, there are some potential environmental concerns for a number of these technologies. These environmental concerns include the potential contribution of these technologies to

**Figure 1.25.** The Bay Front Generating Station in Ashland, Wisconsin is a 28 MW facility that was converted from burning coal to burning waste wood in the early 2010s. A pile of wood fuel is located to the left of the facility. This image (Chris857 2018) has been obtained by the author from the Wikimedia website where it was made available under a CC BY-SA 4.0 licence, https://creativecommons.org/licenses/by-sa/4.0/deed.en. It is included within this chapter on that basis. It is attributed to Chris857.

global greenhouse gas emissions. For example, as noted above, potential deforestation can contribute to greenhouse gases by eliminating a component of natural carbon sequestration. It is also possible that carbon dioxide or other greenhouse gases such as methane may be emitted during the manufacture of infrastructure, and this contribution should be amortized over the lifetime of the facility. In order to fully understand the greenhouse gas implications of different technologies, it is necessary to undertake a life-cycle analysis. This analysis evaluates the greenhouse gas contribution during all stages of infrastructure construction, system operation, and eventual end-of-life processing. In order to account for the fact that different aspects of the system life may produce different greenhouse gases, the results of this analysis are often given in terms of the equivalent concentration of carbon dioxide in accordance with the relative infrared absorption of each gas, as given in table 1.7.

Table 1.12 shows the results for greenhouse gas emissions per unit electrical energy generated as determined by life-cycle analysis for different electricity generation technologies. As different facilities utilizing the same energy technology may be of different designs and sizes and may be at different geographical locations, there is a range of carbon dioxide emissions associated with each technology. The table gives minimum, median, and maximum values obtained from the life-cycle analyses.

It is clear from the data in the table that all 'carbon-free' electricity sources are not created equal from an environmental standpoint, although most have a greenhouse gas footprint that is smaller than that of either natural gas or coal by an order of magnitude. Two notable exceptions to this trend are biofuels and the maximum

**Table 1.12.** Life-cycle analysis of equivalent $CO_2$ greenhouse gas emissions for different electricity-producing technologies. Data adapted from Krey *et al* (2014) and Schlömer *et al* (2014).

| | $CO_2$ emissions (g $kWh^{-1}$) | | |
|---|---|---|---|
| Energy source | Minimum | Median | Maximum |
| Hydroelectric | 1 | 24 | 2200 |
| Wind (onshore) | 7 | 11 | 56 |
| Wind (offshore) | 8 | 12 | 35 |
| Solar (photovoltaic) | 18 | 48 | 180 |
| Biofuels | 130 | 230 | 420 |
| Geothermal | 6 | 38 | 79 |
| Nuclear fission | 3.7 | 12 | 110 |
| Natural gas | 410 | 490 | 650 |
| Coal | 740 | 820 | 910 |

amount of greenhouse gases that can be produced by hydroelectric energy. Biofuels can represent a very diverse collection of different technologies, and fuels can be produced from very different feedstocks using very different technologies. A detailed analysis of different biofuels has been presented by Scharlemann and Laurence (2008). While hydroelectric facilities are generally very environmentally friendly, the discussion above has pointed out that high head facilities with large reservoirs, particularly those in tropical regions, may be a source of excessive atmospheric greenhouse gas contributions. This situation is responsible for the extreme maximum value given in the table, which shows that the adverse environmental effects of such installations can exceed those of conventional fossil fuel use. It might be noted that many of the early hydroelectric facilities were constructed at a time when the addition of significant amounts of economical electrical generating capacity was a more important consideration than environmental sustainability. It is significant that the median value of the greenhouse gas emissions for hydroelectric facilities falls well within the range of values for other renewable energy resources, which shows that hydroelectric power is, for the most part, a suitable source of renewable and sustainable electricity.

A related concept is that of embodied energy (see, for example, Thomitzek *et al* (2019)), which is the total energy required to produce a manufactured item. This includes the energy utilized in the mining and production of the component materials, as well as the energy involved in the manufacturing process itself. This concept is useful in understanding the overall energy needs for different technologies and is a useful tool for material selection and for improving manufacturing techniques. In the case of transportation, embodied energy is one factor in a comparison of the overall environmental impact of different technologies.

### 1.6.7 Utilization of carbon-free electricity

It is clear that the use of electricity generated by carbon-free technologies has increased considerably in recent years. However, it still remains a small fraction of

**Table 1.13.** Typical capacity factors for some energy technologies. The data are average values for the United States for 2019 and are from the U.S. EIA (2020a).

| Energy source | Capacity factor (%) |
|---|---|
| Hydroelectric | 41.2 |
| Wind | 34.3 |
| Solar | 24.1 |
| Biofuels | 61.0 |
| Geothermal | 69.6 |
| Nuclear fission | 93.4 |

that produced from fossil fuels. In evaluating the importance of alternative energy sources for the generation of electricity, it is important to understand the distinction between installed capacity and actual electricity generated. This is because all forms of electricity production do not produce electricity at the maximum capacity all of the time. This is particularly the case for many types of renewable energy. The ratio of the actual energy output from a device to the energy that would be produced if it operated continuously at its maximum output is referred to as the capacity factor (often quoted as a percentage). Table 1.13 gives some typical capacity factors for some carbon-free electricity generation technologies in the United States. Thus, in assessing the overall ability of an energy source to contribute to our energy needs, the capacity factor as well as the total installed capacity must be considered. Some comments, as given below, will help to put this information in context.

Although we might think of hydroelectric facility output as a constant quantity, there are actually substantial seasonal fluctuations in hydroelectric output due to fluctuations in river flow rate. At times of reduced flow, the generator facility will provides less than the maximum output. The maximum capacity of the facility can be designed to optimize the actual output, but there will always be times of less than maximum capacity, leading to a capacity factor of less than 100%. These fluctuations in river flow are more significant for run-of-the-river power stations because they have no reservoir, and this leads to a typically lower capacity factor for these facilities. As a general rule, less than half of the installed capacity of hydroelectric facilities translates into actual energy produced.

Wind velocities typically show both daily as well as seasonal variability. This leads to variability in the generator output and capacity factors of less than 100%. It is interesting to note that wind power is proportional to the cube of the wind velocity, thus it is important to design the wind turbine so that it is efficient during periods of large wind velocity, as this is the time when the most wind energy is available. The range of operating velocities for the turbine must match the wind conditions at the site in order to optimize the capacity factor. Generally, wind turbines positioned offshore have better capacity factors because the wind velocities are more consistent than they are on land and the turbine can be designed to take advantage of the prevailing wind conditions.

Photovoltaics utilize solar energy that has considerable temporal variations. There are several factors that affect the power output of a photovoltaic panel. These include:

- the latitude;
- the day of the year;
- the time of day;
- the orientation of the panel.

These factors determine the angle of incidence of sunlight on the panel as a function of time. In addition, local weather conditions can influence the amount of sunlight that is incident on the panel. Tracking collectors that follow the Sun to ensure that sunlight is incident normal to the surface of the panel improve the capacity factor but at the expense of significant additional cost and design complexity.

As illustrated in table 1.13, nuclear energy shows, by far, the greatest capacity factor. In this case the loss of capacity results largely from periods of maintenance and refueling.

The relative importance of different electricity generation technologies is, perhaps, better represented by the actual amount of electrical energy generated over the course of a year, rather than the installed capacity. The electrical energy generated is a measure of both the installed capacity and the average capacity factor. Table 1.14 shows the percentages of electricity that were generated worldwide in 2017 by different generating technologies. This table shows that fossil fuels are clearly the leading source of electricity worldwide. The relative percentages for renewable electricity and nuclear are the result of their installed capacities and their capacity factors, as described above.

It is important for the development and incorporation of new electricity generation technologies that they are, at least in the long run, economically viable. Based on an analysis of low-carbon technologies, the cost per unit electricity generated has been estimated by the United States Energy Information Administration. The results of their estimates for new facilities entering service in

**Table 1.14.** Percentages of world electricity generated by different sources in 2018. 'Other' includes tidal and solar thermal. Calculated from information available at www.iea.org/data-and-statistics?country=WORLD&fuel=Electricity%20and%20heat&indicator=ElecGenByFuel.

| Energy source | Fraction of electricity (%) |
|---|---|
| Fossil fuels | 63.9 |
| Hydroelectric | 16.2 |
| Nuclear | 10.1 |
| Wind | 4.8 |
| Solar (photovoltaics) | 2.1 |
| Biofuels | 1.9 |
| Geothermal | 0.3 |
| Other | 0.2 |

**Table 1.15.** Estimated average cost of electricity in the United States produced by various methods for new facilities entering service in 2025 (in 2019 USD). Government tax incentives are not included. Data adapted from EIA (2020b, 2020c).

| Source | Electricity cost ($ $kWh^{-1}$) |
|---|---|
| Solar (photovoltaics) | 0.036 |
| Geothermal | 0.037 |
| Wind (onshore) | 0.040 |
| Hydroelectric | 0.053 |
| Natural gas | 0.067 |
| Coal (advanced) | 0.076 |
| Nuclear fission (advanced) | 0.082 |
| Biofuels | 0.095 |
| Wind (offshore) | 0.122 |

2025 are summarized in table 1.15. Solar photovoltaics and wind (at least onshore) are seen in the table to be among the most economical electricity-producing technologies. Thus, the two renewable technologies that have shown significant growth in recent years are likely to continue to show growth.

### 1.6.8 The need for increased energy storage capacity

It is clear from the previous discussion that both wind and solar energy exhibit fluctuations on both a daily timescale and longer timescales, such as seasonal variations. Similarly, hydroelectric energy shows some seasonal fluctuations. It is clear that the widespread use of these energy sources requires careful consideration.

On a seasonal scale, average wind velocities tend to be lower in the summer (in the Northern Hemisphere) and higher in the winter. The opposite is true for solar energy. In many cases, hydroelectric energy tends to peak in late spring when river flows are at their largest due to runoff from melting snow. Thus, these natural seasonal fluctuations in renewable electricity tend to even out some of the seasonal fluctuations in capacity if a suitable mix of different energy sources is implemented.

It is, however, the short-term variations in renewable electricity that require the greatest consideration in order to fulfill our electricity needs and, in particular, to satisfy baseload requirements. Three factors also exacerbate this situation: first, as noted above, an increase in total energy consumption due to population increase and increased industrialization; second, the shift away from fossil fuels and towards renewable energy sources that will be necessitated by resource limitations and environmental factors; finally, the likelihood of required increases in grid capacity in order to accommodate changing transportation needs, as discussed in detail in chapter 6.

All of these factors point to the need to develop energy storage facilities in parallel with the development of renewable energy resources. The major grid-scale electrical storage approaches are pumped hydroelectric storage and battery storage.

Considerable pumped hydroelectric storage is already in place, and in recent years there has been a considerable expansion of battery electrical storage worldwide. Other approaches that might be applicable to some specific situations include compressed-air energy storage, superconducting magnets, and hydrogen. Compressed-air energy storage requires specific locations where substantial underground chambers are available to store compressed air; at present there are two such major facilities, one in Germany and one in the United States. Both facilities utilize the combustion of natural gas to heat the air as it expands during energy recovery and are, therefore, not carbon-free technologies. The use of superconducting magnets is becoming common as a means of grid stabilization and small-scale grid storage. Future developments in high-temperature superconducting magnet technology may make this approach compatible with large-scale grid storage. Hydrogen may be appropriate for specific applications such as electricity storage for remote locations. The use of hydrogen as an energy storage medium is discussed in detail in chapter 3. Dunlap (2020b, 2020c) recently provided a review of energy storage technologies.

## 1.7 Atmospheric carbon removal

There are two approaches that we can take in order to reduce greenhouse gas concentrations in the atmosphere. The best approach would be to eliminate the sources of anthropogenic greenhouse gases by eliminating the use of fossil fuels. This requires the total replacement of our fossil fuel energy infrastructure with carbon-free energy technologies.

Until fossil fuel use is eliminated entirely, it will be possible to capture carbon dioxide and remove it from the atmosphere. Thus far, this strategy has primarily been used to capture and sequester carbon dioxide that is produced by fossil fuel (e.g., coal-fired) power plants, or other facilities such as geothermal wells, where substantial amounts of carbon dioxide are released at a single location. More recently, some facilities have been constructed for the direct air capture of carbon dioxide. In this case, carbon dioxide that is not associated with any specific source is removed from the atmosphere. Carbon dioxide that is captured can either be sequestered by one of the methods described below or it can be utilized for the production of synthetic fuels, such as e-methanol, as described in section 3.10. In the latter case, the process returns the carbon to the atmosphere when the synthetic fuel is burned. Thus, the process is overall carbon-neutral and there is no net change to the atmospheric carbon concentration. The most important carbon sequestration methods are described below.

### 1.7.1 Ocean storage

The longevity of ocean storage is uncertain. Studies have shown that carbon dioxide dissolved in the oceans may diffuse back into the atmosphere on a timescale of a few hundred years. In addition, carbon dioxide dissolved in seawater lowers the pH, leading to ocean acidification and the accompanying environmental consequences as described above.

### 1.7.2 Underground storage

This approach requires the identification of appropriate underground caverns. One possibility is depleted oil and natural gas wells. At present, carbon dioxide is routinely pumped into existing oil wells. This provides storage for some greenhouse gas emissions but is mainly done in order to increase the efficiency of oil extraction. The longevity of carbon dioxide storage underground requires detailed investigation.

### 1.7.3 Chemical storage

Carbon dioxide may be stored in a solid form using reactions such as

$$CO_2 + CaO \rightarrow CaCO_3$$

and

$$CO_2 + MgO \rightarrow MgCO_3.$$

At present, world carbon dioxide emissions related to energy production are about 33 Gt (or $3.3\times10^{13}$ kg) per year. To sequester all of this carbon dioxide would require about 42 Gt ($1.3\times10^{10}$ $m^3$) of CaO or about 30 Gt ($8.4\times10^9$ $m^3$) of MgO. This means that about 10 $km^3$ of material would have to be processed and stored. Only about one third of carbon emissions are from power stations, so from a practical standpoint about 3 $km^3$ of storage material would be needed to cover this component of carbon emissions. This is still a huge and expensive undertaking. An additional complicating factor is that the reactions above do not proceed fast enough to be useful at room temperature. The sequestering material must be heated to fairly high temperatures, thereby increasing the complexity and cost of the process and also requiring the input of additional energy.

The final approach to dealing with atmospheric greenhouse gases is to capture them after they are released to the atmosphere. There are several potentially viable approaches for removing greenhouse gases, particularly carbon dioxide, which are already present in the atmosphere. These methods have recently been reviewed by Geden and Schenuit (2020) and are discussed briefly below.

### 1.7.4 Reforestation or afforestation

Reforestation refers to the growth of new trees in areas where the number of trees has diminished, while afforestation refers to establishing trees in areas where they previously did not grow. This may be done for the purpose of sequestering carbon dioxide from the atmosphere. Estimates of the effectiveness of this approach suggest that extensive reforestation could sequester up to 3.6 Gt $CO_2$ per year.

### 1.7.5 Humus

Humus is the component of soil that is produced by the decomposition of plant material by microorganisms. Humus is the major organic fraction of soil that is without the distinct cellular structure of plant matter. Humus is capable of storing significant quantities of carbon, but this carbon is typically released during cultivation using modern agricultural techniques. The development of appropriate agricultural methods could allow humus to retain 2 Gt–5 Gt $CO_2$ per year.

### 1.7.6 Biochar

Biochar is a high-carbon charcoal that is produced by the pyrolysis of biomass. Pyrolysis is the heating of a material in a low-oxygen or oxygen-free environment and this process produces a mixture of biochar, bio-oil, and syngas. The ratio of the three products can be varied by adjusting the heating temperature. Temperatures in the range of about 400 °C to 500 °C optimize biochar production. Syngas, which is a mixture of mainly hydrogen and carbon monoxide, may be used as a fuel for the production process. Biochar has been used for many years to increase the agricultural productivity of soil. It is also effective at sequestering carbon dioxide from the atmosphere, and recent experiments have shown that the widespread use of biochar in agriculture could remove 0.5 Gt–2 Gt of $CO_2$ per year.

The amount of carbon dioxide that can be removed from the atmosphere by these methods can be put in context by comparing it with the current world annual carbon dioxide emissions of 33 Gt $CO_2$. Thus, the amount of carbon dioxide that can be removed from the atmosphere by the methods described above would amount to a maximum of only about one third of our current emissions. It is also important to see how these emissions and the possibility of mitigating them compare with the total atmospheric carbon that is consistent with minimizing the effects of global climate change. The Paris Agreement (see UNFCCC (2015)) specified that global temperatures should not increase by more than 2 °C above pre-industrialized levels and, if possible, that the increase should be less than 1.5 °C. There have been numerous studies of the relationship between total carbon dioxide in the atmosphere and global temperatures in order to estimate the remaining atmospheric carbon budget that would likely limit global temperature increases to 1.5 °C. Kriegler *et al* (2018) have recently reviewed the literature on this topic. Remaining carbon dioxide budgets (adjusted to 2020) range from −248 Gt $CO_2$ to 752 Gt $CO_2$ in various recent studies. The negative value indicates that the total atmospheric carbon dioxide has already exceeded the amount that is consistent with the requirements of the Paris Agreement. The mean value resulting from these studies gives a remaining carbon dioxide budget of around 234 Gt $CO_2$ (adjusted to 2020), corresponding to about seven years at the current rate of emission. This is consistent with the estimated remaining carbon dioxide budget of of 202 Gt $CO_2$ (adjusted to 2020) provided by the Intergovernmental Panel on Climate Change (IPCC 2014). Clearly, continuing along our current path of extensive fossil fuel use beyond about 2030 will make it difficult to maintain atmospheric carbon dioxide at levels consistent with minimizing the adverse effects of climate change.

## 1.8 The way forward

It is clear from the above discussion that the longevity of fossil fuels is limited, but that it is essential for environmental reasons to transition to low-carbon energy sources as soon as possible. Since our overall energy consumption is expected to increase, as shown in figure 1.4, any decrease in fossil fuel energy must be more than compensated for by the implementation of alternative energy resources. Since oil, natural gas, and coal serve different purposes to some extent, their impacts on alternative energy technology development should be viewed somewhat independently. Oil, for example, is used extensively for transportation, while coal is used primarily for electricity generation, and natural gas is used for electricity generation as well as residential heating and industrial processes.

The relative importance of dealing with different fossil fuels from an environmental standpoint is summarized by the information in table 1.16. In this analysis, we have taken the idealized compositions of coal, gasoline, and natural gas to be C (carbon), $C_8H_{18}$ (octane) and $CH_4$ (methane), respectively. These fuels contain additional components, but this simple assumption gives us considerable insight into their importance for greenhouse gas emissions. The combustion processes for these three fuels are as follows:

$$\text{carbon:}\quad C + O_2 \rightarrow CO_2$$

$$\text{octane:}\quad 2C_8H_{18} + 25O_2 \rightarrow 16CO_2 + 18H_2O$$

$$\text{methane:}\quad CH_4 + 2O_2 \rightarrow CO_2 + 2H_2O.$$

The table shows the energy content of each fuel and the amount of carbon dioxide emitted per unit energy generated. Clearly, coal produces the greatest greenhouse gas emissions per unit energy, followed by gasoline and then natural gas. There is a simple intuitive way of looking at this fact. The energy generated by the combustion of these different fuels can be thought of as the sum of the energy generated by the oxidation of the carbon component plus the energy generated by the oxidation of the hydrogen component. The oxidation of carbon produces carbon dioxide, while the oxidation of hydrogen produces water. Therefore, the amount of carbon dioxide produced per unit energy is inversely related to the hydrogen-to-carbon ratio of the fuel.

As shown in table 1.16, the combustion of coal produces about 2.3 times as much carbon dioxide per unit energy generated as natural gas. This certainly provides motivation to phase out the use of coal sooner than natural gas. This has been the selling point for many governments in favor of the development of natural gas infrastructure to replace coal infrastructure. Certainly, from a greenhouse gas

**Table 1.16.** Carbon dioxide emissions per unit energy for different fossil fuels. Energy content from Dunlap (2019).

| Fuel | Ideal composition | H/C ratio | Energy content (MJ $kg^{-1}$) | Molecular mass (g $mol^{-1}$) | (kg $CO_2$)/(MJ) |
|---|---|---|---|---|---|
| Coal | C | 0:1 | 32.8 | 12 | 0.112 |
| Gasoline | $C_8H_{18}$ | 2.25:1 | 46.8 | 114 | 0.066 |
| Natural gas | $CH_4$ | 4:1 | 55.8 | 16 | 0.049 |

standpoint, this is advantageous but not as advantageous as moving to carbon-free energy sources, such as solar, wind, or even nuclear. Attempts to rationalize the continued extensive use of coal by promoting 'clean coal' technology may reduce particulate matter and sulfur dioxide pollution but have no effect on greenhouse gas emissions and are motivated by economic rather than environmental concerns.

The use of carbon-free energy sources for transportation is essential in order to eliminate the adverse effects of global climate change. This could include the use of renewable sources or nuclear power for electricity generation as well as the direct use of biofuels. However, as pointed out in chapter 6, a detailed analysis of environmental factors, as well as an evaluation of the efficiency of different transportation technologies, is essential when choosing the most viable approach.

## References

BP 2020 *Statistical Review of World Energy*, 69th edn (London: BP). Available online at: www.bp.com/en/global/corporate/energy-economics/statistical-review-of-world-energy.html

BP 2021 *Statistical Review of World Energy*, 70th edn (London: BP). Available online at: www.bp.com/content/dam/bp/business-sites/en/global/corporate/pdfs/energy-economics/statistical-review/bp-stats-review-2021-full-report.pdf

Chris857 2018 File:Bay Front Generating Station - Ashland, Wisconsin.jpg https://commons.wikimedia.org/wiki/File:Bay_Front_Generating_Station_-_Ashland,_Wisconsin.jpg

Delphi234 2014 File:Global Wind Power Cumulative Capacity.svg https://commons.wikimedia.org/wiki/File:Global_Wind_Power_Cumulative_Capacity.svg

Dunlap R A 2019 *Sustainable Energy* 2nd edn (Boston, MA: Cengage)

Dunlap R A 2020a *Renewable Energy Volume 1: Requirements and Sources* (San Rafael, CA: Morgan & Claypool Publishers)

Dunlap R A 2020b *Renewable Energy Volume 2: Mechanical and Thermal Energy Storage Methods* (San Rafael, CA: Morgan & Claypool Publishers)

Dunlap R A 2020c *Renewable Energy Volume 3: Electrical, Magnetic and Chemical Energy Storage Methods* (San Rafael, CA: Morgan & Claypool Publishers)

Dunlap R A 2021 *Energy from Nuclear Fission* (Bristol: IOP Publishing)

EIA (U.S. Energy Information Administration) 2019 *Monthly Energy Review (April 2019) DOE/EIA-0035(2019/4)*. Available online at: www.eia.gov/totalenergy/data/monthly/archive/00351904.pdf

EIA (U.S. Energy Information Administration) 2020a *Electric Power Monthly with Data for August 2020*. Available online at: www.eia.gov/electricity/monthly/current_month/october2020.pdf

EIA (U.S. Energy Information Administration) 2020b *Levelized Cost and Levelized Avoided Cost of New Generation Resources in the Annual Energy Outlook 2020*. Available online at: www.eia.gov/outlooks/aeo/pdf/electricity_generation.pdf

EIA (U.S. Energy Information Administration) 2020c *Annual Energy Outlook 2020 with Projections to 2050*. Available online at: www.eia.gov/outlooks/aeo/pdf/AEO2020%20Full%20Report.pdf

EPA 2012 *Climate Change Indicators in the United States, 2012* 2nd edn EPA 430-R-12-004

EPA 2022 United States Environmental Protection Agency: Sources of Greenhouse Gas Emissions https://www.epa.gov/ghgemissions/sources-greenhouse-gas-emissions

Fintan264 2009 File:Teesside wte.JPG https://commons.wikimedia.org/wiki/File:Teesside_wte.JPG

Geden O and Schenuit F 2020 Unconventional mitigation – carbon dioxide removal as a new approach in EU climate policy *SWP Research Paper 8* (Berlin: Stiftung Wissenschaft und Politik)

Höök M 2010 Coal and oil: the dark monarchs of global energy: understanding supply and extraction patterns and their importance for future production *Doctoral thesis* Uppsala University http://urn.kb.se/resolve?urn=urn:nbn:se:uu:diva-129295

IAEA (International Atomic Energy Agency) 2020 *Power reactor information system—Nuclear Power Capacity Trend.* Available online at: https://pris.iaea.org/PRIS/WorldStatistics/WorldTrendNuclearPowerCapacity.aspx

IEA 2020 *Statistics Report—Coal Information Overview.* Available online at: https://iea.org/reports/coal-information-overview

IEA 2022 *World Energy Outlook 2021.* Available online at: https://iea.blob.core.windows.net/assets/4ed140c1-c3f3-4fd9-acae-789a4e14a23c/WorldEnergyOutlook2021.pdf

IPCC (Intergovernmental Panel on Climate Change) 2000 *IPCC Special Report of Emission Scenarios—Summary for Policy Makers* (Geneva: IPCC) https://www.ipcc.ch/site/assets/uploads/2018/03/sres-en.pdf

IPCC (Intergovernmental Panel on Climate Change) 2014 *Climate Change 2014: Synthesis Report. Contribution of Working Groups I, II and III to the Fifth Assessment Report of the Intergovernmental Panel on Climate Change* ed Core Writing Team, R K Pachauri and L A Meyer (Geneva: IPCC)), 151 Available online at: www.ipcc.ch/site/assets/uploads/2018/05/SYR_AR5_FINAL_full_wcover.pdf

IRENA (International Renewable Agency) 2020 *Renewable Capacity Statistics 2020* (Abu Dhabi: IRENA) Available online at: www.irena.org/publications/2020/Mar/Renewable-Capacity-Statistics-2020

Krey V *et al* 2014 Annex II: metrics and methodology *Climate Change 2014: Mitigation of Climate Change. Contribution of Working Group III to the Fifth Assessment Report of the Intergovernmental Panel on Climate Change* ed O Edenhofer *et al* (Cambridge: Cambridge University Press) Available online at: www.ipcc.ch/site/assets/uploads/2018/02/ipcc_wg3_ar5_annex-ii.pdf

Kriegler E, Luderer G, Bauer N, Baumstark L, Fujimori S, Popp A, Rogelj J, Strefler J and van Vuuren D P 2018 Pathways limiting warming to 1.5 °C: a tale of turning around in no time? *Phil. Trans. R. Soc.* A **376** 20160457

Lindsey R 2022 NOAA Climate.gov: If carbon dioxide hits a new high every year, why isn't every year hotter than the last? www.climate.gov/news-features/climate-qa/if-carbon-dioxide-hits-new-high-every-year-why-isn%E2%80%99t-every-year-hotter-last

Maggio G and Cacciola G 2012 'When will oil, natural gas and coal peak?' *Fuel* **98** 111–23

NASA 2005 National Aeronautics and Space Administration, Goddard Institute for Space Studies. Science Briefs. Trial of the Century: Co-Conspirators Convicted www.giss.nasa.gov/research/briefs/2005_hansen_11

NOAA 2022 NOAA Global Monitoring Laboratory: Trends in Atmospheric Methane www.esrl.noaa.gov/gmd/ccgg/trends_ch4/

Plazak 2013 File:Top 5 Hydropower-Producing Countries.png https://commons.wikimedia.org/wiki/File:Top_5_Hydropower-Producing_Countries.png

Plazak 2015 File:Hubbert Upper-Bound Peak 1956.png https://commons.wikimedia.org/wiki/File:Hubbert_Upper-Bound_Peak_1956.png

Rfassbind 2014a File:PV cume semi log chart 2014 estimate.svg https://commons.wikimedia.org/wiki/File:PV_cume_semi_log_chart_2014_estimate.svg

Rfassbind 2014b File:Global Market Share by PV Technology from 1990 to 2013.svg https://commons.wikimedia.org/wiki/File:Global_Market_Share_by_PV_Technology_from_1990_to_2013.svg

Rfassbind 2015 File:Price history of silicon PV cells since 1977.svg https://commons.wikimedia.org/wiki/File:Price_history_of_silicon_PV_cells_since_1977.svg

Ritchie H 2020 'What are the safest and cleanest sources of energy?' *Our World in Data* 10 February 2020. Available online at: https://ourworldindata.org/safest-sources-of-energy

Ritchie H, Roser M and Rosado P 2022 Hannah Ritchie, Max Roser and Pablo Rosado (2022) - "Energy" https://ourworldindata.org/energy

Rohde R A 2007 File:Nuclear Power History.png https://commons.wikimedia.org/wiki/File:Nuclear_Power_History.png

Rubber Duck 2015 File:Hubbert curve.svg https://commons.wikimedia.org/wiki/File:Hubbert_curve.svg

Rutledge D 2011 Outlook for Long-Term Fossil-Fuel CO2 Emissions www.esrl.noaa.gov/gmd/publications/annual_meetings/2011/slides/106-110421-A.pdf

Scharlemann J P W and Laurence W F 2008 How green are biofuels? *Science* **319** 43–4

Schlömer S *et al* 2014 Annex III: technology-specific cost and performance parameters *Climate Change 2014: Mitigation of Climate Change. Contribution of Working Group III to the Fifth Assessment Report of the Intergovernmental Panel on Climate Change* ed Edenhofer *et al* (Cambridge: Cambridge University Press) Available online at: www.ipcc.ch/site/assets/uploads/2018/02/ipcc_wg3_ar5_annex-iii.pdf

Sorrell S, Speirs J, Bentley R, Brandt A and Miller R 2009 *An Assessment of the Evidence for a Near-Term Peak in Global Oil Production* (London: UK Energy Research Centre) https://ukerc.rl.ac.uk/UCAT/cgi-bin/ucat_query.pl?GoButton=DisplayLanding&ucatID=299

Sovacool B K 2009 Contextualizing avian mortality: a preliminary appraisal of bird and bat fatalities from wind, fossil-fuel, and nuclear electricity *Energy Policy* **37** 2241–8

TammyLe 2019 File:Bird's eye view of installation works at PV Arrays.png https://commons.wikimedia.org/wiki/File:Bird%27s_eye_view_of_installation_works_at_PV_Arrays.png

Thomitzek M, Cerdas F, Thiede S and Herrmann C 2019 Cradle-to-gate analysis of the embodied energy in lithium ion batteries *Proc. CIRP* **80** 304–9

UNFCCC 2015 The full text of the Paris Agreement may be found on the United Nations website at: https://unfccc.int/files/meetings/paris_nov_2015/application/pdf/paris_agreement_english_.pdf

United Nations 2019 *World Population Prospects 2019* (New York: United Nations Department of Economic and Social Affairs) Available online at: https://population.un.org/wpp/Graphs/Probabilistic/POP/TOT/900

U.S. Fish and Wildlife Services 2020 *Threats to Birds Migratory Bird Mortality—Questions and Answers*. Available online at: www.fws.gov/birds/bird-enthusiasts/threats-to-birds.php

Ypna 2013 File:Nuclear power plant construction.jpg https://commons.wikimedia.org/wiki/File:Nuclear_power_plant_construction.jpg

**IOP** Publishing

Transportation Technologies for a Sustainable Future
Renewable energy options for road, rail, marine and air transportation
**Richard A Dunlap**

# Chapter 2

## Battery electric vehicles

### 2.1 Introduction

Batteries are electrochemical devices that store energy and can be used to power a variety of electrical devices such as flashlights, notebook computers, and cell phones. The first battery was constructed in 1800 by Alessandro Volta (1745–1827). The device that Volta constructed consisted of a stack of copper and zinc plates immersed in a brine solution and was capable of producing an electric current for an extended period of time. Around the end of the 19th century the dry cell battery was developed, and this greatly increased the usefulness of batteries for portable applications. In this chapter we consider the application of batteries as a power source for road vehicles. The present treatment deals primarily with light-duty passenger vehicles; however, two-wheeled vehicles (e.g. motorcycles) as well as heavy-duty road vehicles (trucks and buses) are also discussed.

### 2.2 Primary and secondary batteries

There are many types and sizes of battery. One of the most fundamental means of categorizing different batteries is to classify them as primary batteries or secondary batteries. Primary batteries are what we usually refer to as non-rechargeable batteries and secondary batteries are what we normally refer to as rechargeable batteries. Primary batteries, such as most alkaline batteries, are in common use for personal electric and electronic devices such as flashlights, quartz watches, voice recorders, and calculators. Secondary batteries, such as many lithium-ion batteries and lead–acid batteries, are in common use for applications such as cell phones, notebook computers, and for starting gasoline and diesel vehicles. Primary batteries are not appropriate for battery electric vehicles (BEVs), and the reasons for this can be easily demonstrated. First, it would be inconvenient to have to change a vehicle's batteries when they became depleted of charge. However, a more significant reason is related to cost. We can make a rough estimate of the cost of powering an electric vehicle with primary batteries.

doi:10.1088/978-0-7503-5306-9ch2  

We can take alkaline D-cells as an example. These batteries have a capacity of 12 Ah and a voltage of about 1.5 volts, corresponding to a stored energy of 0.018 kWh. An overview of the physics of batteries that explains the significance and relationship of various battery quantities is given in appendix E. If an alkaline D-cell costs (say) 1 USD, then the cost of energy is 56 USD per kWh. This may be compared to a typical cost of (say) 0.13 USD per kWh for electricity purchased from a public utility. That is, electricity provided by an alkaline battery costs almost 500 times as much as the electricity provided by the public utility. This simple example makes it easy to see that primary batteries may be a convenient means of providing energy for small portable electrical devices, but the actual cost of that energy is extremely high. We can see how this relates to the use of primary batteries for energy storage for an electric vehicle. Using the energy cost for the D-cell along with the energy requirements for a passenger vehicle, as discussed below in section 2.5, this amounts to about $9 per km, or about 50–100 times what we would consider acceptable on the basis of current gasoline costs.

The above example illustrates the need for secondary batteries for electric vehicle use. Not only are they more convenient, but the fact that they can be recharged many times also makes them economically viable.

## 2.3 Battery chemistry

Batteries utilize the transport of charge between electrodes to store energy while charging and deliver energy while discharging. There are a number of different battery chemistries, and these utilize different ions to carry charge between the electrodes. The details of the charge and discharge processes are described in detail in the next section for one type of battery, the lithium-ion battery. In this section we look at some of the different battery chemistries that may be of relevance to the development of BEVs. As noted in the previous section, secondary or rechargeable batteries are necessary for the construction of a viable electric vehicle. Three different battery chemistries have been used for electric vehicles. These are lead–acid batteries, nickel–metal hydride batteries, and lithium-ion batteries. The major considerations for the suitability of a battery for an electric vehicle are its ability to produce sufficient power and to store sufficient energy. The first requirement determines how much power the vehicle has and (for example) the available acceleration. The second requirement determines the range of the vehicle. Of course, the total battery power and energy can be increased by increasing the number of batteries. Ideally, however, we would like to minimize the battery weight (and/or volume) while maintaining reasonable vehicle operating characteristics. Therefore, the power and energy storage per unit weight and per unit volume are important considerations when designing an electric vehicle and deciding on the type of battery to be utilized.

Table 2.1 gives a comparison of typical battery characteristics for lead–acid, nickel–metal hydride, and lithium-ion batteries. Specific energy and specific power refer to the energy and power, respectively, per unit mass of the battery. Energy density refers to the energy per unit volume. It is clear from the table that lithium-ion batteries are far superior to the other secondary batteries in the table in

**Table 2.1.** Typical values of the characteristics of some relevant secondary battery chemistries.

| Chemistry | Nominal cell voltage (V) | Specific energy (Wh $kg^{-1}$) | Energy density (Wh $L^{-1}$) | Specific power (W $kg^{-1}$) |
|---|---|---|---|---|
| Lead–acid (Pb–acid) | 2.1 | 30 | 80 | 180 |
| Nickel–metal hydride (NiMH) | 1.2 | 70 | 150 | 400 |
| Lithium-ion (Li-ion) | 3.8 | 160 | 300 | 300 |

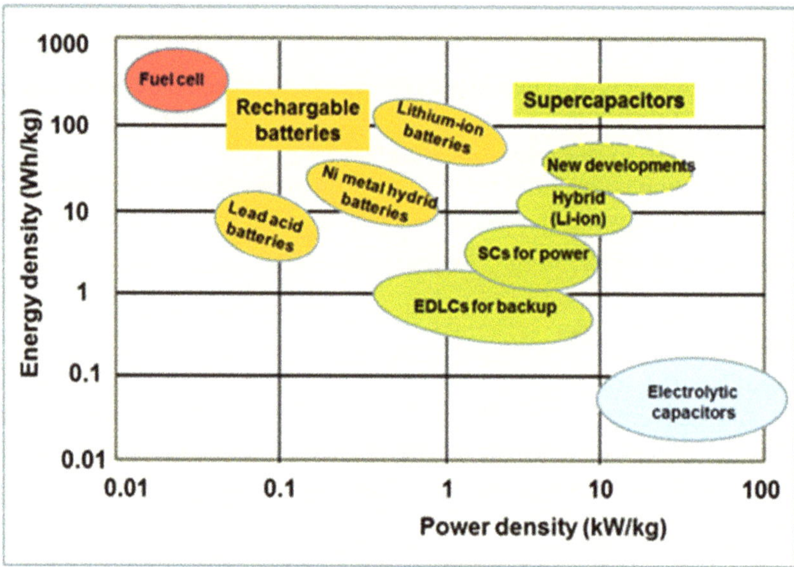


**Figure 2.1.** Ragone plot showing the relationship between specific energy (vertical axis) and specific power (horizontal axis) for different battery chemistries, supercapacitors (as discussed later in this chapter), and fuel cells (as discussed in the next chapter). This image (Elcap 2013a) has been obtained by the author from the Wikimedia website, where it is stated to have been released into the public domain. It is included within this chapter on that basis.

terms of energy per unit mass as well as energy per unit volume. However, lithium-ion batteries typically fall short of nickel–metal hydride batteries in terms of power per unit mass. As demonstrated in section 2.5, the specific energy is the more relevant parameter in terms of satisfying design criteria for a BEV.

It should be noted that the values in the table are typical values and depend considerably on the details of the battery design and the materials used for the battery construction. In order to better understand the relationship between different battery characteristics, we can construct a so-called Ragone (pronounced rə-GOH-nee) plot or diagram, as shown in figure 2.1. This plot is named after American metallurgist David V Ragone (b. 1930) and shows the relationship between specific power and specific energy. As seen in the figure, it can conveniently describe the characteristics of different battery chemistries. It can also be used to illustrate the characteristics of other energy storage media, as discussed in later chapters of this book. The figure shows the significant range of values that the energy and power characteristics of different types of battery can exhibit. This large range is particularly notable for the specific power, as shown on the horizontal axis of the figure. The figure also shows another common battery feature, that is, for a particular battery chemistry, there is typically a trade-off between energy storage capability and power availability.

The different characteristics of different battery chemistries make them suitable for different applications. A large specific energy is certainly a characteristic that is important for many applications, including BEVs. However, other considerations also influence the choice of the most appropriate battery for each use. For example, lead–acid batteries, despite their low energy density, are a cost-effective and reliable choice for applications such as starting internal combustion engines, stationary backup power, and electrical storage for renewable energy sources such as wind.

Nickel–metal hydride batteries are in common use for a variety of small electronic devices; for example, cordless telephones frequently use nickel–metal hydride batteries. They are available in several of the same standard sizes as primary alkaline batteries, which makes them a rechargeable option for many devices that normally take non-rechargeable batteries. Their somewhat lower voltage, however, may be a factor for some applications. Prior to the development of suitable cost-effective lithium-ion batteries, nickel–metal hydride batteries played a major role in the development of BEVs. Their place in BEV history is described in detail in section 2.6. In the past, nickel–metal hydride batteries were the standard energy storage medium for hybrid vehicles. For reasons described in section 2.10, they still remain a suitable choice for many hybrid vehicle applications.

Lithium-ion batteries, because of their superior specific energy, have replaced nickel–metal hydride batteries for many hybrid vehicle applications and they are currently used in the vast majority of BEV applications. However, lithium-ion batteries are more expensive than other rechargeable batteries and cost considerations may outweigh superior energy density for applications in which battery weight and/or volume is not a crucial factor.

## 2.4 Basic battery science

Today, the term 'battery' is used to refer to a unit containing one or more electrochemical cells in a suitable package with connections to allow the cell to be used to power a device. An electrochemical cell, sometimes referred to as a galvanic cell or voltaic cell (named after Luigi Galvani (1737–98) or Alessandro Volta, respectively) is an electrochemical unit with two electrodes separated by an electrolyte. The electrochemical cell is sometimes thought of as two half cells, each consisting of an electrode–electrolyte interface.

The operation of a typical secondary electrochemical cell can be understood by looking at the processes that occur while a cell is charging and discharging. Figure 2.2 shows the charging and discharging of a generic electrochemical cell. During the charge process a current source is attached to the cell, as shown in the

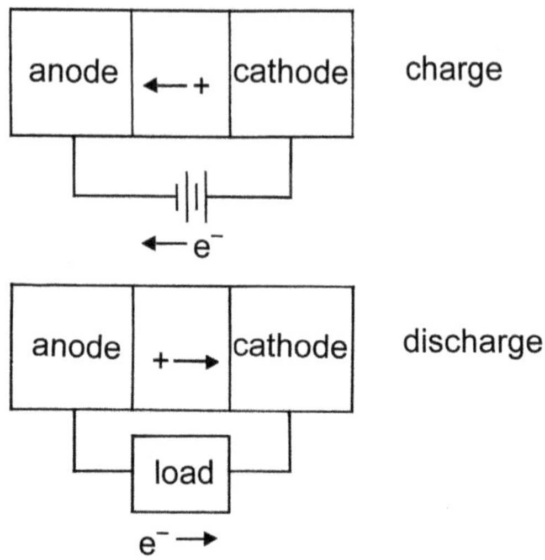


**Figure 2.2.** Charge flows in a generic secondary electrochemical cell as it is charged (top) and discharged (bottom). The negative electrode is the anode, and the positive electrode is the cathode. These definitions of anode and cathode are based on the direction of flow of the positive ions through the electrolyte and are opposite to the definitions of anode and cathode that are used for many other electronic devices in which the current is viewed as a flow of negatively charged electrons.

top part of the figure. The current source transports electrons from the positive electrode (referred to as the cathode) through the external circuit to the negative electrode (referred to as the anode). Positive ions, which are formed in the cathode by the removal of electrons, travel through the electrolyte to the anode, where they recombine with the electrons that traveled through the external circuit.

During the discharge process, a resistive load, i.e. an electrical appliance such as a bulb, heater, motor, etc. is attached across the electrodes of the cell, as shown in the bottom part of figure 2.2. During this process the electrons flowing through the external circuit are able to provide energy to the load and convert electrical energy into a form suitable for our needs, such as illuminating a light bulb, rotating a motor, or producing heat or a magnetic field.

The reason that electrical energy is stored in the battery during the charge process and recovered during the discharge process is because the overall energy of the cell is greater when the ions are combined with the electrons in the anode than when they are combined with electrons in the cathode. This change in energy occurs because the chemical compounds that are formed from the ions when they are in anode have a greater chemical potential than the compounds that are formed from the ions when they are in the cathode.

The general features, as described above, can perhaps be clarified by looking at a specific example. Since lithium-ion batteries are, at present, the most useful type of battery for the development of BEVs, we consider their operation in some detail. A variety of different materials are used for the construction of lithium-ion cells, depending on the specific design and intended application of the

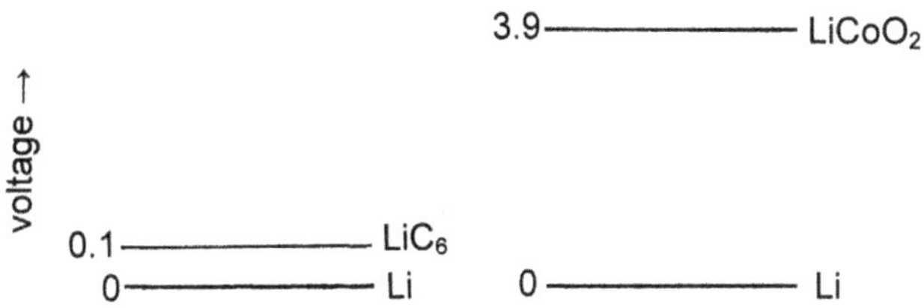


**Figure 2.3.** Chemical potential of $LiC_6$ (left) and lithium in $LiCoO_2$ (right) relative to lithium in the form of elemental lithium metal.

battery, and we consider some typical lithium-ion cell materials in the example below.

A common material that is used for the anode in a lithium-ion cell is carbon, generally in the form of layered graphite. This can readily accommodate the lithium ions between the carbon layers. A common material that is used for the cathode is a lithium metal oxide, such as $LiCoO_2$. Using these two materials as examples of lithium-ion battery electrodes, we can look in detail at how the charge and discharge processes work.

The chemical potentials of lithium in carbon ($LiC_6$) and $LiCoO_2$ are both more positive than that of elemental lithium, as shown in figure 2.3. The voltage of the cell is determined by the potential difference between the positive and negative electrodes. As seen in the figure, the potential difference between the two electrodes in the lithium-ion cell, for the materials as described above, is 3.9 V – 0.1 V = 3.8 V (recall table 2.1) and this is the voltage of the cell.

During the charging and discharging processes, lithium ions are transported between the cathode and the anode through the electrolyte, as shown in figure 2.4. The reactions that take place during the charge and discharge processes at the cathode and anode are

$$LiCoO_2 \leftrightarrow CoO_2 + Li^+ + e^- \quad \text{(cathode)} \tag{2.1}$$

and

$$C_6 + Li^+ + e^- \leftrightarrow LiC_6 \quad \text{(anode)}, \tag{2.2}$$

where, as noted above, the lithium ions travel through the electrolyte and the electrons (which were removed from the lithium atom to create the positively charged lithium ions) travel through the external circuit. In equations (2.1) and (2.2), the arrow points from left to right during the the charging process and from right to left during the discharging process. These two equations represent the reactions of the two half cells and can be combined to give the overall reaction for the complete electrochemical cell (or battery) while it is charging and discharging, as follows:

$$LiCoO_2 + C_6 \leftrightarrow CoO_2 + LiC_6.$$

Batteries that undergo these types of reaction during the charging and discharging processes are sometimes referred to as 'rocking chair' batteries, as the lithium ions travel (or rock) back and forth during cycling.

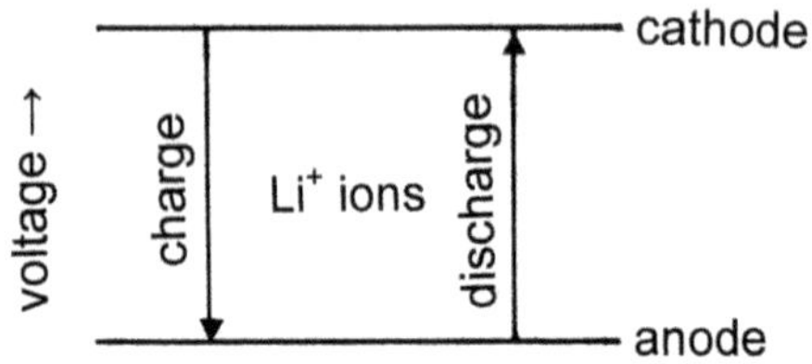


**Figure 2.4.** The flow of $Li^+$ ions through the electrolyte during the processes of charging and discharging a lithium-ion battery.

The general design of a lithium-ion battery is shown in figure 2.5. As shown in the diagram, the electrodes (as described above) are connected to conductive metal foils (called current collectors). The current collectors used are often aluminum for the cathode and copper for the anode. The electrodes are attached to the current collectors using a suitable organic binder (such as polyvinylidene fluoride). The binder is mixed with a conductive material (such as carbon) to electrically connect the electrode and current collector. The electrolyte consists of a lithium-containing salt (such as LiClO or LiPF) in a non-aqueous solution. Organic liquids such as diethyl carbonate or ethylene carbonate are typically used as the solvent. A separator, such as polypropylene or polyethylene, is used to electrically insulate the two electrode materials. The separator is a microporous film that allows the lithium ions to pass from one side to the other.

It should be noted that rechargeable battery electrodes deteriorate over time. The transfer of ions (such as lithium) in and out of the electrode material results in the formation and propagation of cracks in the material, with the result that portions of the electrode become electrically disconnected, which reduces battery capacity. Much research work has been undertaken over the years in developing new electrode materials and preparation processes to increase specific energy and to minimize electrode degradation during the charge/discharge cycle and thereby increase cycle life.

Lithium-ion batteries come in a variety of different geometries including cylindrical cells (commonly used in flashlights) and rectangular cells or flat packs (commonly used in cell phones). Battery packs for electric vehicles consist of a large number of individual batteries. For example, the 85 kWh Tesla battery pack (see the details of Tesla vehicles below) utilizes 7104 individual 18 650 lithium-ion batteries. The standard numbering scheme for cylindrical lithium batteries indicates the battery size. In this case, 18 650 refers to a battery that is 18 mm in diameter and 65.0 mm in length.

## 2.5 Battery electric vehicle characteristics

Ideally, we would like to have an electric vehicle that has characteristics similar to those of a fossil-fuel internal combustion engine vehicle. The gasoline-powered vehicles available today cover a wide range of categories, ranging from commuter vehicles designed for city use to high-performance sports cars. Today's available models of BEV also cover a wide range of categories, many of which are outlined in

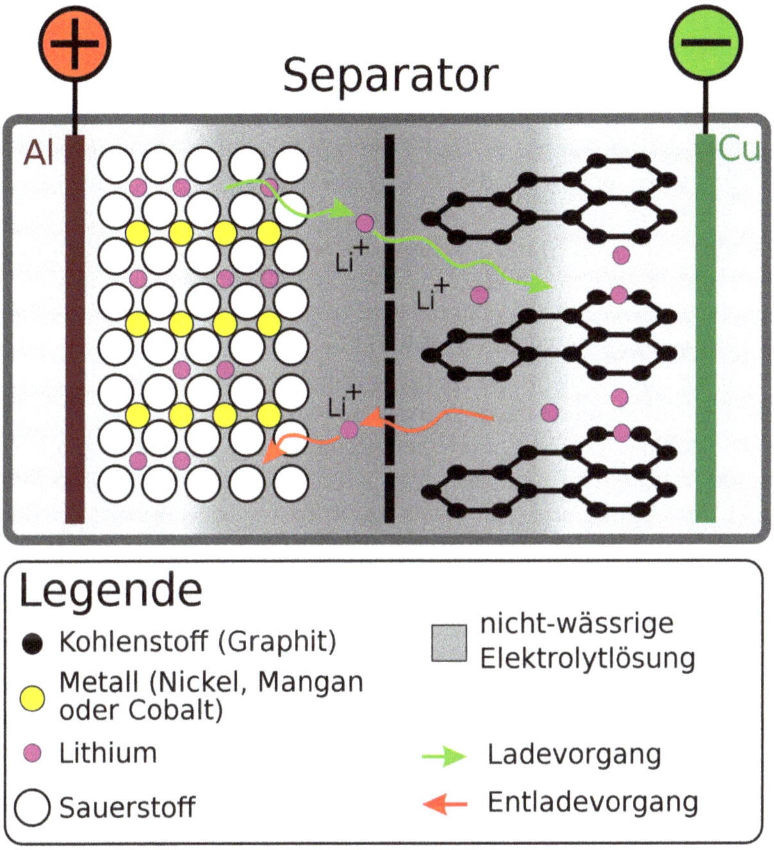


**Figure 2.5.** Schematic diagram of the construction of a typical lithium-ion battery. Kohlenstoff (graphit) = carbon (graphite); Metall (Nickel, Mangan oder Cobalt) = metal (nickel, manganese, or cobalt); Sauerstoff = oxygen; nicht-wässrige Elektrolytlösung = non-aqueous electrolytic solution; Ladevorgang = charge; Entladevorgang = discharge. This image (Cepheiden 2020) has been obtained by the author from the Wikimedia website where it was made available under a CC BY-SA 2.0 DE licence, https://creativecommons.org/licenses/by-sa/2.0/de/deed.en. It is included within this chapter on that basis. It is attributed to Cepheiden.

section 2.7. For the purpose of understanding the relationship between battery properties and electric vehicle characteristics, we consider the design of an electric vehicle that may serve as a replacement for a traditional internal combustion engine family sedan, such as a four-cylinder Toyota Camry.

According to the United States Environmental Protection Agency (EPA), a typical midsize gasoline-powered family sedan has a fuel economy of about 32 mpg (combined city/highway). In the United States, fuel economy is expressed in miles per gallon. In many other countries (such as Canada), fuel consumption is used and is expressed in liters per 100 kilometers. In the case of the midsize vehicle, fuel consumption is typically around 7.5 L/100 km. For the present purpose of comparing a gasoline-powered vehicle to a BEV, we do not consider hybrid vehicles. These are discussed in section 2.10.

From appendix D, it can be seen that the energy content of gasoline is 34.8 MJ $L^{-1}$. Since, as noted in chapter 1, a gasoline internal combustion engine is about 20% efficient, about $(0.2)\times(34.8\ \text{MJ L}^{-1}) = 7\ \text{MJ L}^{-1}$ is available for vehicle propulsion. Using the fuel consumption for the midsize gasoline vehicle, we can easily calculate the energy needed to transport the vehicle a distance of 1 km. This calculation gives $(7.5\ \text{L}/100\ \text{km}) \times (7\ \text{MJ L}^{-1}) \approx 0.5\ \text{MJ km}^{-1}$. Using the conversion factor in appendix C, this may be expressed as 140 Wh $km^{-1}$. As a typical battery/motor drive system in a BEV might be 90% efficient, the required battery energy storage capacity for an electric vehicle is $(140\ \text{Wh km}^{-1})/(0.9) = 156\ \text{Wh km}^{-1}$.

Now that we know how much energy is required to transport a typical vehicle a given distance, we can relate the performance of a BEV to the battery characteristics. Returning to table 2.1, we see that the specific energy of a lithium-ion battery is typically 160 Wh $kg^{-1}$. This means that a BEV requires about $(156\ \text{Wh km}^{-1})/(160\ \text{Wh kg}^{-1}) \approx 1.0\ \text{kg km}^{-1}$. So, each kg of lithium-ion batteries provides about 1 km of vehicle range. We can compare this simple calculation with an actual production BEV. As an example, based on EPA published information, the Tesla Model 3 has a battery-weight-to-range ratio of 0.9 kg $km^{-1}$. Of course, just as gasoline fuel economy can vary considerably for different types of vehicle and different types of driving, the range of a BEV also depends on these factors. It should also be noted that temperature plays an important role in the operation of a BEV for two reasons. First, the specific energy of most batteries is a function of temperature, since batteries supply less energy at low temperatures than they do at high temperatures. Second, air conditioning and heating, which are needed at high and low ambient temperatures, respectively, require energy, and this energy must be supplied by the batteries. While the need for air conditioning may be at least partly compensated for by the additional battery capacity at high temperatures, electric vehicles operating at low ambient temperatures suffer from the combined effects of reduced battery capacity and heating requirements. Some electric vehicle manufactures, such as Nissan and Jaguar, utilize a heat pump to provide heating when the outside temperature is low, rather than using simple resistance heating (Cooley 2021).

A typical midsize gasoline-powered passenger vehicle has a fuel capacity of about 15 U.S. gallons (about 57 L) and a corresponding range of $(15\ \text{gal})\times(32\ \text{mpg}) = 480\ \text{m} \approx 780\ \text{km}$. The range of a BEV is limited by the capacity of an acceptably sized battery. On the one hand, a large range is desirable, but on the other hand minimizing battery weight and cost is also desirable. Current technology suggests that a suitable BEV that is the best compromise for a midsize family vehicle might utilize about 500 kg of batteries, which is consistent with current high-end Tesla electric vehicles. This would give a typical range of about 500 km, somewhat less than the range of the gasoline-powered vehicle.

In addition to vehicle range, the battery size also determines the available power. Using the specific power for lithium-ion batteries from table 2.1 of 300 W $kg^{-1}$ gives an estimate of the total power available from 500 kg of batteries, which is 150 kW.

It is important to note that this is the maximum power that can be provided by the batteries. The actual peak power supplied to the wheels is determined by the properties of the electric motor and the control electronics and may be somewhat less that the maximum that can be supplied by the battery.

We can compare these rough estimates of vehicle characteristics to those of some actual midsize passenger vehicles. Table 2.2 shows several features that are typical in a comparison of current model gasoline and BEVs. Electric vehicles tend to have more available power and less range than similar gasoline vehicles. This feature is a direct result of the properties of lithium-ion batteries. Specifically, the ratio of specific power to specific energy determines the relative power-to-range characteristics of the vehicle.

When discussing vehicle performance, peak power is often used as the defining measure. This is particularly the case for automobile marketing, in which peak power (in kW or hp) is generally promoted as the best indicator of performance. However, torque (commonly measured in N·m or ft-lb) is also an important factor in vehicle performance. Although power and torque are both related to engine size, in fact, it is engine torque that determines the force between the drive wheels and the road and is directly related to the available acceleration. High torque at low engine speed is important for initial vehicle acceleration, and it is in this respect that the difference between an internal combustion engine and a battery-operated electric motor is most significant.

Figure 2.6 shows the relationship between power and torque for an internal combustion engine as a function of engine speed. Both power and torque increase with engine speed up to a maximum (in this case) at about 5000 rpm for power and about 3500 rpm for torque. Figure 2.7 shows a similar graph for an electric motor. It is clear that an electric motor with the same peak power as an internal combustion engine produces more torque. Most importantly, though, the electric motor produces its maximum torque on start-up. This torque characteristic of electric motors results in some of the very impressive acceleration statistics that are published for BEVs. It is also beneficial for the performance of battery electric/gasoline hybrids when starting in the electric mode.

Table 2.2 also shows that electric vehicles are typically more expensive than similarly sized gasoline vehicles. From a price standpoint, the Tesla Model 3 is more comparable to the BMW than to the Toyota or Honda. This difference in price is, to

**Table 2.2.** Characteristics of some current (2021) compact and midsize vehicles. Specifications are for base-model vehicles.

| Vehicle | Energy source | Weight (kg) | Power (kW) | Range (km) | Fuel consumption (L/100 km) | Base price (USD) |
|---|---|---|---|---|---|---|
| Toyota Camry | gasoline | 1495 | 151 | 811 | 7.4 | 24 970 |
| Honda Accord | gasoline | 1433 | 143 | 737 | 7.6 | 24 970 |
| BMW 330i | gasoline | 1711 | 190 | 711 | 8.3 | 41 250 |
| Tesla Model 3 | Li-ion batteries | 1612 | 211 | 423 | 1.7 equivalent | 37 990 |

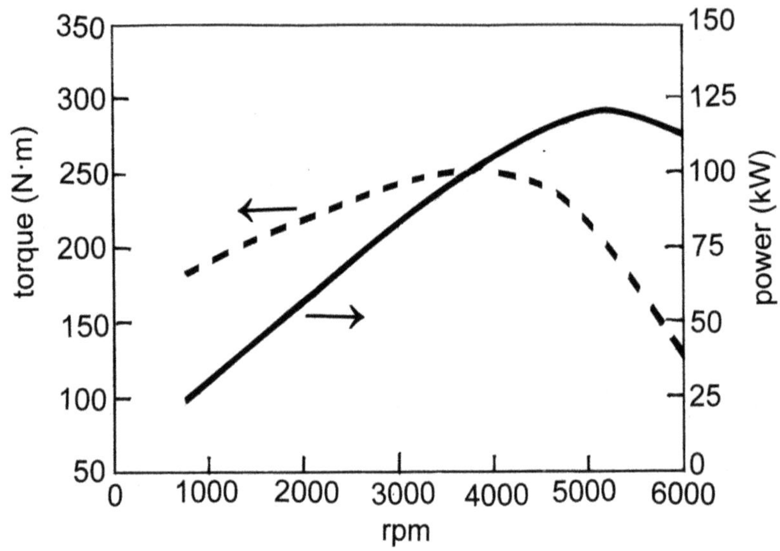


**Figure 2.6.** Typical torque (broken line) and power (solid line) curves as a function of engine speed for an internal combustion engine with a peak power of about 125 kW.

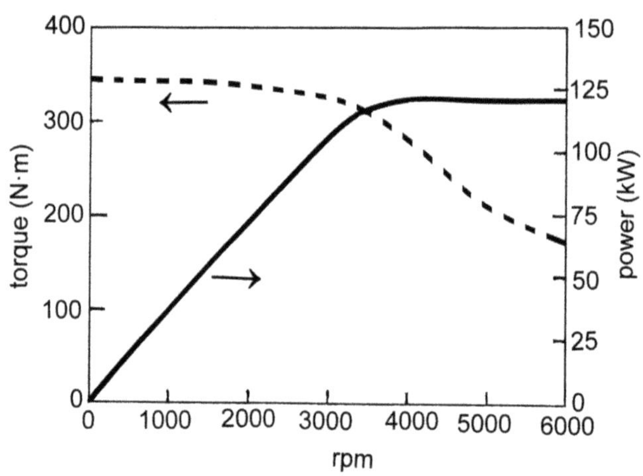


**Figure 2.7.** Typical torque (broken line) and power (solid line) curves as a function of engine speed for an electric motor with a peak power of about 125 kW.

some extent, offset by several factors. First, in many jurisdictions, government subsidies are available for the purchase of electric vehicles. Second, operating costs are lower for electric vehicles than for gasoline vehicles. This is a direct result of the efficiency of the propulsion system. A BEV is about 90% efficient at converting stored electrical energy in the battery to mechanical energy at the vehicle's wheels. By comparison, a gasoline vehicle is about 20% efficient at converting chemical energy in gasoline to mechanical energy at the vehicle's wheels. Thus, to travel the same distance, the energy provided by a battery in an electric vehicle only needs to be about 100 × (20/90) ≈ 22% of the energy provided by the gasoline in an internal combustion vehicle. This is reflected in the fuel consumption shown in table 2.2; here, the equivalent fuel consumption (based on an energy analysis) as quoted by the manufacturer of the Tesla is 1.7 L/100 km compared to an average of 7.5 L/100 km for the gasoline vehicles.

Obtaining a quantitative fuel cost analysis for a BEV and comparing it with the operating costs of a gasoline vehicle is not straightforward. The reason for this is that the costs of electricity and gasoline are not well defined. The cost of electricity is relatively stable in time but varies considerably at different locations. For example, even within the United States, recent residential electricity prices range from about 0.08 USD per kWh in the Southeast to around 0.37 USD per kWh in Hawaii. This obviously translates to a factor of four or more in operating costs for an electric vehicle, depending on location. However, for the purpose of establishing a typical operating cost of the electric vehicle, we consider an average electric cost of 0.13 USD per kWh, which is typical of much of the continental United Sates. Using the energy per km as calculated previously, this gives an average operating cost for fuel of (0.156 kWh $km^{-1}$) × (\$0.13/kWh) = 0.020 USD $km^{-1}$. Establishing a similar cost for a gasoline vehicle for comparison purposes is even more uncertain. The cost of gasoline not only varies considerably from one location to another but also fluctuates in time on a relatively short timescale. Using a gasoline cost of 4.00 USD per U.S. gallon, which has been typical for much of the United States during 2022, gives a fuel cost of (\$4.00/gal)/(32 mi $gal^{-1}$ × 1.6 km $mi^{-1}$) = 0.078 USD/$km^{-1}$. Thus, in this particular example, the fuel cost for the electric vehicle is about one quarter of that for the gasoline vehicle. In addition, the maintenance costs may, in general, be lower for electric vehicles than for internal combustion engine vehicles, but this depends greatly on vehicle design and use.

Another characteristic of BEVs that has been a concern for their widespread acceptance is charging time. In recent years, technical developments have resulted in substantial reductions in charging time. Part of this improvement has been due to the continuing advancement of battery technology. However, a major factor contributing to more rapid charging has been the development and availability of high-power charging stations. The Tesla Model 3 is a good example of the effects of charging power. A home electrical system, which is a convenient means of recharging an electric vehicle overnight, typically provides about 7.4 kW of power. Under these conditions, recharging a Tesla Model 3 Long Range version requires about 12 h (from 0% to 100% capacity). By comparison, the Tesla Supercharger (for Tesla use only), which is available at some public locations, can recharge a Model 3 (from 10% to 80% capacity) in 22 min. The availability of suitable BEV infrastructure, as discussed below in section 2.8, is an essential component of the future widespread implementation of BEVs.

## 2.6 The history of the battery electric vehicle

### 2.6.1 The development of road vehicles

We tend to think of the development of BEVs as a modern endeavor. However, BEVs have been in existence for well over a century and actually played a very important role in the widespread use of personal vehicles, particularly in North America, during the early 20th century. It is relevant to an understanding of the development of electric vehicles to begin with a brief history of powered road vehicles.

The desire for a practical source of power for the propulsion of vehicles, as well as the availability of mechanical energy for manufacturing, resulted in the development of three devices: the steam engine, the electric motor, and the internal combustion engine. The steam engine was the first of these devices to be developed. The first practical steam engine was constructed by the British inventor Thomas Newcomen (1664–1729) around 1712. The electric motor followed, with the first experimental device being constructed around 1828 by Hungarian physicist Ányos Jedlik (1800–95). The first practical electric motor followed a few years later, in 1832, and was constructed by the British scientist William Sturgeon (1783–1850). The internal combustion engine was the last of the three devices to be developed. The Belgian engineer Étienne Lenoir (1822–1900) created the first practical design around 1860.

Early devices, particularly steam engines, tended to be quite large and were best suited for stationary industrial applications. The development of motorized road vehicles not only required the availability of a suitable engine, but also the parallel development of associated steering, suspension, and braking systems. In addition, the availability of adequate road surfaces was necessary for their widespread use. An additional problem for electric vehicles was the availability of a suitable power source. The invention of the rechargeable lead–acid battery by the French physicist Gaston Planté (1834–89) in 1859 provided such a power source.

Much of the development of motorized road vehicles during the second half of the 19th century was slowed by social attitudes. The perceived dangers of motorized vehicles resulted in the implementation of strict laws governing their use in many jurisdictions. In fact, in 1865, the United Kingdom enacted legislation that required many motorized road vehicles to be preceded by a person waving a red flag and blowing a horn to alert the public. However, by the 1890s, the convenience of motorized transportation resulted in the widespread popularization of these devices. In fact, by 1900, steam-powered road vehicles, as well as electric and internal combustion automobiles, were quite common. In the early 20th century, about 40% of road vehicles in the United States were steam powered, about 38% were electric and the remaining 20% were gasoline powered. All three approaches had their own drawbacks. Steam vehicles required an extended period of time to heat the boiler before they could be operated. Electric vehicles had limited range and limited charging infrastructure (as discussed below). Gasoline vehicles were noisy, produced pollution, and were difficult to start, as the engine had to be hand-cranked. The development of the muffler (to reduce noise) and the invention of the electric starter (to eliminate hand-cranking) did much to advance the standing of the internal combustion engine vehicle. These developments, along with the mass production of the Ford Model T and the growing need for motorists to travel greater distances (at least in North America), led to the prevalence of gasoline vehicles by the end of the 1910s. Here, we look specifically at the history of electric vehicles.

### 2.6.2 Early electric vehicles

Perhaps the earliest electric road vehicle was the Flocken Elektrowagen, as shown in figure 2.8. This vehicle was constructed by Andreas Flocken (1845–1913) of Coburg, Germany in 1888. In the early years of automotive history, electric vehicles achieved a number of successes. In fact, the first road vehicle to achieve a speed in excess of 100 km $h^{-1}$ was La Jamais Contente (shown in figure 2.9). This vehicle was powered by two 25 kW direct drive electric motors and weighed 1450 kg. Driven by the Belgian driver Camille Jenatzy, La Jamais Contente established a land speed record of 105.9 km $h^{-1}$ at Achères, Yvelines near Paris, France on 29 April 1899.

In the early years of the 20th century, electric vehicles achieved considerable commercial success. The Detroit Electric, as shown in figure 2.10, manufactured by the Anderson Electric Car Company in Detroit, Michigan, was a popular electric vehicle during that era. Between 1907 and 1939 the Anderson Electric Car Company manufactured about 13 000 electric road vehicles.

By the 1920s many of the earlier disadvantages of internal combustion vehicles no longer existed and driver expectations for the range of personal vehicles had changed. As a result, internal combustion vehicles became the dominant type of

**Figure 2.8.** An 1888 Flocken Elektrowagen, reconstructed in 2011. This image (Haag Franz 2011) has been obtained by the author from the Wikimedia website where it was made available under a CC BY-SA 3.0 licence, https://creativecommons.org/licenses/by-sa/3.0/deed.en. It is included within this chapter on that basis. It is attributed to Franz Haag.

**Figure 2.9.** La Jamais Contente. This image (Afernand74 2011) has been obtained by the author from the Wikimedia website where it was made available under a CC BY-SA 3.0 licence, https://creativecommons.org/licenses/by-sa/3.0/deed.en. It is included within this chapter on that basis. It is attributed to Afernand74.

road vehicle and electric vehicles (as well as steam-powered vehicles) essentially disappeared from the roads.

### 2.6.3 The Henney Kilowatt

In the late 1950s there was a revival of interest in electric vehicles. The first commercial electric vehicle from this modern era was the Henney Kilowatt (figure 2.11). This vehicle was produced in the United States, with much of the bodywork based on the French Renault Dauphine. The Kilowatt used lead–acid batteries and had a 5.2 kW electric motor (about 7 hp). It had a range of up to 100 km and a top speed of 97 km $h^{-1}$. At a price of about 3600 USD, it was substantially more expensive than contemporaneous gasoline vehicles, which had much superior performance. Consequently, the Henney Kilowatt was not a commercial success and was discontinued after only 47 units were sold between 1959 and 1960.

### 2.6.4 The General Motors EV1

Research into electric vehicle development continued from the 1960s to the 1990s. A substantial number of automobile companies, as well as other businesses and organizations, were involved in electric vehicle development. In the 1970s, the global oil crisis motivated this work, and by the late 1980s, concerns over air pollution in cities as well as a growing awareness of global climate change caused by

**Figure 2.10.** A 1911 Detroit Electric. This image (Cullen328 2016) has been obtained by the author from the Wikimedia website where it was made available under a CC BY-SA 3.0 licence, https://creativecommons.org/licenses/by-sa/3.0/deed.en. It is included within this chapter on that basis. It is attributed to Cullen328.

carbon emissions added to the motivation for electric vehicle development. During this time a number of experimental vehicles were produced; some were purpose-designed vehicles, but many were based on existing gasoline vehicle platforms. Very little commercial success was achieved during this period because of the high cost and short range of the vehicles.

The first modern attempt to mass produce electric vehicles came in the late 1990s and was largely the result of the clean air legislation passed in California in the early 1990s, which required major automobile manufacturers to develop zero-emission vehicles in order to continue marketing their fossil-fuel vehicles in the state. During the 1990s, Chrysler, Ford, Honda, Nissan, General Motors, and Toyota were all involved in the development of limited-production electric vehicles. The General Motors EV1 and the Toyota RAV4 EV were the most notable vehicles to emerge from this era of electric vehicle development and are described briefly below.

The General Motors EV1, shown in figure 2.12, was a purpose-designed BEV produced from 1996 to 1999. EV1 vehicle specifications changed over time; the most notable change was the replacement of the lead–acid batteries used in early versions of the vehicle with NiMH batteries in later versions. Some of the specifications of the EV1 are shown in table 2.3a.

**Figure 2.11.** A 1960 Henney Kilowatt. This image (DRoberson 2005) has been obtained by the author from the Wikimedia website, where it is stated to have been released into the public domain. It is included within this chapter on that basis.

**Figure 2.12.** A General Motors EV1, produced from 1996 to 1999. This image (Rowen R 2002) has been obtained by the author from the Wikimedia website where it was made available under a CC BY-SA 2.0 licence, https://creativecommons.org/licenses/by-sa/2.0/deed.en. It is included within this chapter on that basis. It is attributed to RightBrainPhotography (Rick Rowen).

**Table 2.3a.** Specifications of early and late model General Motors EV1s.

| Property | Early model | Late model |
|---|---|---|
| Weight (kg) | 1400 | 1319 |
| Range (km)* | 89 | 169 |
| Batteries | Lead-acid | NiMH |
| Battery capacity (kWh) | 16.5 | 26.4 |
| Maximum speed (km $h^{-1}$)** | 128 | 128 |
| Power (kW) | 102 | 102 |

* Based on 2019 EPA evaluation.
** Electronically limited.

**Table 2.3b.** Specifications of the first-generation Toyota RAV4 EV.

| Property | Value |
|---|---|
| Weight (kg) | 1560 |
| Range (km) | 153 |
| Batteries | NiMH |
| Battery capacity (kWh) | 27.4 |
| Maximum speed (km $h^{-1}$)* | 126 |
| Power (kW) | 50 |

* Electronically limited.

The EV1 was never offered for sale to the public but was offered for lease at selected locations in California, Arizona, and Georgia. The monthly lease price was between $399 and $549 and was based on an assumed selling price of $34 000. Estimates of the actual cost to General Motors to produce an EV1 were in the range of $80 000 to $100 000.

Over the years 1996–1999, 1117 EV1s were produced. At the end of their lease periods, they were repossessed by General Motors and were not offered to former lessees for purchase (as is typical for most automobile leasing programs). Most EV1s were crushed (see figure 2.13) or shredded. About 40 ex-lease EV1s were donated to museums or educational institutions, but virtually all had their drivetrains disabled so they could no longer be driven.

General Motors' approach to the EV1 has often been seen as less than enthusiastic. It has even been speculated that the company used the lack of financial success of the EV1 project and its ultimate demise, along with the destruction of the vehicles, as evidence that that it was not possible for automobile manufacturers to comply with California's zero-emission vehicle legislation.

### 2.6.5 The Toyota RAV4 EV (first-generation)

The Toyota RAV4 EV, as shown in figure 2.14, was produced from 1997 to 2003. A substantially redesigned second-generation RAV4 EV was produced from 2012 to 2014 in collaboration with Tesla Motors. The present discussion refers to the first-generation vehicle, and table 2.3b summarizes the specifications of this vehicle.

**Figure 2.13.** Crushed General Motors EV1s. This image (Plug In America 2010) has been obtained by the author from the Wikimedia website where it was made available under a CC BY-SA 2.0 licence, https://creativecommons.org/licenses/by-sa/2.0/deed.en. It is included within this chapter on that basis. It is attributed to Plug In America.

**Figure 2.14.** A Toyota RAV4 EV produced between 1997 and 2003. This image (Ortiz R MD 2012b) has been obtained by the author from the Wikimedia website where it was made available under a CC BY-SA 3.0 licence, https://creativecommons.org/licenses/by-sa/3.0/deed.en. It is included within this chapter on that basis. It is attributed to Mariordo - Mario Roberto Durán Ortiz.

Unlike the General Motors EV1, the Toyota RAV4 EV was based on an existing gasoline-powered vehicle and utilized many of the same basic components. The RAV4 EV was first made available in 1997 for lease to businesses in California. In March 2002, Toyota made the vehicle available for lease or purchase by individuals. The manufacturer's suggested retail price for the RAV4 EV was $42 000. In total, 1484 vehicles were leased or sold, all in California.

Although a small number of BEVs were commercially produced prior to the end of the 20th century, automobile manufacturers were primarily motivated to pursue development in this field in order to comply with government regulations (primarily in the state of California), rather than for the purpose of economic reward. This approach was the result of continued low petroleum prices in the United States and the corresponding minimal interest in expensive and inconvenient electric vehicles taken by the American public.

After the turn of the century, interest in BEVs grew. This was due to a number of factors, as follows:

- increasing gasoline prices in many parts of the world;
- a growing awareness of the hazards associated with long-term fossil-fuel use;
- developments in lithium-ion battery technology and accompanying lower costs;
- improvements in microelectronic control devices;
- improved economic viability for manufacturers to produce electric vehicles;
- the realization by automobile manufacturers of the long-term benefits of electric vehicle development.

It was therefore in the early 2000s that automobile manufacturers undertook serious efforts to develop practical and cost-effective BEVs for consumer use. Several approaches to producing BEVs that could effectively fit into the existing automobile market included:

- vehicles designed for urban use;
- vehicles designed to replace traditional internal combustion engine vehicles;
- vehicles designed for high-end performance.

We look at the development and resulting specifications of three BEVs that were developed and marketed between about 2000 and 2010 that nicely exemplify these three categories, namely, the Reva, the Nissan Leaf, and the Tesla Roadster.

### 2.6.6 The REVA/REVAi

Many of the BEVs developed in the early 2000s were designed for urban use. These vehicles, sometimes referred to as *neighborhood electric vehicles* (NEVs), had minimal power (and top speed) and relatively short ranges. They were primarily marketed in Europe, Asia, and South America and because of their limited power were exempt from certain automobile regulations in many countries.

The Reva Electric Car Company (which became Mahindra Electric Mobility Limited in 2010) is an Indian company based in Bangalore founded in 1994 for the purpose of developing an electric road vehicle. The REVA (an acronym for 'Revolutionary Electric Vehicle Alternative') was first sold in 2001. It was replaced with the REVAi (see figure 2.15) in 2008. Table 2.4a provides the basic specifications of the REVAi. It is interesting to note that the REVA and REVAi utilized lead–acid batteries, which accounted for 270 kg (or about 40%) of the vehicle's weight. At the time, this was adequate for the vehicle's design objective, which was a very small urban vehicle with limited power and range. More than half of the REVA and REVAi vehicles that were produced during the 2000s were sold in India. The major export market was the United Kingdom, where it was marketed as the REVA G-Wiz and where the base price was around £10 000. In 2009, Reva marketed a

**Figure 2.15.** This image (Reva Norge 2009) has been obtained by the author from the Wikimedia website where it was made available under a CC BY-SA 2.0 licence, https://creativecommons.org/licenses/by-sa/2.0/deed.en. It is included within this chapter on that basis. It is attributed to RevaNorge.

**Table 2.4a.** Specifications of the REVAi.

| Property | Value |
|---|---|
| Weight (kg) | 665 |
| Range (km) | 80 |
| Batteries | Lead-acid |
| Battery capacity (kWh) | 9.6 |
| Maximum speed (km $h^{-1}$) | 80 |
| Power (kW) | 13 |

**Table 2.4b.** Specifications of the 2011 Nissan Leaf.

| Property | Value |
|---|---|
| Weight (kg) | 1521 |
| Range (km) | 117 |
| Batteries | Lithium-ion |
| Battery capacity (kWh) | 24 |
| Maximum speed (km $h^{-1}$) | 150 |
| Power (kW) | 80 |

lithium-ion battery version of the REVA, the REVA L-ion. The implementation of lithium-ion batteries reduced the vehicle weight to 565 kg and increased the range to 120 km. Since the company became the Mahindra Electric Mobility Company, it has marketed a lithium-ion battery vehicle, the Mahindra e2o.

### 2.6.7 The Nissan Leaf

The Nissan Leaf was introduced in 2010 (model year 2011) and was the first mass-produced BEV aimed at a mainstream market (see figure 2.16). It was a five-door hatchback with overall dimensions similar to those of contemporaneous compact internal combustion engine vehicles. Its batteries were charged through an electrical socket located on the front of the vehicle, as shown in figure 2.17. The specifications of the first-generation vehicle as introduced in 2010 are shown in table 2.4b. Its driving characteristics were designed to be similar to those of compact vehicles, such as the Toyota Corolla, except for the limited range. Later first-generation Leafs had the range increased to 172 km, largely due to an increased capacity battery (30 kWh). A second-generation Leaf was introduced in 2017 with significantly improved power and range. A major concern with first-generation Leafs was the lack of options (particularly outside of Japan) for refurbishing or replacing batteries that had reached the end of their life. Over 500 000 Nissan Leafs have been sold since its inception and it was the best-selling electric vehicle until its annual sales were surpassed by the Tesla Model 3 in 2020.

### 2.6.8 The Tesla Roadster

The first-generation Tesla Roadster was developed in the 2000s and utilized the greater power that is typically available for BEVs (compared to gasoline internal combustion engine vehicles) to produce a sports car, for which performance was the primary consideration and cost was somewhat secondary. The first-generation Tesla Roadster is shown in figure 2.18 and was produced from 2008 to 2012. It was the first serial production BEV to utilize lithium-ion batteries and was based on a Lotus Elise gasoline vehicle platform. Lotus supplied gliders (i.e. complete vehicles without the powertrain), into which Tesla installed the battery electric drive system. The specifications of the base-model first-generation Tesla Roadster are given in table 2.5.

The Tesla Roadster had much superior performance and range compared with the Nissan Leaf. However, the Tesla Roadster was aimed at a much more limited

**Figure 2.16.** A first-generation Nissan Leaf manufactured between 2010 and 2017 (image: Richard A Dunlap).

**Figure 2.17.** Charging socket on the front of a Nissan Leaf (image: Richard A Dunlap).

**Figure 2.18.** A first-generation Tesla Roadster. This image (IFCAR 2011) has been obtained by the author from the Wikimedia website, where it is stated to have been released into the public domain. It is included within this chapter on that basis..

**Table 2.5.** Specifications of the 2008 Tesla Roadster.

| Property | Value |
|---|---|
| Weight (kg) | 1305 |
| Range (km) | 393 |
| Batteries | Lithium-ion |
| Battery capacity (kWh) | 53 |
| Maximum speed (km $h^{-1}$) | 201 |
| Power (kW) | 185 |

market, that is, high-end performance sports cars. This difference was clearly reflected in the sales price (roughly three times that of the Leaf) and the total production, approximately 2450 units over five years. However, the technology developed for the Tesla Roadster had a significant impact on the subsequent development of BEVs, as described below.

## 2.7 Current battery electric vehicles

During the past decade there has been significant development and marketing of BEVs. At present there are more than 50 different makes of battery electric road vehicle that are available to the general public worldwide. These vehicles more or less follow the trends established in previous years for the different categories of BEV. Specifically, most of the currently available vehicles fall into two categories:

NEVs and highway-capable electric vehicles. The following overview considers a few of the many BEVs currently available in these two categories.

### 2.7.1 Neighborhood electric vehicles

NEVs, as they are commonly known in North America, or quadricycles, as they are known in Europe, are small, lightweight electric vehicles designed for urban use. They follow the same lines as the Reva concept described above. European quadricycles are categorized as either light quadricycles or heavy quadricycles. This categorization is based largely on top speed, where 45 km $h^{-1}$ is the dividing value between the two categories.

NEVs are legal in certain jurisdictions, and their licensing and use are governed by the specific vehicle properties. In general, since they are designed for urban use, the available power, top speed, and range are not major factors. In addition, recharge time is not necessarily an important consideration, as they are most often recharged overnight.

Two notable NEVs are the Citroën Ami and the Renault Twizy, both of which have enjoyed a degree of popularity in Europe. These are shown in figures 2.19 and 2.20, respectively, and a comparison of their specifications is given in table 2.6.

**Figure 2.19.** A Citroën Ami. This image (Ibex73 2020) has been obtained by the author from the Wikimedia website where it was made available under a CC BY-SA 4.0 licence, https://creativecommons.org/licenses/by-sa/4.0/deed.en. It is included within this chapter on that basis. It is attributed to Ibex73.

**Figure 2.20.** A Renault Twizy. This image (Jwh 2013) has been obtained by the author from the Wikimedia website where it was made available under a CC BY-SA 3.0 LU licence, https://creativecommons.org/licenses/by-sa/3.0/lu/deed.en. It is included within this chapter on that basis. It is attributed to Jwh.

**Table 2.6.** A comparison of the specifications of the Citroën Ami and Renault Twizy Urban 80 NEVs. The Citroën has been available since 2020, and the Renault has been available since 2012.

| Maker | Model | Classification | Weight (kg) | Power (kW) | Top speed (km $h^{-1}$) | Battery capacity (kWh) | Range (km) |
|---|---|---|---|---|---|---|---|
| Citroën | Ami | Light quadricycle | 485 | 6 | 45 | 5.5 | 75 |
| Renault | Twizy Urban 80 | Heavy quadricycle | 474 | 13 | 80 | 6.1 | 90 |

### 2.7.2 Highway-capable electric vehicles

Highway-capable electric vehicles are larger than NEVs and are more compatible with the typical concept of an automobile. They generally have top speeds in excess of 100 km $h^{-1}$. Numerous examples of highway-capable BEVs are available worldwide, and these cover a wide range of styles and costs. Table 2.7 provides information on a number of popular vehicles that are currently available. The

**Table 2.7.** Specifications of some currently available highway-capable BEVs, in order of increasing electric motor power. In many cases, the top speed is electronically limited. The range figure provided is the estimate provided by the US EPA except where noted, as follows: (1) Worldwide Harmonized Light Vehicle Test Procedure (WLTP), (2) manufacturer's data, (3) New European Driving Cycle (NEDC). Note: EPA ranges tend to be lower than many other estimates, particularly those provided by manufacturers.

| Make | Model | Category | First year available | Weight (kg) | Power (kW) | Top speed (km $h^{-1}$) | Battery capacity (kWh) | Range (km) |
|---|---|---|---|---|---|---|---|---|
| Smart | EQ forfour | City car | 2007 | 900 | 55 | 125 | 17.6 | 93 |
| Renault | Zoe | Supermini | 2012 | 1468 | 100 | 135 | 52 | 395(1) |
| BMW | i3 | Supermini luxury | 2013 | 1195 | 130 | 150 | 42.2 | 246 |
| Chevrolet | Bolt EV | Subcompact | 2016 | 1616 | 149 | 150 | 66 | 383 |
| Volkswagen | ID.3 | Subcompact | 2019 | 1600 | 150 | 160 | 77 | 550(1) |
| Nissan | Leaf | Compact | 2010 | 1640 | 160 | 150 | 62 | 364 |
| BYD | e6 | Compact crossover | 2009 | 2020 | 200 | 160 | 80 | 400(2) |
| Polestar | 2 | Compact executive | 2020 | 2140 | 300 | 205 | 78 | 375 |
| Xpeng | P7 | Midsize | 2020 | 2060 | 316 | 170 | 80.9 | 562(3) |
| Tesla | Model 3 | Compact executive | 2017 | 1847 | 340 | 261 | 82 | 504 |
| Ford | Mustang Mach-E | Compact crossover | 2021 | 1993 | 358 | 198 | 88 | 402 |
| Tesla | Model S | Full-size luxury | 2012 | 2250 | 580 | 249 | 100 | 647 |

specifications (or names) of vehicles have often changed over the years, and the information in the table is for the most recent (as of 2021) version. Also, in many cases, manufacturers offer more than one battery and motor option for a certain vehicle model. The following table gives the specifications for the version with the greatest electric motor power and/or the greatest range. Representative vehicles are shown in figures 2.21–2.25.

It is clear that the Smart EQ forfour is intended for short trips because of its range and motor power. This is a direct result of the fairly low battery capacity. All other vehicles in the table have motor powers that are similar to, or greater than, the typical family-size gasoline-powered passenger automobiles shown in table 2.2. However, as expected for BEVs, the range is typically around 400–500 km, compared to the generally greater range (700–800 km) of gasoline vehicles.

### 2.7.3 Two-wheeled battery electric vehicles

Two-wheeled BEVs begin at the low-power end of the spectrum with what are now referred to as e-bikes. These basically consist of a bicycle with an added electric motor. They can be propelled by pedaling or by the electric motor, or by some combination of both, depending on the particular design. In a sense, as they use two

**Figure 2.21.** Smart EQ forfour. This image (Migl A 2019) has been obtained by the author from the Wikimedia website where it was made available under a CC BY-SA 4.0 licence, https://creativecommons.org/licenses/by-sa/4.0/deed.en. It is included within this chapter on that basis. It is attributed to Alexander Migl.

**Figure 2.22.** Renault Zoe. This image (DaMastah 2020) has been obtained by the author from the Wikimedia website where it was made available under a CC BY-SA 4.0 licence, https://creativecommons.org/licenses/by-sa/4.0/deed.en. It is included within this chapter on that basis. It is attributed to DaMastah.

**Figure 2.23.** Polestar 2. This image (Migl A 2019a) has been obtained by the author from the Wikimedia website where it was made available under a CC BY-SA 4.0 licence, https://creativecommons.org/licenses/by-sa/4.0/deed.en. It is included within this chapter on that basis. It is attributed to Alexander Migl.

**Figure 2.24.** Xpeng P7. This image (Jengtingchen 2021) has been obtained by the author from the Wikimedia website where it was made available under a CC BY-SA 4.0 licence, https://creativecommons.org/licenses/by-sa/4.0/deed.en. It is included within this chapter on that basis. It is attributed to Jengtingchen.

**Figure 2.25.** Ford Mustang Mach-E (GT version). This image (Migl A 2021) has been obtained by the author from the Wikimedia website where it was made available under a CC BY-SA 4.0 licence, https://creativecommons.org/licenses/by-sa/4.0/deed.en. It is included within this chapter on that basis. It is attributed to Alexander Migl.

propulsion systems, they can be referred to as electric–human hybrids. E-bikes typically have battery capacities in the range of 0.3 kWh to 0.5 kWh and motors with maximum outputs of 200 W to 500 W.

Two-wheeled BEVs intended to replace gasoline-powered two-wheeled vehicles typically fall into two categories: scooters and motorcycles. The distinction between these two categories is not always clear. Typically, scooters have less power and smaller wheel diameters. However, the usual distinguishing feature of scooters is that they have a step-through frame. At present, there are a large number of different battery electric scooters and motorcycles available. Some of these are manufactured by companies that are traditionally associated with gasoline-powered two-wheeled vehicles, while others are manufactured by companies that specialize in electric vehicles. Table 2.8 gives the specifications of some currently available battery electric scooters and motorcycles. Figure 2.26 shows a typical electric scooter, and figure 2.27 shows a typical electric motorcycle. Electric scooters such as the Honda and Vespa have low top speeds, and are, therefore, not highway capable. They also have limited range and are primarily designed for city transportation. Battery electric motorcycles have characteristics that are similar to those of their gasoline-powered counterparts and are designed for highway use and long-distance travel. An interesting aspect of electric two-wheeled vehicles is the energy required per kilometer of travel, as shown in the table. This is calculated as the simple ratio of total battery capacity divided by total range (and converted into MJ km$^{-1}$). Scooters require less energy than motorcycles, largely because they are lighter and travel at lower speeds. The values in the table can be compared to the energy per kilometer as calculated previously for passenger automobiles, which is about 0.5 MJ km$^{-1}$.

**Table 2.8.** Specifications of some two-wheeled BEVs (i.e. scooters and motorcycles). Vehicles are listed in order of increasing top speed.

| Make | Model | Weight (kg) | Power (kW) | Battery capacity (kWh) | Range (km) | Top speed (km $h^{-1}$) | Energy per km (MJ $km^{-1}$) |
|---|---|---|---|---|---|---|---|
| Honda | PCX electric | 144 | 4.2 | 2.06 | 41 | 60 | 0.18 |
| Vespa | Elettrica | 130 | 4 | 3.95 | 100 | 70 | 0.14 |
| Alrendo | TS Bravo | 249 | 58 | 14.3 | 250 | 134 | 0.21 |
| Harley-Davidson | LiveWire | 250 | 78 | 15.5 | 234 | 178 | 0.24 |
| Zero | SR/F | 220 | 82 | 14.4 | 260 | 200 | 0.20 |

**Figure 2.26.** Vespa elettrica battery electric scooter. This image (Triplecaña 2019) has been obtained by the author from the Wikimedia website where it was made available under a CC BY-SA 4.0 licence, https://creativecommons.org/licenses/by-sa/4.0/deed.en. It is included within this chapter on that basis. It is attributed to Triplecaña.

### 2.7.4 Battery electric trucks and buses

The requirements of battery electric commercial vehicles, such as trucks and buses, are somewhat different than for light-duty road vehicles. In the case of trucks, sufficient power is needed because of the significant vehicle weight. In addition, a sufficiently long range is needed for many trucking applications. In contrast to the case of automobiles, in the case of trucking, battery weight and volume are less

**Figure 2.27.** Alrendo TS Bravo battery electric motorcycle. This image (Cozmcrae 2020) has been obtained by the author from the Wikimedia website where it was made available under a CC BY-SA 4.0 licence, https://creativecommons.org/licenses/by-sa/4.0/deed.en. It is included within this chapter on that basis. It is attributed to Cozmcrae.

important factors in the vehicle design. As noted previously for battery electric light-duty vehicles, batteries with an energy storage capacity that is sufficient to provide a suitable range are typically able to provide more than ample power. In fact, this available power is convenient for hill climbing with a heavy load and is an important advantage of electric trucks compared to their diesel-powered counterparts.

Battery electric trucks are at an earlier stage of development and implementation than battery electric automobiles. However, a number of manufacturers are now in the process of making battery electric trucks available. Figure 2.28 shows a typical example, the Tesla Semi.

Battery electric buses designed for long-distance travel share many of the same challenges as battery electric trucks. Electric buses for city use have a greater number of design options. The shorter required range and the potential ability to use external power sources can reduce the battery capacity requirements or eliminate the need for batteries altogether, as discussed further below in the sections on electric roads and supercapacitors. In any case, battery electric buses have become popular in recent years. An example is illustrated in figure 2.29. As of 2019, more than 425 000 battery

**Figure 2.28.** Battery electric Tesla Semi. Reprinted from Jurvetson (2019) Flickr (CC-BY 2.0).

**Figure 2.29.** Battery electric Proterra BE35 bus operated by San Joaquin Regional Transit District (San Joaquin, California) at a fast-charging station. This is one of about 300 battery electric buses in use in the United States. This image (SanJoaquinRTD 2013) has been obtained by the author from the Wikimedia website where it was made available under a CC BY-SA 3.0 licence, https://creativecommons.org/licenses/by-sa/3.0/deed.en. It is included within this chapter on that basis. It is attributed to SanJoaquinRTD.

electric buses are on the roads worldwide. Ninety-nine percent of these are operated in China, where they make up about 17% of the total number of buses.

## 2.8 Battery electric vehicle infrastructure

In recent years, our awareness of the need to deal with global climate change has motivated the development and marketing of BEVs. Figure 2.30 shows the substantial increase in the number of registered battery electric road vehicles during the past decade, although the current number of BEVs (around seven million) is still only a very small fraction of the estimated 1.4 billion road vehicles worldwide.

For any transportation technology to be successfully implemented, it is essential that the appropriate infrastructure should be in place. Since gasoline and diesel-powered vehicles first became commonplace, more than a century ago, an extensive infrastructure of fueling stations has developed. The successful implementation of electric vehicle technology requires an analogous infrastructure to be established. At present, the established approach to BEV infrastructure is the use of charging stations. Charging stations generally fall into three categories: private slow chargers, public slow chargers, and public fast chargers. We look briefly at each of these categories followed by a rough evaluation of charging times under different circumstances.

### 2.8.1 Private slow chargers

Private slow chargers operate on AC voltage and are usually facilities that are provided by the vehicle owner using their home electrical system. They may also be facilities provided by companies for employees. Since batteries must be charged with

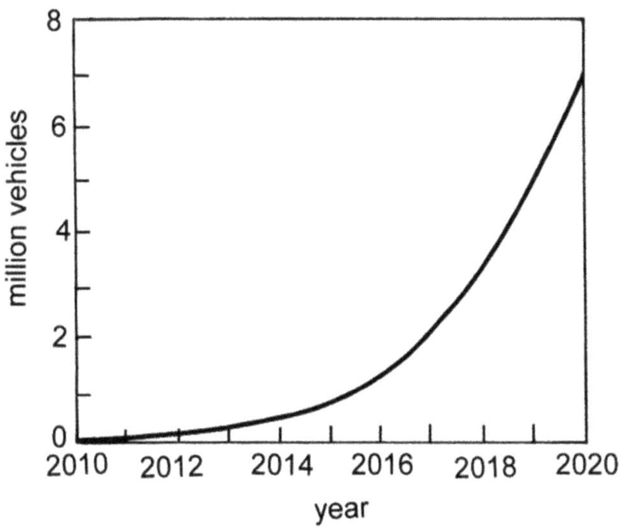


**Figure 2.30.** Number of BEVs worldwide as a function of time between 2010 and 2020. Note that many sources of similar information report the number of BEVs and plug-in hybrids (see section 2.11) combined.

DC current, home private chargers that produce AC current utilize electronics that are integrated into the vehicle (i.e. an AC/DC converter or rectifier) to change the AC current to DC in order to supply power to the battery. The simplest chargers in North America operate on 120 V residential electric power and plug into a standard three-prong grid-connected electrical socket. They typically provide up to about 2 kW and are sometimes referred to as Level 1 chargers. So-called Level 2 chargers operate at 208–240 VAC and can be installed by an electrician in virtually any residential electrical system. They can typically provide up to about 20 kW. The number of private slow chargers is similar to the number of BEVs on the road, as the majority of electric vehicle owners utilize the option of home charging.

### 2.8.2 Public slow chargers

Public slow chargers are similar to Level 2 private chargers except that they are available for public use (generally for payment). They are typically AC chargers and are limited to a power output of 22 kW. As of 2019, there were about 600 000 public slow chargers available worldwide.

### 2.8.3 Public fast chargers

Public fast chargers provide DC current to the vehicle; therefore, the vehicle's internal electronics that converts AC to DC is bypassed. DC fast chargers may be Level 1, providing up to 80 kW of power, or Level 2, providing up to 400 kW of power. As of 2019, there were about 250 000 public fast chargers in the world. The majority (about 80%) were in China. With the increase in the number of BEVs (and plug-in hybrids, as discussed later in the chapter), the number of public chargers, particularly fast chargers, has grown substantially in recent years. Figure 2.31 shows a typical public charger station. Many governments incentivize the establishment of public charging stations in order to promote the production and sale of battery vehicles.

The different chargers as described above are distinguished by different plug configurations at the vehicle end of the charger cable, as illustrated in figure 2.32.

### 2.8.4 Charging time

The length of time that is required to charge an electric vehicle battery depends on several factors. These include the following:

- the capacity of the battery;
- the depth of discharge, i.e. the amount of remaining charge;
- whether the charging current is AC or DC;
- the power provided by the charger;
- the specific design of the vehicle.

Much progress has been made in recent years in improving charging times for BEVs. This has primarily been due to improvements in lithium-ion battery technology. In addition, the increased availability of public fast chargers has greatly increased the attractiveness of electric vehicle ownership. Although there is some

**Figure 2.31.** Public charger station in Caraquet, New Brunswick, Canada operated by NB Power showing a fast charger (larger orange unit on the left) and a slow charger (smaller orange unit on the right). Image: Richard A Dunlap.

variability in charge times between different vehicles, some general approximate charging times for highway-capable vehicles are summarized in table 2.9. Charging from 0% to 100% at a Level 2 private charging station typically takes 6–8 h. This is convenient as a means of charging overnight at home. Similar times for public slow chargers often make this charging option unacceptably long. Public fast chargers are more compatible with charging while traveling. Charge times are often much shorter when charging to (say) 80% of capacity than when charging to 100% capacity. Charge times are typically reduced further for charging from 30% to 80% rather than charging a fully discharged battery. Many manufacturers indicate that partial recharging takes on the order of 30 min. For example, Tesla claims that a 50% charge can be reached in 30 min (for an additional range of about 250 km) using a DC fast charger, which supplies up to 150 kW, for the Tesla Model 3.

Despite the improvements in battery technology in recent years, charging time remains a concern for the general widespread use of BEVs, as does the availability of public DC fast chargers. Two possible alternatives to traditional BEV utilization have been considered. These are battery swapping and the electric road, as discussed below.

| charger level | connector designation | pin configuration |
|---|---|---|
| level 1 | J1772 connector | 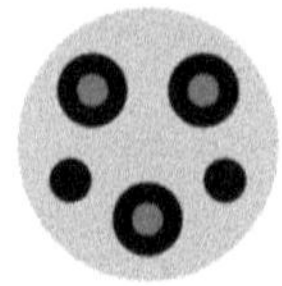 |
| level 2 | J1772 connector | 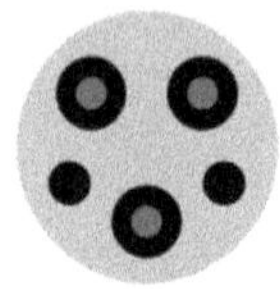 |
| DC fast charger | CCS connector |  |
| | CHAdeMO connector | 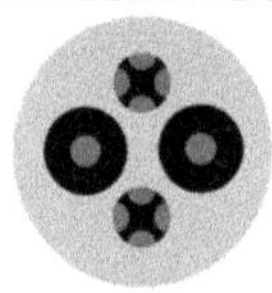 |
| | Tesla connector |  |

**Figure 2.32.** Standard electric vehicle charge plugs (vehicle end of cable) for different levels of charger. Adapted from the U.S. Department of Energy (2021) document 'Developing infrastructure to charge plug-in electric vehicles.'

**Table 2.9.** Typical additional vehicle range as a function of charging time for different charger levels. Data taken from the U.S. Department of Energy (2021) document 'Developing infrastructure to charge plug-in electric vehicles,' available at https://afdc.energy.gov/fuels/electricity_infrastructure.html.

| Charger level | Additional range |
|---|---|
| Level 1 | 3–8 km $h^{-1}$ of charge |
| Level 2 | 16–32 km $h^{-1}$ of charge |
| DC fast charger | 100–130 km per 20 min |

### 2.8.5 Swappable batteries

In the late 19th and early 20th centuries when electric vehicles enjoyed considerable popularity, charging was an important concern. During these years, many homes in North America did not have electric power. In addition, there was no established public charging infrastructure. A solution to this problem that was prevalent from the early 1900s until the mid-1920s was a battery exchange service. This would typically involve a customer purchasing an electric vehicle without a battery and paying a monthly service fee to cover the replacement of discharged batteries with charged ones.

The first modern version of the battery exchange concept was started in 2007 by the American company Better Place. This company operated BEV charging stations and developed a limited network of battery-swapping stations. Since this concept requires the battery and vehicle design to be compatible, the Better Place network was specifically designed to be compatible with the Renault Fluence Z.E., a vehicle equipped with swappable battery technology that has a battery-swapping time of about 5 min. Beginning in 2011, the Better Place battery-swapping network operated in Israel and Denmark. About 1000 vehicles were involved in the network in Israel and about 400 in Denmark. The commercial viability and practicality of such a program has not been demonstrated and Better Place filed for bankruptcy in 2013. The successful implementation of this approach would require vehicle designs to facilitate battery removal and installation and some degree of compatibility between swapping station battery inventory and vehicle battery requirements. This approach does, however, have some potential advantages for the consumer. It provides a viable option for electric car owners who have minimal home charging options and for those who are reluctant to purchase an electric vehicle because of recharging times. In addition, the cost of replacing aging batteries is amortized over time, so that the vehicle owner is not burdened with a large one-time battery replacement cost.

The possibility of battery swapping for battery electric scooters and motorcycles faces fewer difficulties than for automobiles. The major differences in battery requirements for two-wheeled electric vehicles compared to automobiles are the smaller battery capacity and the (typically) greater accessibility of the battery. In some cases, batteries are modular and consist of more than one smaller power pack. In many cases, a battery swap can be done by one person. This possibility means that home charging can provide a spare charged battery ready for swapping. The possibility of commercial battery-swapping stations has been considered by a number of electric scooter and motorcycle manufacturers. Several companies, including Honda, Kawasaki, Suzuki, and Yamaha have agreed to a plan to standardize battery design in order to ensure interchangeability.

### 2.8.6 Electric road

An electric road is a road which incorporates technology to provide electric power to vehicles which travel on it. There are three basic designs of electric road: overhead power lines, ground-level conductive rails, and inductive coils.

Overhead power lines, sometimes called catenary lines because of the shape of the suspended cable, have been used since the 19th century to provide electric power to buses, sometimes referred to as trolleybuses or trams. Power is supplied to the bus through trolley poles which are attached to the top of the bus and make electrical contact with the overhead power cables.

Overhead power lines are in common use for city buses, as shown in figure 2.33, but are generally considered unsuitable for other types of vehicle or for travel outside urban areas.

Ground-level conductive rails that provide electric power for vehicular use have been implemented for railway use since the 1880s. This arrangement, sometimes referred to as the third rail system, is not suitable for road transportation. However, in more recent years, there has been interest in the development of electric roads that incorporate conductive current-carrying rails embedded in the road surface that would be suitable for road vehicles of all sizes. An arm extending from underneath the vehicle and contacting the conductive rails would carry current to the vehicle for propulsion. While railways which utilize ground-level conductive rails can incorporate fencing to ensure public safety, electrical conductors embedded in road surfaces,

**Figure 2.33.** A bus (sometimes known as a trolleybus) with overhead electrical connections operating in Cambridge, Massachusetts. This image (Reinhold A 2004) has been obtained by the author from the Wikimedia website where it was made available under the GNU Free Documentation licence, https://en.wikipedia.org/wiki/en:GNU_Free_Documentation_License. It is included within this chapter on that basis. It is attributed to ArnoldReinhold.

particularly in city environments, must take more innovative approaches to eliminating pedestrian exposure to high voltages. A common solution to this problem is to only energize short (1–2 m) sections of the conductive rails that are underneath the vehicle as it passes over them and to de-energize the rails after the vehicle passes. Each section of the conductor can be energized by a signal from an appropriate vehicle, and this would ensure the safety of pedestrians. This approach would also provide a means of identifying appropriate vehicles and billing customers for the electricity used.

The final alternative for electric roads is the use of inductive coupling as a means of supplying electricity to a vehicle. This is sometimes referred to as dynamic charging. In this case, electromagnets producing AC magnetic fields are embedded in the roadway. Coils inside a vehicle are energized by inductive coupling to the AC magnetic field and provide power to the vehicle. Mechanical coupling to the vehicle, as used by the embedded rail approach, is not needed, although a similar approach can be taken in terms of energizing the coils when a vehicle is overhead as well as for customer billing. This approach is appropriate for passenger vehicles as well as larger commercial vehicles.

In the past 20 years or so, there have been investigations of possible ground-level conductive rails and dynamic charging systems in a number of countries. Typically, short (<2 km) sections of roadway are constructed to test possible designs. Dynamic charging is certainly the most attractive approach to an electric road, and Sweden, China, France, Germany, Israel, Japan, South Korea, the United Kingdom, and the United States have investigated this possibility. Much of this work has dealt with commercial vehicles, particularly city buses. In general, infrastructure costs for an electric road system are quite high, and detailed Swedish studies have suggested that dynamic charging systems are about 2.8 times as costly as embedded rails. The utilization of an electric road is most cost-effective in urban areas where there is high traffic density. It is less certain whether this approach could be made viable in rural areas.

A major benefit of electric roads is their effect on vehicle design requirements. Because batteries are charged more or less continuously while driving, lower-capacity batteries are sufficient to fulfill the vehicle's needs. This has several obvious advantages. First, vehicle costs can be reduced, and second, battery requirements become less demanding, as power is a more important consideration than range. The use of supercapacitors, as discussed in section 2.10, in conjunction with smaller batteries can be an effective means of increasing power. Also, since the batteries are charged while driving, the charging time for static external charging is not as important and the driving range can, in principle, be unlimited. The reduction in battery requirements for dynamically charged vehicles can mitigate the demands on the Earth's lithium supplies to a significant extent, as discussed in section 2.9.

While both swappable batteries and electric roadways are potentially attractive options for future battery electric road vehicles, both of these approaches would require the implementation of extensive additional infrastructure. So, for the near future at least, the development of a BEV infrastructure will depend primarily on the improvement of vehicle technology and the implementation of a more extensive public charger network with particular emphasis on public fast chargers. In addition, the development of batteries with higher specific energy, faster charging times, longer cycle life, and lower cost will help to facilitate the widespread adoption of BEVs. The next two sections discuss some of the trends in ongoing battery research that may contribute to this goal, both in the short term and the long term. A related technology, i.e. supercapacitors, is also outlined.

## 2.9 More about lithium batteries

### 2.9.1 Types of lithium-ion battery

Lithium-ion is, at present, the established standard battery chemistry for electric vehicles. As discussed in section 2.4, typical lithium-ion batteries utilize a carbon anode and a cathode that is based on $LiCoO_2$. Much research has been undertaken in recent years in order to optimize the composition of materials used in lithium-ion batteries, with specific emphasis on the material used for the cathode. A number of factors have gone into the analysis of different materials. These factors include specific energy, cost, and cycle life, as well as consideration of battery safety. Most current battery technologies use cathode materials in which other elements are substituted for some of the cobalt in $LiCoO_2$. The most common cathode material used is $LiNi_xMn_yCo_zO_2$, where $x + y + z \approx 1$. These batteries are referred to as NMC batteries. A major benefit of substituting other metals for cobalt, and thereby reducing the amount of cobalt that is needed for the battery, is the fact that cobalt is expensive. The market prices for some relevant elements as of March 2021 are shown in table 2.10. The prices of some of the elements in the table, e.g. Li, tend to be fairly volatile. It should also be noted that cobalt is toxic and batteries which contain this element that are not properly recycled or disposed of can result in adverse environmental consequences. So, it is clear that eliminating cobalt (as long as battery performance does not suffer significantly) is advantageous. NMC batteries are commonly used for BEV applications, including the BMW i3, Chevrolet Bolt, Nissan Leaf, Renault Zoe, and Volkswagen ID.3.

Perhaps the most notable exceptions to the use of NMC for BEVs are the vehicles made by Tesla. These use cathodes comprised of $LiNi_xCo_yAl_zO_2$, where $x + y + z \approx 1$. These materials are referred to as NCA. Again, these materials have the advantage that they eliminate some of the expensive and environmentally hazardous cobalt from the cathode.

**Table 2.10.** Market prices of some elements as of March 2021. Note that lithium prices are often quoted for lithium carbonate ($Li_2CO_3$) as opposed to elemental lithium.

| Element | Price (USD/kg) |
| --- | --- |
| Lithium | \$74 |
| Manganese | \$3 |
| Cobalt | \$53 |
| Nickel | \$16 |

Another possible cathode material that completely eliminates cobalt is lithium iron phosphate, which has the composition $LiFePO_4$. This is usually referred to as LFP. This also eliminates the nickel that is present in both NMC and NCA. Although nickel is less expensive than cobalt, as noted in table 2.10, it is still substantially more costly than iron. In addition to cost and environmental factors, LFP batteries have some other advantages over NMC batteries. Among these are a more constant voltage output over the battery's lifetime, higher discharge rates, and greater thermal stability. These features provide simplified voltage-regulating electronics, greater power availability, and less chance of fire or explosion, respectively. However, LFP batteries have some disadvantages. Principal among these is the need to avoid fully discharging the battery. While the battery has a nominal output of 3.2 V, discharging it to less than 2.5 V can cause damage to the cell. Thus, appropriate electronic controls must be put in place to avoid this situation. Much recent research has been directed towards developing a commercially viable lithium iron phosphate rechargeable battery that can be useful in vehicular applications, which offers hope that this technology will be available sometime in the foreseeable future.

An additional potential improvement to lithium-ion batteries addresses the choice of anode material. While carbon-based anode materials are commonly used in commercial lithium-ion batteries, there has been recent interest in the possible use of nanocrystalline lithium titanate, $Li_2TiO_3$, (known as LTO) as an anode material. This material has an effective surface area of about 100 square meters per gram, or about 30 times that of typical carbon anode materials. This very large anode surface area allows reactions to occur very rapidly, leading to quick discharges and large corresponding power outputs, as well as fast recharge times. LTO batteries also have long cycle lives compared to many other lithium-ion batteries. The major drawback of these batteries is the chemical potential associated with the anode. Referring to figure 2.3, the LTO anode is at a much more positive potential than that of $LiC_6$, meaning that the difference between the cathode and anode potentials is smaller than in carbon anode batteries. The corresponding cell voltage is around 2.4 V compared to around 3.8 V in most current lithium-ion batteries (see table 2.1). As a result of this lower voltage, the specific energy associated with LTO cells is less than those of other lithium-ion batteries (see appendix E for a discussion of the relationship of energy and voltage). Nevertheless, their rapid charge capabilities have made them attractive for a number of applications, and they have been used in the Honda Fit EV and some Mitsubishi i-MiEV BEVs.

### 2.9.2 Lithium–sulfur batteries

Lithium–sulfur batteries are an alternative to traditional lithium-ion batteries. These batteries utilize a cathode that is comprised of sulfur and an anode that is comprised of lithium metal or a lithium alloy. This battery chemistry is of particular interest because the low densities of lithium (534 kg $m^{-3}$) and sulfur (~2000 kg $m^{-3}$) result in a very high specific energy. Lithium–sulfur batteries have an advantage over lithium-ion batteries because that they do not utilize expensive cobalt, or even nickel.

A generic diagram of a lithium–sulfur cell is shown in figure 2.34. A electrolyte commonly used in these cells is a solution of lithium bis(trifluoromethanesulfonyl) imide, $LiC_2F_6NO_4S_2$ (often known as LiTFSI) in dioxolane and dimethyl ether and a common separator material is microporous polypropylene. Figure 2.34 shows the discharge cycle of the lithium–sulfur cell. During the charge cycle, the lithium ions travel in the opposite direction. The half-cell reactions at the cathode and anode are

$$Li_2S \leftrightarrow S + 2Li^+ + 2e^- \quad \text{(cathode)} \tag{2.3}$$

and

$$2Li^+ + 2e^- \leftrightarrow 2Li \quad \text{(anode)},$$

where the arrow points from left to right during charging and from right to left during discharging. Combining the two half-cell reactions leads to the overall reaction during charging (arrow to the right) and discharging (arrow to the left), as follows:

$$Li_2S \leftrightarrow 2Li + S.$$

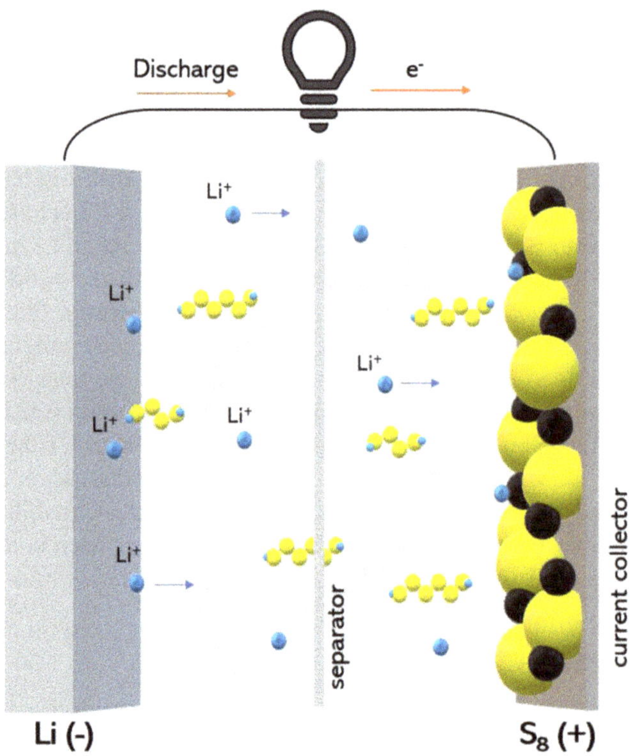


**Figure 2.34.** Operating principle of the lithium–sulfur battery during discharge showing the Li anode, the S cathode, and the flow of $Li^+$. This image (Egibe 2019) has been obtained by the author from the Wikimedia website where it was made available under a CC BY-SA 4.0 licence, https://creativecommons.org/licenses/by-sa/4.0/deed.en. It is included within this chapter on that basis. It is attributed to Egibe.

Interest in lithium–sulfur batteries dates back to the 1960s (see Nole and Moss (1970)), when primary cells operating on this principle were first constructed. By the late 1980s, the operation of a lithium–sulfur secondary battery had been demonstrated (Peled *et al* 1989). Although there is significant commercial interest in the development of lithium–sulfur batteries, progress towards this goal has been slow. Two major challenges face the development of a viable lithium–sulfur battery: cathode design and the so-called polysulfide shuttle. Cathode design is a concern for two reasons. First, unlike materials such as lithium alloys that are typically used for cathodes in conventional lithium-ion batteries, sulfur has virtually no electrical conductivity, making it difficult to transport lithium ions into and out of the cathode; second, the sulfur cathode experiences significant volume changes during the charge/discharge cycle, leading to electrode degradation, as noted previously.

A common approach to the problem of the low conductivity of the sulfur cathode is the incorporation of carbon into the electrode. This may be done, for example, by the simple process of carbon coating the sulfur cathode. The problem with this approach stems from the second concern noted above. Significant expansion and contraction of the sulfur cathode during lithium loading and unloading can result in the electrical disconnection of the cathode from the carbon coating, leading to battery failure. One approach to dealing with this problem is to construct a composite cathode using sulfur with carbon nanofibers. This can avoid the problems associated with the electrical disconnection caused by cathode volume changes, but can also significantly increase battery construction costs.

The second challenge, as noted above, concerns the polysulfide shuttle. When the reaction of lithium with sulfur occurs at the cathode during discharge (equation (2.3)), lithium polysulfides ($Li_2S_x$ with $6 \leqslant x \leqslant 8$) can also be formed. These lithium polysulfides are highly soluble in the electrolytes that are typically used in these cells. In what is referred to as the *shuttle effect*, the polysulfide molecules that are released into the electrolyte solution from the cathode travel to the anode, as shown in figure 2.34. At the anode, they are reduced, thereby depositing sulfur on the anode, and then travel back to the cathode, where they pick up more sulfur. This process is repeated and results in the contamination of the anode with sulfur and the degradation of cell performance.

Although much progress has been made in resolving these issues with lithium–sulfur batteries, more research is necessary before they will be able to compete with more traditional lithium-ion batteries from a commercial standpoint.

### 2.9.3 Lithium and other battery resources

Extensive research in recent years on materials and designs for lithium batteries has made enormous progress in increasing specific energy, decreasing cost, reducing the charging time, and improving safety. It has been as a result of these improvements that BEVs have achieved the level of success that they have. Ongoing research will undoubtedly lead to further improvements in lithium-ion battery properties. However, despite the advantage of reducing or eliminating expensive elements such as cobalt and possibly nickel from the battery electrode materials, these batteries still utilize chemistry that requires the presence of lithium. As noted in table 2.10, lithium itself is an expensive element. This cost arises as a result of the limited supply of lithium that can be conveniently and economically extracted from nature. It is, therefore, of interest to better understand the extent of the available lithium resources.

Although lithium is a relatively common element in the Earth's crust, it is generally not easy to extract. Most lithium occurs in deposits where it exists in very low concentrations. It is, therefore, difficult and expensive to utilize. It is not straightforward to estimate the amount of lithium that is available that is commercially viable. The existing lithium deposits that are currently mined commercially are estimated to contain about $1.35\times10^{10}$ kg of lithium. It is believed that this corresponds to about one third of all terrestrial lithium that could be utilized.

We need to consider how this amount of available lithium compares to the needs of the BEV market. The amount of lithium that is required to produce the batteries that are necessary for the construction of one BEV is quite variable. It depends significantly on the battery capacity, which is largely determined by the purpose of the vehicle. However, as a first estimate, it is reasonable to assume that about 10 kg of lithium is required per vehicle. Using the amount of lithium associated with existing mining operations, this would correspond to 1.35 billion vehicles, which is about the same as the estimate of the current number of road vehicles worldwide. So, converting all existing road vehicles to BEVs that utilize lithium-ion batteries would consume the total lithium resources in existing mines. It is also important to consider the timescale for this development. The current world production of lithium amounts to about $5.2 \times 10^7$ kg per year (for 2019). To produce the amount of lithium necessary for 1.35 billion road vehicles at current production rates would require about 250 years. Clearly, phasing out fossil-fuel use for transportation in favor of lithium-ion BEVs on a timescale that is compatible with mitigating the effects of climate change would require a substantial increase in lithium production rates.

The viability of this approach seems rather marginal, but there are a few important factors to consider. On the positive side, lithium battery technology is likely to improve in the future, and less lithium will be needed to provide the same capacity. In addition, many vehicles will likely fall into the NEV category and will require batteries with less capacity and less lithium than the amount estimated above. Added to this is the fact that it is estimated that there are another $2.7 \times 10^{10}$ kg of terrestrial lithium resources to be developed. However, there are some very significant factors on the negative side. First, the number of vehicles worldwide is likely to increase in the future, as transportation demands in developing countries will undoubtedly increase. Second, as vehicles are retired, it will be necessary to recycle virtually all of the lithium for use in new vehicles. Most important, however, is the fact that lithium has other uses in addition to its use in BEVs. At present, about two thirds of all lithium that is mined is used for batteries. The remainder is mostly used for the manufacture of glass, ceramics, polymers, and grease. The lithium-containing batteries that are produced at present are partly used in vehicles, but are also used for other applications, such as electronics and tools. Most important, however, is the present and future need for batteries for grid power backup. As more renewable energy, such as wind and solar, is brought online to replace fossil-fuel-generated energy, the need for grid backup will increase.

Batteries are one of the most versatile and dependable means of storing electrical energy for grid integration. Another potential future use for lithium is as a breeding fuel for nuclear fusion reactors. At present, the possibility of implementing this technology in the future is uncertain, but if it does come to fruition, it will also compete for the world's lithium resources (see Dunlap (2021)).

The above assessment suggests that lithium resources must be utilized carefully and may be a limiting factor for future sustainable energy development. There are two possible approaches to alleviating this potential difficulty: locate more lithium resources or find an alternative battery chemistry. The first of these approaches is discussed below and the second is discussed in the following section.

A major source of lithium is the world's oceans. Seawater contains between 0.14 ppm and 0.25 ppm of lithium. Since it is estimated that the oceans contain $1.34 \times 10^{18}$ m$^3$ of water, the total amount of lithium present is about $2.6 \times 10^{14}$ kg. This is nearly 10 000 times the total estimated terrestrial lithium. While it would certainly not be possible to extract all of the lithium that is contained in the oceans, even a small fraction (say) 0.1% could increase the world's lithium resources by a factor of ten over terrestrial resources. Although the technology for lithium extraction from seawater has been demonstrated (see Yu *et al* (2020)), it remains to be determined whether this approach could be scaled up and made economically viable.

As suggested in table 2.10, cobalt, in addition to lithium, is an element with limited availability. Cobalt is used primarily for the manufacture of rechargeable lithium-ion batteries (although not only for electric vehicle applications), permanent magnets, and high-strength alloys. Most cobalt is a by-product of nickel or copper mining. At present, the known cobalt resources worldwide are about $7.6 \times 10^9$ kg (U.S. Geological Survey 2022). Almost half of these resources are in the Democratic Republic of the Congo. In 2021 the total annual cobalt production was $1.7 \times 10^8$ kg, of which $1.2 \times 10^8$ kg was from the Democratic Republic of the Congo.

It is clear that the continued use of cobalt at the present rate would lead to the depletion of known cobalt resources in about 40 years. Certainly, the probable increase in BEV use will deplete these resources even sooner (e.g. Kane (2019)). There are several approaches that can be taken to help to mitigate cobalt resource limitations for electric vehicles:

- Locate additional cobalt resources. The United States Geological Survey has reported that more than $1.2 \times 10^{11}$ kg exists in polymetallic nodules and crusts on the ocean floor. The commercial viability of this cobalt is not known.
- Recycle material from batteries at the end of their life cycle. This, as noted above, is an important consideration for lithium, but it is also an important factor for cobalt, not only for resource utilization but also for health and environmental reasons.
- Develop secondary batteries (lithium or other chemistries) that do not require cobalt-containing cathodes.

Certainly, the last option is the most desirable from cost and environmental standpoints.

## 2.10 Advanced batteries and related technologies

A review of BEV development over the years shows that the development of high-capacity lithium-ion batteries has been key to the success of these vehicles, and it is certain that the development of BEV technology in the foreseeable future will continue to depend on lithium-ion batteries. As noted in the previous section, however, the potential drawbacks of lithium-ion battery technology include the cost of materials, waste recycling, long-term sustainability of lithium resources, and substantial recharging times. In recent years, there has been extensive research with the aim of addressing these issues, and some of this work has focused on the development of low-cost, high-capacity rechargeable batteries based on chemistries which utilize elements other than lithium. Such chemistries are often based on elements that share chemical properties with lithium. Specifically, for example, sodium is in the same column of the periodic table as lithium. Another approach that may be advantageous to reducing recharging times is the use of batteries with liquid electrodes. In this section, we look at some of the possible alternatives to lithium-ion batteries for electric vehicles that are based on these concepts. In addition, the possible role of supercapacitors in future vehicle design is discussed.

### 2.10.1 Sodium batteries

One alternative to lithium battery chemistry is the use of sodium to carry charge between the electrodes. A neutral lithium atom has three electrons, two of which form the closed shell characteristic of helium and one which is a loosely bound valence electron. This electronic structure is the reason lithium is a very active element that is easily ionized. A neutral sodium atom has eleven electrons. Ten of these electrons form the closed shell characteristic of neon and the eleventh electron acts very much like the third electron in lithium. Thus, we would expect sodium to behave very much like lithium and form similar compounds. This is largely true. However, from the battery chemistry standpoint, there is one very important difference; sodium atoms (and sodium ions) are very much larger than lithium atoms (and lithium ions). Specifically, the $Na^+$ ion has an ionic radius of 0.102 nm compared to 0.076 nm for the $Li^+$ ion. Therefore, compared to lithium ions, it is more difficult for sodium ions to enter into the structure of an electrode material and bind with the electrode atoms without causing damage to the crystal structure. Two approaches for rechargeable batteries based on sodium chemistry that have shown some promising results are the sodium–sulfur battery and the sodium-ion (Na-ion) battery. These are discussed briefly below.

#### *2.10.1.1 Sodium–sulfur batteries*

In a sodium–sulfur battery, as illustrated in figure 2.35, the anode consists of liquid sodium, while the cathode consists of liquid sulfur. Electrical connections are made to the liquid sodium at the center of the cylindrical cell by an inner metal can (the metal insert in the diagram) and to the liquid sulfur by an outer metal can (the cell container in the diagram).

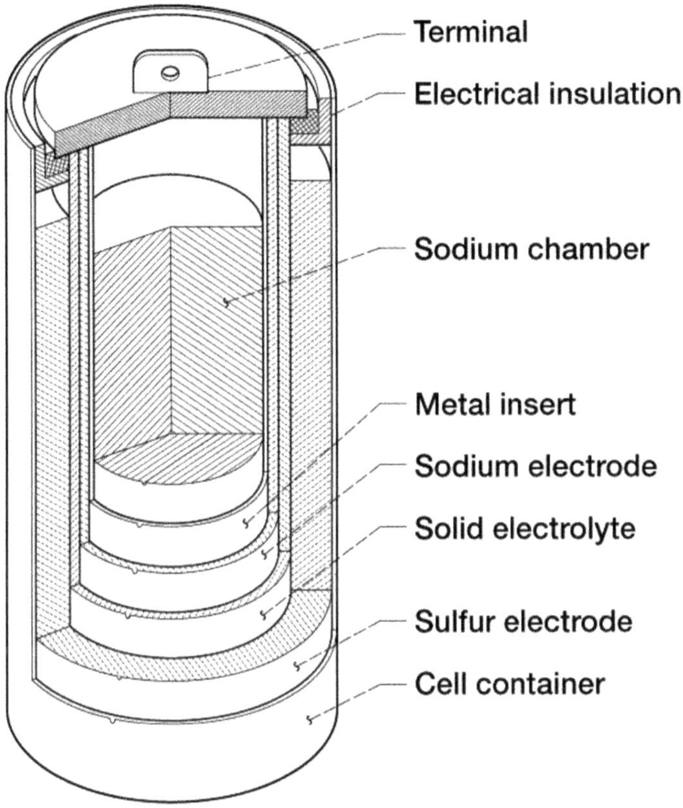


**Figure 2.35.** The structure of a sodium–sulfur battery. This image (NASA John Glenn Research Center 2011) has been obtained by the author from the Wikimedia website, where it is stated to have been released into the public domain. It is included within this chapter on that basis. Credit: NASA John Glenn Research Centre.

The liquid electrodes are separated by a β-alumina solid electrolyte (generally abbreviated as BASE), which is shown as the solid electrolyte in the figure. β-alumina is related to the common form of aluminum oxide, α-alumina ($Al_2O_3$), and has the general formula $Na_{1+x}Al_{11}O_{17+x/2}$, where $x$ is typically around 0.3.

Solid electrolytes in batteries are made from materials that are selectively permeable to certain ions, in this case, $Na^+$ ions. During discharge the sodium ions travel between the anode and the cathode through the solid electrolyte, while the electrons liberated from the sodium atoms travel through the external circuit where they can do work. This process leads to the following overall reaction during the discharging process:

$$2Na + 4S \rightarrow Na_2S_4.$$

This reaction is reversible, leading to the opposite reaction during the charging process. The solid electrolyte is relatively impermeable to electrons, and this results in a very small self-discharge rate in the charged cell.

Sodium–sulfur batteries have a number of advantages over other battery chemistries. These include high efficiency for energy storage, a long cycle life, and the fact that they are constructed from plentiful low-cost materials. Unfortunately, there are also some disadvantages to sodium–sulfur batteries. Since the sodium and sulfur electrodes must be kept in the molten state, the battery must be operated at an elevated temperature. This high-temperature operation is also necessary because the β-alumina solid electrolyte only achieves the proper electrical properties above about 250 °C. Therefore, sodium–sulfur batteries are generally operated at 300 °C–350 °C and require an external source of heat. During continuous charging/discharging cycles, the self-heating that results from currents passing through the electrolyte helps to raise the temperature of the battery, thereby reducing, at least to some extent, the necessary external heating requirements. This effect is most beneficial for larger facilities that incorporate a significant number of cells in close proximity. The design of the battery must also consider the chemical properties of the cell materials. Specifically, the sodium polysulfide ($Na_2S_4$) that is formed in the cathode during the discharging process is highly corrosive. Therefore, the cell container must be made from a material that is highly corrosion resistant, typically molybdenum or chromium.

Because of the characteristics of sodium–sulfur batteries, they are most efficient for large installations such as facilities for electric grid storage that are commonly associated with intermittent renewable energy sources such as wind or solar power (Spataru and Bouffaron 2016). In fact, they have been very successful for such applications. The battery grid storage facility at the Buzen Substation of the Kyushu Electric Power Company in Buzen City, Fukuoka Prefecture, Japan uses sodium–sulfur batteries. This facility has an electrical storage capacity of 300 MWh and can deliver up to 50 MW of power for up to 6 h. However, further technical developments would be necessary to make sodium–sulfur batteries suitable for BEVs.

#### *2.10.1.2 Sodium-ion batteries*

Sodium-ion batteries are analogous to lithium-ion batteries. In sodium-ion batteries, positive $Na^+$ ions take the place of positive $Li^+$ ions, carrying the charge from the cathode to the anode during the charging process and from the anode to the cathode during the discharging process (Hwang *et al* 2017). The electrode materials must be chosen to be compatible with the insertion and extraction of sodium ions during cycling. As noted in section 2.4, graphite is a common material for the anodes of lithium-ion batteries. Although sodium reacts chemically with carbon in a manner that is appropriate for the anode material of a sodium-ion battery, the sodium ions are too large to conveniently fit between the carbon layers in graphite. Instead of graphite, sodium-ion batteries often use a disordered amorphous form of carbon, known as hard carbon, for the anode material. This form of carbon has more open space in the structure than graphite and allows the sodium ions to enter into the structure when the cell is charged. Research into other anode materials for sodium-ion batteries, such as metallic alloys and metal oxides, has met with limited success. The large size of the sodium ion introduces excessive strain into the material during the charging and discharging processes, leading to a limited cycle life. In addition, compared to anode materials that have been investigated in recent years, carbon provides a higher operating voltage.

During the past decade or so, the development of appropriate cathode materials has been a major focus of sodium-ion battery research (see Skundin *et al* (2018)). As in the case of lithium-ion cells, sodium-ion cells have functioned well using transition-metal oxide cathodes, i.e. sodium transition-metal oxides. Layered materials which provide structures that can readily accommodate sodium have met with the most success and also provide a suitable operating voltage of about 3.2 V relative to a carbon anode. An important aspect of the research into these materials has been the optimization of the energy storage capacity and cycle life while utilizing materials that eliminate, or at least minimize, expensive and rare elements such as cobalt, nickel, chromium, and vanadium (see Yabuuchi and Komaba (2014)). At present, however, the most successful cathode materials, in terms of capacity and cycle life, contain at least some nickel. A recent economic evaluation of different battery chemistries including traditional lithium-ion batteries (utilizing NMC-based cathodes), lithium-ion batteries (utilizing LFP cathodes), and sodium transition-metal oxide-based sodium-ion batteries has been written by Peters *et al* (2019). The conclusions of this report are summarized in figure 2.36.

Another class of material that has met with some success as sodium-ion battery cathodes is the polyanions. While these materials typically have slightly lower capacities than those of oxide-based cathode materials, the strong bonding that exists between atoms of the polyanion better resists degradation during cycling and has been found to improve cycle life. Polyanions include compounds such as the rhombohedral $Na_2MnFe(CN)_6$, which has shown good capacity and cycle life as a sodium-ion battery cathode and which utilizes only common inexpensive materials.

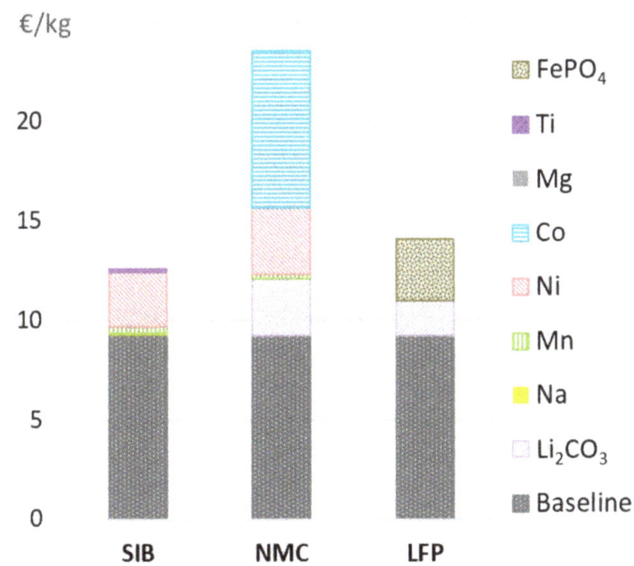


**Figure 2.36.** Comparison of active cathode material costs in Euros per kg for different battery types. SIB = sodium-ion battery, NMC = nickel manganese cobalt battery, LFP = lithium iron phosphate battery. Reproduced from Peters *et al* (2019). CC BY 4.0.

While research and development into sodium-ion batteries is in its very early stages compared to commercially available lithium-ion batteries, it is clear that they offer economic advantages even when they use nickel-containing cathodes. In addition, the early results for nickel-free cathode materials have shown encouraging results. In any case, the possibility of utilizing a sodium-based battery chemistry, rather than a lithium-based one, has the advantage of eliminating the long-term dependence of transportation technologies on limited terrestrial lithium resources.

### 2.10.2 Magnesium batteries

Magnesium is in the adjacent column to lithium and sodium in the periodic table and therefore has two outer electrons rather than one. It is possible that magnesium ions could take the place of lithium or sodium ions in a battery and the abundance of magnesium in the Earth's crust certainly makes this element an attractive and inexpensive alternative to lithium for battery chemistry. In fact, magnesium is superior to lithium in some ways, although not in others. Specifically, its energy per unit mass is about half that of lithium, but its energy per unit volume is about 50% greater than that of lithium. Thus, for automobile applications, for example, magnesium-ion batteries would occupy less space, but have greater mass than lithium-ion batteries of the same capacity.

Primary magnesium-based batteries were developed in the early part of the 20th century and were commercially available from the 1940s until the mid-1980s, when they were overtaken by lithium-based batteries. Secondary magnesium batteries, however, have not yet been developed to the point where they are commercially viable. Rechargeable magnesium-ion battery research is in its fairly early stages compared to lithium-ion and sodium-ion battery research, and while some progress has been made in developing a practical battery design, much more research is still needed. The problems facing magnesium battery development are summarized below.

The successful design of a secondary battery depends on the identification of suitable materials for the anode, cathode, and electrolyte. Although much research has been done in recent years using a variety of different approaches and materials, the choice of suitable materials for magnesium-ion batteries is not so clear (see e.g. Nguyen and Song (2017)). The larger size of the magnesium ion, compared to the lithium ion, along with its greater charge (+2 in the electrolyte solution, compared to +1 for lithium), are certainly complicating factors. The ionic transport through the electrolyte and the insertion of the ions into the electrodes typically occur more slowly for magnesium than for lithium. In addition, electrode damage caused by ion insertion is a significant concern. Experimental magnesium-ion cells have generally suffered from poor cyclability, low capacity (compared to the theoretical maximum), low operating voltage, and poor performance at ambient temperature. However, continuing research may lead to improvements that will make batteries based on this chemistry suitable alternatives to lithium-ion batteries for electric vehicle applications.

### 2.10.3 Aluminum–air batteries

Aluminum–air batteries utilize the reaction of an aluminum electrode with oxygen from the air. The general design of an aluminum–air battery is shown in figure 2.37. Typically, the anode is aluminum, the electrolyte is an aqueous solution, and the separator is polytetrafluoroethylene (PTFE), which is the basis for Teflon. A suitable catalytic material, as described below, is used for the cathode.

The half-cell reaction at the aluminum anode during the discharging process is

$$Al + 3OH^- \rightarrow Al(OH)_3 + 3e^-$$

and the half-cell reaction at the cathode during the discharging process is

$$O_2 + 2H_2O + 4e^- \rightarrow 4OH^-,$$

leading to the following overall reaction:

$$4Al + 3O_2 + 6H_2O \rightarrow 4Al(OH)_3.$$

The typical properties of an aluminum–air battery are a cell voltage of 1.2 V, a specific power of 200 W $kg^{-1}$, and a specific energy of 1300 Wh $kg^{-1}$. These properties can be compared with those for other battery chemistries given in table 2.1. It is the specific power (about eight times that of a lithium-ion battery) along with the abundance and low cost of aluminum (~$3/kg) that make the aluminum–air cell very attractive for electric vehicle use. In particular, the specific energy can provide sufficient capacity in a typical vehicle configuration to yield a range in the vicinity of 2000 km to 4000 km.

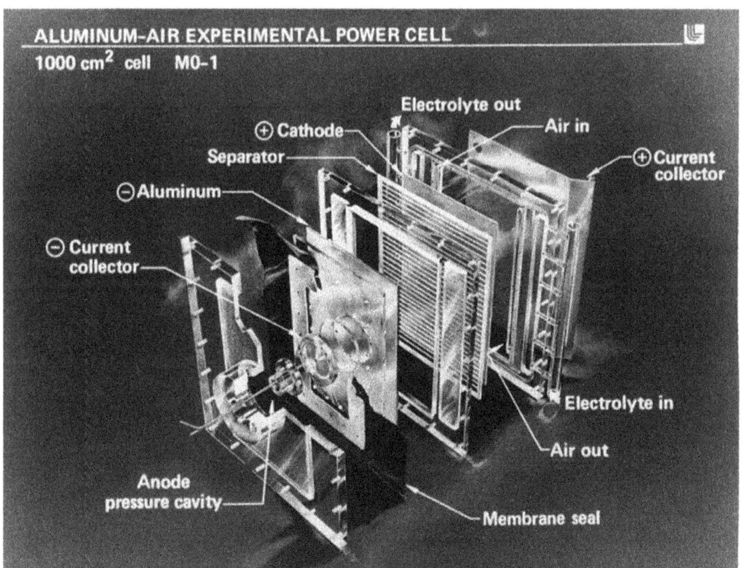


**Figure 2.37.** Diagram of an aluminum–air battery. Reprinted from U.S. Department of Energy (2014) Flickr (public domain).

Although the features described above would seem to make aluminum–air batteries highly attractive for battery electric applications, there are two major drawbacks to this technology. First, at present, suitable cathode catalytic materials tend to be very expensive. Traditionally, silver (current price ~750 USD $kg^{-1}$) has commonly been used, although future research may lead to more economic alternatives. Second, and perhaps most significantly, aluminum–air batteries are primary (i.e. non-rechargeable) batteries. The discharge process oxidizes the aluminum anode, and this is inherently an irreversible process. The cost of primary batteries, as noted at the beginning of this chapter, is a certainly a concern for the use of aluminum–air batteries. The low cost of the aluminum anode material can offset this concern to some extent, but the potential high cost of suitable cathode materials must be factored into any overall economic assessment. Although aluminum–air batteries are not rechargeable in the conventional sense, it is not necessary to replace the entire battery. These batteries can be mechanically recharged by replacing only the oxidized aluminum anodes. If batteries and vehicles were designed to facilitate this procedure, the long range noted above might be sufficient, at some point in the future, to outweigh the disadvantages of this technology. Thus far, most aluminum–air battery use has been for military applications, although the manufacturer Eviation (see section 5.4) has considered the possibility of utilizing them in future designs of its battery electric aircraft.

### 2.10.4 Flow batteries

Flow batteries utilize liquid anodes and cathodes. This battery design has clear advantages over solid-electrode battery designs in terms of recharging time. Instead of moving ions from one electrode to another during recharging, it is only necessary to pump depleted electrode solutions out of the battery and replace them with fresh solutions.

The general diagram of a flow battery is shown in figure 2.38. Electrochemical energy is stored in the cell as a result of the difference in the chemical potential of the ions on the two sides of the cell. The ions are in solution in the liquid electrolytes that make up the positive and negative electrodes of the cell. These electrolytes are analogous to the anode and cathode of a conventional electrochemical cell, but are referred to as the anolyte and catholyte, respectively. The two electrolytes are separated by an ion-selective membrane that allows specific species of ions to flow between the catholyte and the anolyte during charge, and vice versa during discharge, while electrons flow through the external circuit. The electrolytes are stored in tanks, and fresh electrolyte is circulated through the cell by pumps in order to replenish the concentrations of the respective anolyte and catholyte ions on each side of the separator. In general, the electrical storage capacity of the cell is a function of the volume of electrolytes stored in the tanks, as this represents the amount of electrochemical energy stored by the ions. The power available from the cell is a function of the area of the separator, as this determines the rate at which ions can be exchanged between the anolyte and the catholyte. This is a convenient situation for many potential applications because (unlike traditional solid-state batteries) capacity and power are independent design factors.

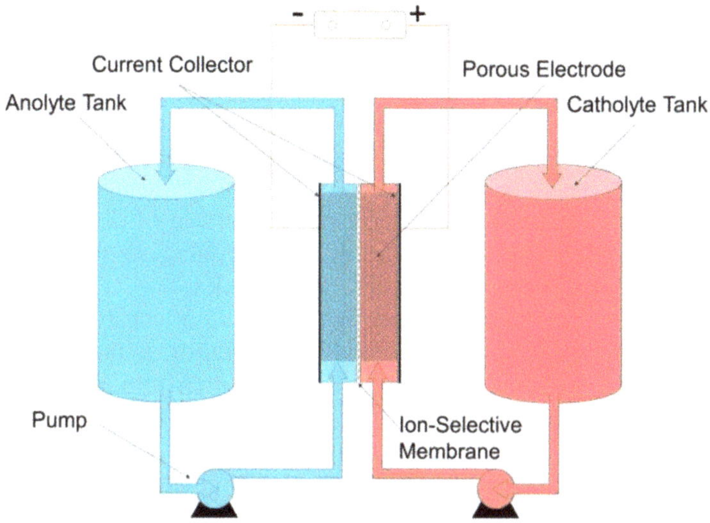


**Figure 2.38.** Diagram of a flow battery. Reproduced from Qi and Koenig (2017). CC BY 4.0.

Vanadium redox flow batteries (also known as vanadium redox batteries or vanadium flow batteries) are the most common type of flow battery and their operation is used here as an example of how this category of rechargeable battery functions (Doetsch and Burfeind 2016). The operation of a vanadium redox flow battery is a result of the specific properties of vanadium ions in aqueous solutions, in which the vanadium ions can exist in four different valence states. In the anolyte solution, vanadium can exist in the 2+ and 3+ valence states, as $V^{2+}$ and $V^{3+}$ ions in solution, respectively. In the catholyte solution, vanadium can exist in the 4+ and 5+ valence states, as vanadium oxide complexes $VO^{2+}$ and $VO_2^+$, respectively. The reaction in the anolyte is

$$V^{2+} \leftrightarrow V^{3+} + e^-,$$

while in the catholyte the reaction is

$$VO_2^+ + e^- + 2H^+ \leftrightarrow VO^{2+} + H_2O,$$

where the reactions move from left to right during the discharging process and from right to left during the charging process. During discharge, $V^{2+}$ in the anolyte is oxidized to become $V^{3+}$ and releases an electron, which travels through a load in an external circuit. The electron, which arrives at the catholyte, reduces $V^{5+}$ in the form of $VO_2^+$ to produce $V^{4+}$ in the form of $VO^{2+}$. During this process, $H^+$ ions travel through the separator from the anolyte to the catholyte in order to maintain charge neutrality. In the charged state the potential difference between the catholyte and the anolyte is about 1.26 V, and this represents the cell voltage.

Once the cell is discharged it may be recharged in the same way as traditional solid-state secondary batteries by replacing the external load with an external current supply that forces electrons from the catholyte to the anolyte in order to drive the above half-cell reactions from the right to the left. An important potential advantage of a vanadium redox flow battery is that there is another simple way to return the cell to the charged state. Discharged electrolytes can be pumped out of the catholyte and anolyte tanks and replaced with fresh electrolytes with vanadium in the $V^{2+}$ and $V^{5+}$ oxidation states, respectively. In this way, recharging a vanadium redox flow battery can be as simple and as quick as filling up the fuel tank of a gasoline-powered vehicle, and the discharged solutions can be recharged at a location outside the vehicle. Another general advantage of flow batteries is that there are no solid-state electrodes to degrade during cycling, so their cycle life is very long, typically thousands or tens of thousands of cycles.

Vanadium redox flow batteries were first successfully constructed in the 1980s (Rychcik and Skyllas-Kazacos 1988) and are currently a fairly well-developed technology that has been successfully commercialized (Alotto *et al* 2014). Because their storage capacity is determined by the volume of their electrolyte storage tanks, and these can be made very large, flow batteries can have very large capacities. As such, they are ideal for large-scale electrical storage facilities, such as grid storage and backup. In fact, the world's largest electric grid storage facility has recently been completed in January 2021. This is the 800 MWh capacity Dalian Vanadium Redox Flow Battery facility in Liaoning Province, China. It is capable of providing up to 200 MW of power for up to 4 h.

While vanadium redox flow batteries have some characteristics that would make them very attractive for road vehicle use, they unfortunately have very low specific energy capacity. This is not a problem for grid backup facilities, where size is not a consideration, but it is a major disadvantage for vehicle use. A battery of a suitable size to fit in a road vehicle would provide an unreasonably short range.

Other types of flow battery, which have the same advantages as vanadium redox flow batteries but a much higher specific energy, have been developed for possible vehicle use. Some of these have recently been reviewed by Tolmachev *et al* (2015). Current hydrogen–lithium halide flow batteries can have a specific energy capacity that is 30–50 times that of commercial vanadium redox flow batteries and a power per unit separator area that is 10–20 times larger. However, their specific energy is still less than that of lithium-ion batteries and the complexity of their design puts them at a disadvantage for vehicle applications. Further work on other flow battery chemistries may provide a design that is suitable for application to road vehicle energy storage.

### 2.10.5 Supercapacitors

Like batteries, capacitors are devices that store electrical energy. Batteries store energy because of the difference in the chemical potentials of different compounds. Capacitors, on the other hand, store energy in an electric field between positively and negatively charged electrodes. The simplest capacitor consists of two parallel conductive plates, as shown in figure 2.39. The area of each plate is $A$, the distance between them is $d$, and the space between the plates contains a material of permittivity $\varepsilon$. The capacitor is in the charged state when there is a charge of $+Q$ on one plate and the opposite charge $-Q$ on the other plate. In this state, the electric

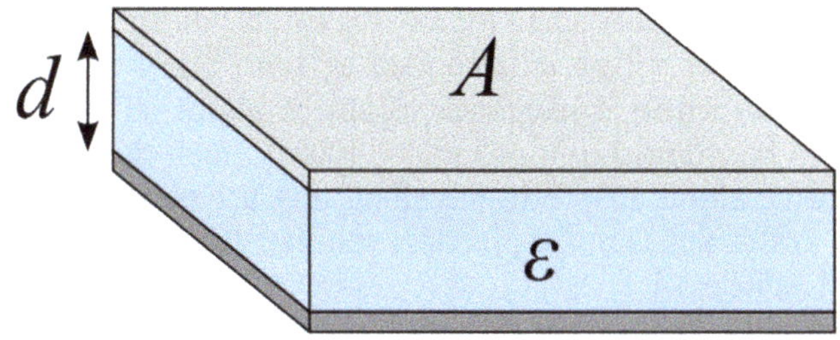


**Figure 2.39.** Diagram of a parallel plate capacitor. This image (Fabian_R 2013) has been obtained by the author from the Wikimedia website where it was made available under the GNU Free Documentation Licence, https://en.wikipedia.org/wiki/en:GNU_Free_Documentation_License. It is included within this chapter on that basis. It is attributed to Fabian_R.

field between the plates and in a direction perpendicular to the plates, $\boldsymbol{E}_z$, is given by Gauss's law, as follows:

$$\boldsymbol{E}_z = \frac{Q}{\varepsilon A}.$$

The voltage between the plates is an integral of the electric field over the distance between the plates:

$$V = \int_0^d \boldsymbol{E}_z dz = \boldsymbol{E}_z d = \frac{Qd}{\varepsilon A}, \tag{2.4}$$

and the capacitance of the capacitor is defined as the ratio of the charge stored on the plates to the voltage difference between the plates:

$$C = \frac{Q}{V}.$$

Therefore, from equation (2.4),

$$C = \frac{\varepsilon A}{d}. \tag{2.5}$$

The energy stored in the capacitor can be determined by considering the movement of a differential charge, $dQ$, across the voltage, $V$, between the plates. This movement of charge requires the following energy:

$$E = \int_0^Q V dQ = \int_0^Q \frac{Q dQ}{C} = \frac{1}{2}\frac{Q^2}{C} = \frac{1}{2}CV^2.$$

Using equations (2.4) and (2.5), this may be written as

$$E = \frac{\varepsilon A V^2}{2d} = \frac{1}{2}\varepsilon A d \boldsymbol{E}_z^2.$$

The maximum energy that can be stored in a capacitor of a particular geometry is determined by the maximum electric field, $\boldsymbol{E}_\mathrm{b}$, that does not result in a breakdown of the dielectric. Unfortunately, the energy storage capacity of a capacitor constructed according to figure 2.39 is very small for any reasonable dimensions. For example, if we consider a capacitor with a surface area of 1 $m^2$ and a distance of 0.01 mm between the plates, then its capacitance would be about $10^{-6}$ Farads (F). If this capacitor were to be charged to a voltage of 1000 V, then the energy stored in the capacitor would be about $10^{-5}$ Wh. Recalling that a typical road vehicle requires about 156 Wh $km^{-1}$, this is not a practical solution to energy storage that can be applied to road vehicles.

In order to produce a capacitor that could make a useful contribution to energy storage for transportation, we would have to significantly decrease the plate spacing or increase at least one of the following:

- the static permittivity of the dielectric material;
- the charge voltage;
- the plate area.

The range of static permeabilities of suitable materials for capacitor construction is not very large. So, while the material can be optimized for the best performance, this has a relatively small effect on the overall capacitance. If the electric field exceeds the breakdown field then electricity flows through the dielectric, causing the capacitor to discharge. Thus, the spacing in a parallel plate capacitor cannot be made arbitrarily small, nor can the voltage be made arbitrary large. The supercapacitor (also referred to as an ultracapacitor) takes a somewhat different approach to increasing the capacitance of a capacitor without significantly increasing its overall size. To understand the operation of the supercapacitor, we begin with a description of some of its basic design characteristics.

A diagram of a generic supercapacitor in the discharged and charged states is shown in figure 2.40. This design is referred to as a double-layer capacitor and consists of two electrodes separated by a dielectric (the electrolyte). As the figure illustrates, an ion-permeable membrane (the separator) divides the two sides of the supercapacitor. The electrolyte contains a random distribution of both positively and negatively charged ions. The supercapacitor is charged by applying positive and negative voltages to the appropriate electrodes. The electric field created between the electrodes causes the ions to migrate through the separator and attracted to the electrode of the opposite charge, as shown in the figure. The figure shows that a thin layer of charge, referred to as the Helmholtz layer, is formed at the surface of each electrode when the supercapacitor is charged. The thickness of the Helmholtz layer is typically of the order of 0.5 nm and this defines the distance over which charge separation occurs. Thus, the capacitance of the supercapacitor is inherently large because of the small distance between the charges in the charged state.

The total capacitance of the supercapacitor is the result of the capacitance of the two Helmholtz layers, $C_+$ and $C_-$, on the positive and negative electrodes, respectively, that are electrically connected in series. This gives the total capacitance, which is

$$C = \frac{C_+C_-}{C_+ + C_-}.$$

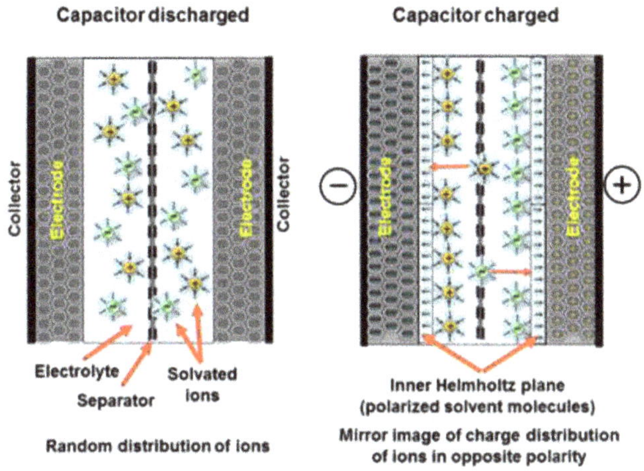


**Figure 2.40.** Discharged and charged supercapacitors. This image (Elcap 2013b) has been obtained by the author from the Wikimedia website, where it is stated to have been released into the public domain. It is included within this chapter on that basis.

In addition to the double-layer capacitance described above, supercapacitors can, depending on their design, have an additional contribution to the capacitance known as the electrochemical pseudocapacitance. Pseudocapacitance occurs when specific components of the electrolyte are de-solvated and contribute ions that are adsorbed onto the surfaces of the electrodes, thereby contributing to the charge accumulation. The relative importance of the traditional double-layer capacitance and the electrochemical pseudocapacitance is related to the specific supercapacitor design.

Although the design of the supercapacitor is somewhat different than that of the simple parallel plate capacitor, it is the proper choice of materials for the electrolyte, separator, and electrodes that optimizes the capacitance of the supercapacitor. We now consider the requirements and appropriate materials for these three supercapacitor components.

*Electrolyte*—For traditional capacitors the dielectric is chosen to be a material with a high static permittivity and a high breakdown electric field. In the case of supercapacitors, the electrolyte must fulfill additional criteria. The electrolyte must be a liquid, as it is necessary for it to contact all of the surface area of the porous electrodes, as discussed below. The dielectric molecules, which are in solution in the liquid electrolyte, must produce an equal number of positively and negatively charged ions. The solvent can be either water or an organic liquid. In water-based electrolytes, ions are formed by the solution of inorganic chemicals such as potassium hydroxide. Organic electrolytes are based on solvents such as acetonitrile, with ions provided by solutes such as ammonium chloride.

*Separator*—The separator electrically isolates the two electrodes but allows positive and negative ions to pass through. A variety of porous insulating materials have been used as separators in supercapacitors, ranging from paper and polymer films to woven ceramic fibers.

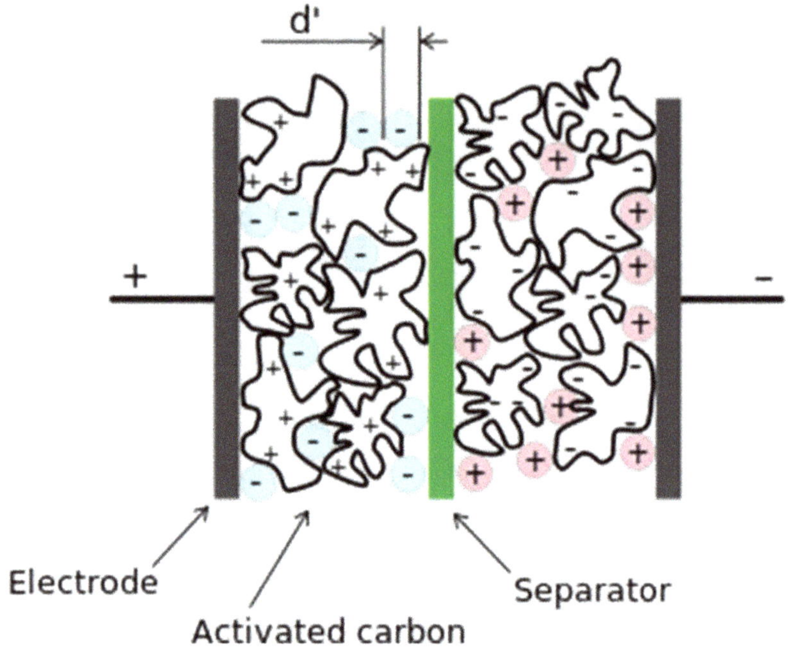


**Figure 2.41.** Diagram of a supercapacitor showing the distribution of charges on the surfaces of the porous electrodes. This image (Zurek 2006) has been obtained by the author from the Wikimedia website where it was made available under the GNU Free Documentation licence, https://en.wikipedia.org/wiki/en:GNU_Free_Documentation_License. It is included within this chapter on that basis. It is attributed to Stan Zurek.

*Electrodes*—As noted above, the surface area of the electrode is an important factor in establishing the high capacitance exhibited by supercapacitors. Figure 2.41 shows a common design for the electrodes of a supercapacitor. Each electrode consists of a metallic conductor that allows the supercapacitor to be connected to an electric circuit. The volumes between the metal electrodes and the separator are mostly filled with activated carbon. This carbon is electrically conductive and very porous. Its porosity means that it has a large surface area per unit mass, up to around 3500 $m^2$ $g^{-1}$ of material. It is necessary for the electrolyte to be in liquid form to penetrate the porous structure of the carbon. When the supercapacitor is charged, ionic charge accumulates on the large surface area associated with the pores, as shown in the figure. Thus, the large capacitance is primarily due to the very large surface area on which the charge collects. Some recently developed forms of nanostructured carbon, such as carbon nanotubes and graphene, can have larger surface areas per unit volume than activated carbon, but for most applications, these materials are not viable from an economic standpoint.

*2.10.5.1 Transportation applications of supercapacitors*

To understand the relationship between supercapacitors and how they might be useful in vehicle propulsion, we can look at their relative placement on a Ragone plot, as shown in figure 2.1. In the figure, the area labeled 'SCs for power' shows the location of supercapacitors that might be relevant for electric vehicle use. These supercapacitors have a specific energy density in the range of about 2 Wh $kg^{-1}$ or typically about only 2% of the specific energy of a lithium-ion battery.

We can look at the relevance of the specific energy of the supercapacitor in the context of the energy requirements of a typical road vehicle. Referring to section 2.5, we recall that the energy requirement for the batteries of a typical BEV is 156 Wh $km^{-1}$. So, if we consider a vehicle utilizing 500 kg of supercapacitor (similar to the weight of batteries in a highway-capable BEV), we see that the supercapacitors have a total energy storage capacity of about (2 Wh $kg^{-1}$) × (500 kg) = 1000 Wh. This would provide a range for the vehicle of (1000 Wh)/(156 Wh $km^{-1}$) = 6.4 km. This is not suitable as a source of energy for a useful road vehicle, as previously described in this chapter. On the other hand, we see from the diagram that the specific power of the supercapacitor is substantially greater than it is for a lithium-ion battery, typically by a factor of about 5–10. This means that the supercapacitor can be discharged very rapidly (compared to the battery) and produce proportionately more power. It can also be charged more rapidly.

These particular properties of supercapacitors can be put to use in the design of a road vehicle. Three approaches to the use of supercapacitors have been taken in recent years, as follows:

*Additional power during acceleration for hybrid vehicles*—This approach has been utilized for high-performance sports cars such as the Lamborghini Sián FKP 37 described in the next section. This same approach has been used for racing cars such as the Toyota TS030 Hybrid.

*Storage of energy generated by regenerative braking*—The Mazda i-eloop system that is incorporated into recent Mazda 6 vehicles utilizes supercapacitors to store electrical energy that is generated during braking. This energy is used to supplement the electrical energy from the vehicle's battery to operate electrical accessories in the vehicle, such as the radio, lights, fans, etc. This reduces the load on the engine, which is used to turn the alternator to generate electricity, thereby improving fuel efficiency. As the energy stored in the supercapacitor does not contribute to the energy needed to drive the wheels, the vehicle can not, as discussed in the next section, properly be described as a hybrid.

*Energy for short-range city buses*—Electrical energy may be stored for city buses by recharging supercapacitors at bus stops in order to provide energy between stops.

## 2.11 Hybrid vehicle technology

In general, the term 'hybrid vehicle' refers to a vehicle that has two (or more) sources of stored energy that can contribute to propulsion. More specifically, the Society of Automotive Engineers defines a hybrid vehicle as a vehicle with more than one energy storage system that can provide propulsive power, either together or independently. This more precise definition is different in some subtle ways from the more common use of the term 'hybrid,' and this will be pointed out for some of the specific examples cited below. Although a number of different possible combinations of energy sources exist, the term 'hybrid vehicle' has most commonly been used in recent years to refer to a vehicle that uses batteries in conjunction with a gasoline internal combustion engine.

The present chapter has outlined some of the technologies related to BEVs and how this approach to road transportation has developed over the years. While it is likely that some form of BEV will account for a significant proportion of road vehicles for some time into the future, today's hybrid vehicles are an important step in the transition to a fossil-fuel-free transportation system. This section of the chapter provides an overview of gasoline–battery hybrid light-duty vehicles, as well as gasoline–supercapacitor hybrid vehicles and diesel–battery hybrid vehicles. Some other possible hybrid energy combinations that do not utilize fossil fuels will be discussed in later chapters of this book.

While some hybrid vehicles were constructed in the early years of road vehicle development, the modern era of hybrid development began in the 1990s as a means of reducing fossil-fuel consumption while maintaining vehicle performance characteristics. The first modern mass-produced hybrid vehicle was the Toyota Prius, as shown in figure 2.42, which was first marketed in Japan in 1997. The first hybrid vehicle to be sold in North America was the Honda Insight in 1999. The Toyota Prius was not sold in North America until 2000. Table 2.11 shows a comparison between the specifications of the first-generation Prius and Insight and similar size

**Figure 2.42.** First-generation Toyota Prius hybrid vehicle. This image (M.rJirapat 2013) has been obtained by the author from the Wikimedia website where it was made available under a CC BY-SA 4.0 licence, https://creativecommons.org/licenses/by-sa/4.0/deed.en. It is included within this chapter on that basis. It is attributed to M.rJirapat.

**Table 2.11.** Specifications of Toyota and Honda first-generation hybrid vehicles available in North America in 2000 and a comparison with similar size entry-level gasoline internal combustion engine vehicles available from these manufacturers at the same time (the Toyota Corolla and the Honda Civic). The specifications of the internal combustion engine vehicles are those of the base model available in North America. The base price does not include any government incentives.

| | | Gasoline engine | | Battery | | Electric | | EPA fuel consumption (L/100 km) | | Base price (USD) |
|---|---|---|---|---|---|---|---|---|---|---|
| Make | Model | Type | Power (kW) | Type | Capacity (kWh) | power (kW) | Weight (kg) | City | Highway | |
| Toyota | Prius | 1.5L I4 | 52 | NiMH | 1.78 | 33 | 1254 | 5.2 | 4.5 | 19 995 |
| Toyota | Corolla | 1.8L I4 | 93 | — | — | — | 1095 | 7.6 | 6.2 | 12 418 |
| Honda | Insight | 1.0L I3 | 50 | NiMH | 0.94 | 10 | 838 | 3.9 | 3.4 | 18 880 |
| Honda | Civic | 1.6L I4 | 79 | — | — | — | 1143 | 7.3 | 6.4 | 10 750 |

(compact) Toyota and Honda internal combustion engine vehicles from the same time. The specifications in the table show that early hybrid vehicles provided substantially improved fuel economy but somewhat less total available power. It is also clear that, at that time, hybrid vehicles were significantly more expensive (in terms of initial purchase price) than their gasoline-only counterparts. It has also been speculated that manufacturers were selling these early hybrid vehicles at a loss in order to make inroads into the emerging market for more environmentally conscious vehicles. From an overall financial standpoint, lower fuel costs can, at some point, outweigh the additional purchase costs for individuals.

Although the term 'hybrid vehicle' (as described in this section) refers to light-duty road vehicles that utilize both a gasoline internal combustion engine and battery-driven electric motors, there are a number of distinctly different technical designs that have been produced over the years and various terminology has subsequently been used to describe these different designs. Here, we look at the details of some of the important categories of hybrid vehicle. It is important, however, to realize that these categories are not always uniquely defined, nor are they always mutually exclusive.

### 2.11.1 Parallel hybrids

Parallel hybrids are the most common form of hybrid available at present. In this configuration, an internal combustion engine and an electric motor can both provide power to the wheels, either individually or together. Both sources of power are coupled to a gearbox, which in turn drives the wheels through an automatically controlled clutch. Either or both power sources can be coupled through the respective clutches to the gearbox. If both power sources are coupled to the gearbox at the same time, then the speeds must be matched. An alternative arrangement that has recently been used is to couple the internal combustion to one axle and the electric motor to the other. This configuration is sometimes referred to as a *through the road hybrid.* The relative proportions of power that can be supplied by the internal combustion engine and the electric motor vary between different hybrid vehicles. Like most types of hybrid, parallel hybrids generally take advantage of regenerative braking to assist in maintaining battery charge.

In the case of the Honda Insight, as illustrated by the specifications in table 2.10, the power available from the electric motor is much smaller than that available from the internal combustion engine. In this case, the vehicle is generally referred to as a *mild hybrid*, and the electric motor is largely used to supplement the internal combustion engine when additional power is needed during acceleration or hill climbing.

A variation on the parallel hybrid is the so-called power-split hybrid, sometimes referred to as a series–parallel hybrid. In this design, the power from the internal combustion engine and the power from the electric motor are divided by a power-splitting device so that the ratio of the power to the wheels can be anywhere from 100% internal combustion engine to 100% electric motor. The Toyota Prius is an example of this type of configuration.

### 2.11.2 Series hybrids

In a true series hybrid, the wheels are driven only by the electric motor and the internal combustion engine is coupled to a generator to provide energy to recharge the batteries. Strictly speaking, a series hybrid of this design is not really a hybrid according to the definition used by the Society of Automotive Engineers, because the internal combustion engine cannot be coupled to the drive wheels. In fact, some manufacturers have avoided referring to series hybrids as hybrids at all. Decoupling the internal combustion engine from the drive wheels has two potential advantages. First, the engine is faced with a fairly constant load and can operate at a speed that is compatible with its maximum efficiency. Second, since the average power utilized by the vehicle is much less than the maximum needed (e.g. for acceleration), the internal combustion engine can be much less powerful than would be otherwise required. A series hybrid drive arrangement has long been used for diesel–electric locomotives and ships but has only been utilized for road vehicles during the past decade or so.

One of the earliest production series hybrid automobiles was the Fisker Karma (see figure 2.43), manufactured from 2011 to 2012. Table 2.12 gives some of the specifications of the Fisker Karma. The Karma was designed as a luxury vehicle and had a base price in (2011) of 95 900 USD. It is clear from the specifications that Fisker emphasized performance, in keeping with the expected market for this vehicle. The specifications also illustrate, on the basis of range, that the Karma relied heavily on the consumption of gasoline. The Karma also incorporated photovoltaic cells to provide an additional source of energy. The roof of the Karma, as shown in figure 2.44, contained a photovoltaic panel that provided electricity to maintain battery charge. The manufacturer claimed that solar energy could provide up to 8 km of additional range per week. Fisker produced around 2500 Karmas between 2011 and 2012 before ending production due to financial difficulties. The rights to the Karma drivetrain technology were sold to the Chinese

Figure 2.43. Fisker Karma series hybrid manufactured from 2011 to 2012. This image (Indianhilbilly 2011) has been obtained by the author from the Wikimedia website where it was made available under a CC BY-SA 3.0 licence, https://creativecommons.org/licenses/by-sa/3.0/deed.en. It is included within this chapter on that basis. It is attributed to Indianhilbilly.

Table 2.12. Specifications of some series hybrids. Specifications are for first-generation vehicles and/or base-model vehicles. Prices are for vehicles sold in the United States and are exclusive of any government incentives.

| | | Gasoline engine | | Battery | | | | EPA range (km) | | |
|---|---|---|---|---|---|---|---|---|---|---|
| Make | Model | Type | Power (kW) | Type | Capacity (kWh) | Electric power (kW) | Weight (kg) | Electric | Electric+ gasoline | Base price (USD) |
| Fisker | Karma (2011) | 2.0L I4 | 194 | Li-ion | 20.1 | 240 | 2400 | 51 | 370 | 95 900 |
| Chevrolet | Volt (2011) | 1.4L I4 | 63 | Li-ion | 16.0 | 166 | 1721 | 56 | 610 | 40 280 |
| BMW | i3 (2014) | 0.65L I2 | 25 | Li-ion | 18.2 | 130 | 1315 | 130 | 241 | 46 125 |

manufacturer Wanxiang Group Corporation, which has marketed an updated version of the Fisker Karma since 2016 under the name Karma Revero with a starting price of around 130 000 USD.

**Figure 2.44.** Solar roof panel on a Fisker Karma. This image (Ajzh2074 2015) has been obtained by the author from the Wikimedia website where it was made available under a CC BY-SA 4.0 licence, https://creativecommons.org/licenses/by-sa/4.0/deed.en. It is included within this chapter on that basis. It is attributed to Ajzh2074.

Perhaps the best-known series hybrid is the Chevrolet Volt, manufactured from 2011 to 2019. The Volt is shown in figure 2.45 and its specifications are given in table 2.12. Like the Fisker Karma, the range of the Volt is dominated by its gasoline supply. However, it is clear that the maximum power available is largely the result of the properties of the electric motor and the power available from the lithium-ion batteries. Strictly speaking, the Volt is not a true series hybrid, as the gasoline internal combustion engine does mechanically connect to the drive wheels at high speed to provide additional power to assist the electric drive. It does, therefore, have some characteristics, under some circumstances, of a parallel hybrid. The Volt was quite successful during its nine years of production, selling about 177 000 units worldwide.

Series hybrids are sometimes referred to as *extended-range electric vehicles.* In the case of the Fisker Karma and the Chevrolet Volt it is clear that the vehicle's range is dominated by the energy stored in its gasoline supply. The term 'extended-range electric vehicle' is actually more appropriate for the BMW i3. While the i3 is basically a BEV, as shown in table 2.7, it is also available, at an additional cost (about 3850 USD), as an extended-range electric vehicle. This version of the i3, shown in figure 2.46, has a 25 kW 647 cc inline two-cylinder engine that BMW uses in its BMW C650 GT maxi-scooter. Additional specifications of this vehicle are

**Figure 2.45.** Chevrolet Volt series hybrid showing the charger cable. This image (IFCAR 2007) has been obtained by the author from the Wikimedia website, where it is stated to have been released into the public domain. It is included within this chapter on that basis.

**Figure 2.46.** BMW i3 with range extender. This image (Tokumeigakarinoaoshima 2014) has been obtained by the author from the Wikimedia website, where it is stated to have been released into the public domain. It is included within this chapter on that basis.

given in table 2.12. The i3 with range extender operates as a BEV until the battery charge drops below a certain level, at which time the gasoline engine starts up and drives a generator to maintain the battery charge level at about 30% for as long as the gasoline supply lasts. This approach to a vehicle that is basically a BEV represents a possible transition to a pure electric vehicle for drivers who obtain comfort from the inclusion of a traditional energy source and the knowledge that they are not only dependent on the range and charge times associated with the batteries. As discussed further in chapter 6, this approach to an extended-range BEV may provide a viable future option if the gasoline is replaced by a low-carbon alternative. The BMW i3 was introduced in 2013 and continues to be available, at least as of 2021. It has been quite successful, with cumulative global sales exceeding 172 000 as of 2019.

### 2.11.3 Plug-in hybrids

Plug-in hybrids may be either parallel hybrids or series hybrids, as described above. They are distinguished from conventional hybrids because the batteries can be charged either by an onboard generator run by the internal combustion engine (as is the case for conventional hybrids) or from an external current source. For most users, the current source is an AC home outlet connected to grid power. Faster charging can be accomplished by a dedicated high-current external DC charger. For AC charging, the vehicle must contain appropriate AC-to-DC converter electronics.

For short trips, the plug-in hybrid can more or less be utilized in the same way as a BEV, i.e. operating on power from the batteries and without the need for the internal combustion engine. If the batteries are recharged from an external source, the vehicle can be considered carbon-free and because they can, in principle, be used without burning gasoline, they are often included in data compilations of electric vehicles. However, some plug-in parallel hybrid configurations utilize the internal combustion engine for additional power during high-demand situations. When the battery is sufficiently depleted during driving, the internal combustion engine starts to recharge the batteries and/or provide mechanical power to the wheels, depending on the drivetrain layout.

Plug-in hybrid vehicles typically have greater battery capacity than similar non-plug-in hybrid vehicles, in order to provide as much emission-free propulsion as possible until the batteries can be recharged from an external current source. Following the same philosophy, the batteries in plug-in hybrid vehicles (compared to non-plug-in vehicles) are designed for deeper cycling, in order to avoid the need to use the internal combustion engine during short trips.

The history of hybrid vehicles, like that of BEVs, begins during the early years of the development of the automobile in the 19th century. One of the early notable plug-in hybrid gasoline–battery road vehicles was the Lohner–Porsche Mixed Hybrid shown in figure 2.47.

The interest in plug-in hybrids was revived around 2000 during the early development and commercialization of BEVs and conventional hybrid vehicles. The first plug-in hybrid that was available for sale was the Chinese-made BYD

**Figure 2.47.** A 1902 Lohner–Porsche Mixed Hybrid vehicle. This was a series plug-in hybrid with independent hub motors in each of the four wheels, as can be seen in the photograph. This image (Lohner Posrche 1902) has been obtained by the author from the Wikimedia website, where it is stated to have been released into the public domain. It is included within this chapter on that basis.

F3DM. This vehicle, as seen in figure 2.48, was made available for fleet sales in 2008 and for sale to the general public in 2010. Series hybrids, such as the Chevrolet Volt, BMW i3 with range extender, and Fisker Karma, are plug-in hybrids. In addition, a number of previously available hybrids, such as the Toyota Prius, have subsequently been made available as plug-in hybrids. The most popular plug-in hybrid vehicles in recent years have been the Chevrolet Volt, the Mitsubishi Outlander PHEV, and the Toyota Prius PHV. As of 2020, there were about three million light-duty plug-in hybrid vehicles worldwide.

### 2.11.4 Diesel–electric hybrids

While some prototype or limited-production light-duty diesel–electric hybrid vehicles have been produced, the largest fraction of diesel–electric hybrid road vehicles have been heavy-duty vehicles, such as buses and trucks, that traditionally use diesel engines. This approach has most commonly been applied to city buses, although some truck prototypes have also been developed. As in the case of light-duty road vehicles, hybrid technology as applied to heavy-duty vehicles falls into two categories: parallel hybrid designs and series hybrid designs. Commercial diesel–battery hybrid buses have been produced in both configurations; examples include a parallel hybrid bus, the Volvo B5LH, and a series hybrid bus, the Alexander Dennis Enviro400 MMC. More than 20 manufacturers produce hybrid buses commercially, and these buses are in common use as city buses in more than 200 urban areas

**Figure 2.48.** BYD F3DM plug-in hybrid being charged. This image (Ortiz 2012a) image has been obtained by the author from the Wikimedia website where it was made available under a CC BY-SA 3.0 licence, https://creativecommons.org/licenses/by-sa/3.0/deed.en. It is included within this chapter on that basis. It is attributed to Mariordo (Mario Roberto Duran Ortiz).

worldwide, mostly in North America and Europe. Series hybrid drivetrain buses, in particular, seem to be advantageous for city transportation, as they allow the traditionally large diesel engine to be replaced with a much smaller engine that runs efficiently at constant speed to maintain battery charge. Diesel engines are particularly suited for constant-load operation and offer high-efficiency operation and excellent reliability. Plug-in hybrid buses are also under development.

Hybrid diesel–electric trucks have not been commercialized to the same extent as hybrid diesel–electric buses and have not found universal acceptance as long-range transport trucks. However, some models, such as the Volvo FE Hybrid and the Mitsubishi Fuso Canter Eco Hybrid, have found applications as package delivery vehicles and garbage collection vehicles.

### 2.11.5 Supercapacitor hybrids

As noted above, supercapacitors do not have sufficient specific energy storage capacity to replace batteries in BEVs. However, their high specific power makes them attractive for certain hybrid vehicle applications, and supercapacitors have been put to use in some specialized vehicles. The Lamborghini Sián FKP 37, shown in figure 2.49, is an interesting example of a hybrid vehicle that employs supercapacitors. The Sián, which was introduced in 2020, is an all-wheel drive vehicle with a 6.5L V-12 gasoline engine that produces 577 kW. A 25 kW electric motor energized by supercapacitors provides additional power as needed. The electric

**Figure 2.49.** Lamborghini Sián FKP 37. This image (Migl A 2019b) has been obtained by the author from the Wikimedia website where it was made available under a CC BY-SA 4.0 licence, https://creativecommons.org/licenses/by-sa/4.0/deed.en. It is included within this chapter on that basis. It is attributed to Alexander Migl.

motor is located within the gearbox, where it can be directly linked to the drive wheels. A comparison between the power provided by the internal combustion engine and that provided by the electric motor clearly indicates that the Sián is a mild parallel hybrid.

Hybrid vehicle technology has been transitional and will continue to be so for some time into the future. Hybrids have been instrumental in reducing carbon emissions from road vehicles, even though they continue to burn fossil fuels. Transitioning to plug-in hybrids and extended-range BEVs may be a logical step in expanding the use of all electric vehicles. Hybrids have also been important platforms for technology development that will, no doubt, benefit future carbon-free transportation technologies, and they have also served to highlight public awareness of the detrimental effects of the continued use of fossil fuels for transportation.

## 2.12 Summary

This chapter has described the development and enormous growth of BEVs in the past two decades or so. This has included both the substantial increase in the number of battery electric and plug-in hybrid vehicles as well as the development of infrastructure such as public charging stations. Nevertheless, electric vehicles still represent less than 1% of all road vehicles worldwide. Public acceptance of this technology, as well as government incentives, suggest that this growth will continue. In the short term, this continued growth requires not only the continued technical development of vehicles, but also the continued expansion of charging infrastructure. It is also important to consider the careful management of lithium resources, as the need for lithium for vehicle batteries will compete with other lithium needs. Lithium production rates will need to be assessed to ensure adequate supply. Recent research into lithium-ion batteries that are more economical and have higher specific energy and shorter recharging times has increased the viability of these batteries for vehicle use. In addition, the development of non-lithium-based battery chemistries has made progress in recent years, and at some point, these batteries are likely to make an impact on electric vehicle designs and will help to alleviate cost and resource availability concerns.

The further integration of BEVs into our transportation infrastructure will also require a consideration of grid generating capacity. Sustainable, i.e. carbon-free, electricity generation must meet the increased demand resulting from the reduction of fossil-fuel energy utilization. This is considered more quantitatively in chapter 6. The eventual development of technologies such as electric roads may be necessary to accommodate all future road transportation needs. Finally, although hybrid vehicles, as discussed in this chapter, still rely on the availability of fossil fuels, the possibilities offered by other hybrid transportation technologies that are environmentally sustainable are discussed in chapters 3 and 4.

## References

Afernand74 2011 Jamais Contente - Compiègne - 2011 https://commons.wikimedia.org/wiki/File:Jamais_Contente_-_Compi%C3%A8gne_-_2011.JPG

Alotto P, Guarnieri M and Moro F 2014 Redox flow batteries for the storage of renewable energy: a review *Renew. Sustain. Energy Rev.* **29** 325–35

Ajzh2074 2015 FiskerKarma 7 https://commons.wikimedia.org/wiki/File:FiskerKarma_7.jpg

Cepheiden 2020 Li-Ion-Zelle (NMC-Carbon, Schema) https://commons.wikimedia.org/wiki/File:Li-Ion-Zelle_(NMC-Carbon,_Schema).svg

Cooley B 2021 'What drops an EV's range? Learn the risk factors' *Road Show* (11 September 2021) available at: www.cnet.com/roadshow/news/ev-range-electric-car-mileage/

Cozmcrae 2020 TS Bravo https://commons.wikimedia.org/wiki/File:TS_Bravo.jpg

Cullen328 2016 1911 Detroit Electric https://commons.wikimedia.org/wiki/File:1911_Detroit_Electric.jpg

DaMastah 2020 Renault Zoe ZE50 White https://commons.wikimedia.org/wiki/File:Renault_Zoe_ZE50_White.jpg

Doetsch C and Burfeind J 2016 Vanadium redox flow batteries ed T M Letcher *Storing Energy—With Special Reference to Renewable Energy Sources* (Amsterdam: Elsevier), ch 12 pp 227–46

DRoberson 2005 Kilowatt https://commons.wikimedia.org/wiki/File:Kilowatt.jpg

Dunlap R A 2021 *Energy from Nuclear Fusion* (Bristol: IOP Publishing)

Egibe 2019 Li-S battery EN https://commons.wikimedia.org/wiki/File:Li-S_battery_EN.png

Elcap 2013a Supercapacitors-vs-batteries-chart https://commons.wikimedia.org/wiki/File:Supercapacitors-vs-batteries-chart.png

Elcap 2013b EDLC-Charge-Distribution https://commons.wikimedia.org/wiki/File:EDLC-Charge-Distribution.png

Fabian_R 2013 Plate CapacitorII https://commons.wikimedia.org/wiki/File:Plate_CapacitorII.svg

Haag Franz 2011 1888 Flocken Elektrowagen https://commons.wikimedia.org/wiki/File:1888_Flocken_Elektrowagen.jpg

Hwang J Y, Myung S T and Sun Y K 2017 Sodium-ion batteries: present and future *Chem. Soc. Rev.* **46** 3529–614

Ibex73 2020 Citroën Ami 2020 (1) https://commons.wikimedia.org/wiki/File:Citro%C3%ABn_Ami_2020_(1).jpg

IFCAR 2007 Chevrolet-Volt-DC https://commons.wikimedia.org/wiki/File:Chevrolet-Volt-DC.jpg

IFCAR 2011 Tesla Roadster -- 02-11-2011 https://commons.wikimedia.org/wiki/File:Tesla_Roadster_--_02-11-2011.jpg

Indianhilbilly 2011 Fisker Karma 2 https://commons.wikimedia.org/wiki/File:Fisker_Karma_2.jpg

Jengtingchen 2021 Xpeng P7 012 https://commons.wikimedia.org/wiki/File:Xpeng_P7_012.jpg

Jurvetson S 2019 The Tesla Semi Truck by Steve Jurvetson https://www.flickr.com/photos/jurvetson/40705940423/

Jwh 2013 Eng Renault Twizy ZE https://commons.wikimedia.org/wiki/File:Eng_Renault_Twizy_ZE.jpg

Kane M 2019 'Quick look at cobalt and nickel usage in xEVs for H1 2019' *InsideEVs* (10 October 2019) https://insideevs.com/news/375146/cobalt-nickel-usage-xevs-h1-2019/

Lohner Porsche 1902 File:Lohner Porsche.jpg https://commons.wikimedia.org/wiki/File:Lohner_Porsche.jpg

Migl A 2019a Polestar 2 Genf 2019 1Y7A6000 https://commons.wikimedia.org/wiki/File:Polestar_2_Genf_2019_1Y7A6000.jpg

Migl A 2019b Lamborghini Sian at IAA 2019 IMG 0332 https://commons.wikimedia.org/wiki/File:Lamborghini_Sian_at_IAA_2019_IMG_0332.jpg

Migl A 2019c Smart EQ forfour at IAA 2019 IMG 0799 https://commons.wikimedia.org/wiki/File:Smart_EQ_forfour_at_IAA_2019_IMG_0799.jpg

Migl A 2021 Ford Mustang Mach-E GT IAA 2021 1X7A0176 https://commons.wikimedia.org/wiki/File:Ford_Mustang_Mach-E_GT_IAA_2021_1X7A0176.jpg

M.rJirapat 2013 Toyota Prius in Toyota Automobile Museum https://commons.wikimedia.org/wiki/File:Toyota_Prius_in_Toyota_Automobile_Museum.jpg

NASA John Glenn Research Center 2011 NaS battery, derivative work: Quibik https://commons.wikimedia.org/wiki/File:NaS_battery.png

Nole D A and Moss V 1970 Battery employing lithium–sulphur electrodes with non-aqueous electrolyte *U.S. Patent* 3532543 https://image-ppubs.uspto.gov/dirsearch-public/print/downloadPdf/3532543

Nguyen D-T and Song S-W 2017 Magnesium stannide as a high-capacity anode material for magnesium-ion batteries *J. Power Sources* **368** 11–7

Ortiz M R D 2012a Bilbao 05 2012 BYD F3DM 2447 https://commons.wikimedia.org/wiki/File:Bilbao_05_2012_BYD_F3DM_2447.JPG

Ortiz R M D 2012b Toyota RAV4 EV WAS 2012 0759 https://commons.wikimedia.org/wiki/File:Toyota_RAV4_EV_WAS_2012_0759.JPG

Peled E, Gorenshtein A, Segal M and Sternberg Y 1989 Rechargeable lithium–sulfur battery *J. Power Sources* **26** 269–71

Peters J F, Peña Cruz A and Weil M 2019 Exploring the economic potential of sodium-ion batteries *Batteries* **5** 10

Plug In America 2010 Ev1 crush5 https://commons.wikimedia.org/wiki/File:Ev1_crush5.jpg

Qi Z and Koenig G M Jr. 2017 Review Article: Flow battery systems with solid electroactive materials *J. Vaccum Sci. Technol. B* **35** 040801

Reinhold A 2004 Trolleybus4120.Harvard.agr https://commons.wikimedia.org/wiki/File:Trolleybus4120.Harvard.agr.JPG

Reva Norge 2009 Reva i silver https://commons.wikimedia.org/wiki/File:Reva_i_silver.jpg

Rowen R 2002 Green EV1 by RightBrainPhotography https://commons.wikimedia.org/wiki/File:Green_EV1.jpg

Rychcik M and Skyllas-Kazacos M 1988 Characteristics of a new all-vanadium redox flow battery *J. Power Sources* **22** 59–67

SanJoaquinRTD 2013 Proterra Electric Bus at Charging Station https://commons.wikimedia.org/wiki/File:Proterra_Electric_Bus_at_Charging_Station.jpg

Skundin A M, Kulova T L and Yaroslavtsev A B 2018 Sodium-ion batteries (a review) *Russ. J. Electrochem.* **54** 113–52

Spataru C and Bouffaron P 2016 Off-grid energy storage ed T M Letcher *Storing Energy—With Special Reference to Renewable Energy Sources* (Amsterdam: Elsevier), ch 22 pp 477–97

Tokumeigakarinoaoshima 2014 BMW i3 Range Extender (1Z06) front https://commons.wikimedia.org/wiki/File:BMW_i3_Range_Extender_(1Z06)_front.JPG

Tolmachev Y V, Piatkivskyi A, Ryzhov V V, Konev D V and Vorotyntsev M A 2015 Energy cycle based on a high specific energy aqueous flow battery and its potential use for fully electric vehicles and for direct solar-to-chemical energy conversion *J. Solid State Electrochem.* **19** 2711–22

Triplecaña 2019 Vespa elettrica 2 https://commons.wikimedia.org/wiki/File:Vespa_elettrica_2.jpg

U.S. Department of Energy 2014 Aluminum–air experimental power cell https://www.flickr.com/photos/departmentofenergy/13471575435/

U.S. Department of Energy 2021 'Developing infrastructure to charge plug-in electric vehicles' available at: https://afdc.energy.gov/fuels/electricity_infrastructure.html

U.S. Geological Survey 2022 'Cobalt' https://pubs.usgs.gov/periodicals/mcs2022/mcs2022-cobalt.pdf

Yabuuchi N and Komaba S 2014 Recent research progress on iron- and manganese-based positive electrode materials for rechargeable sodium batteries *Sci. Technol. Adv. Mater.* **15** 043501

Yu J, Fang D, Zhang H, Leong Z Y, Zhang J, Li X and Yang H Y 2020 Ocean mining: a fluidic electrochemical route for lithium extraction from seawater *ACS Mater. Lett.* **2** 1662–8

Zurek S 2006 Supercapacitor diagram https://commons.wikimedia.org/wiki/File:Supercapacitor_diagram.svg

**IOP** Publishing

Transportation Technologies for a Sustainable Future
Renewable energy options for road, rail, marine and air transportation
**Richard A Dunlap**

# Chapter 3

# Hydrogen and its derivatives

## 3.1 Introduction

Hydrogen is the most abundant element in the Universe, accounting for over 90% of all atoms (or ions) and about 74% of the mass. It is the third most abundant element on the Earth's surface in terms of the number of atoms and the tenth most abundant by mass. There has been much written about the possible contribution of hydrogen to future energy systems. Specifically, the so-called 'hydrogen economy' has been promoted as a replacement for our current fossil-fuel-dominated energy system. It is essential, however, to realize that hydrogen is not an energy source, although it is often referred to as such. Hydrogen is merely an energy storage medium, in the same sense as a rechargeable battery. There are no reserves of free hydrogen on Earth. Virtually all hydrogen atoms available from the Earth's crust or atmosphere are bound to other atoms in the form of compounds. In order to obtain atoms of free hydrogen, the compounds in which it is contained must be broken down, and this requires energy.

As an example of the difference between an energy source and an energy storage medium, it is useful to compare hydrogen to gasoline. Gasoline is not found on Earth in its pure state but must be prepared by processing crude oil. Crude oil, which is extracted from underground wells, is transported from the well site to a refinery, where it is separated into its various constituent hydrocarbons according to their molecular weights. The lightest of these is methane, $CH_4$, which is commonly referred to as natural gas. The heaviest hydrocarbons are materials such as tar, which is used for roofing and road paving purposes. Gasoline falls in between these two extremes and is closer to the light end; its compositions are close to that of octane, $C_8H_{18}$. Once these middle-weight hydrocarbons are extracted from the crude oil, they are transported to facilities such as gasoline stations for distribution to the end users. This entire process uses energy, but this energy is equivalent to perhaps 5% of the energy content of the gasoline that is provided to the consumer. Thus, the vast majority of the energy that is obtained from the environment is available for use. This may also be viewed in terms of the ratio of the energy available from the gasoline to the energy input into processing, which is about 20 to one.

doi:10.1088/978-0-7503-5306-9ch3

The situation described above does not apply to hydrogen. Hydrogen is obtained by the decomposition of hydrogen-containing compounds using an energy input. A simple example of this process is the production of molecular hydrogen, $H_2$, by the decomposition of water, $H_2O$, using electrolysis. This is described in detail later in the chapter. The hydrogen thus formed can be stored and later used as a fuel by recombining it with oxygen to produce water and energy in the form of electricity (using a fuel cell) or heat (by combustion). Thus, the overall process of making hydrogen and using it as a fuel can be represented as follows:

$$2H_2O + \text{energy in} \rightarrow 2H_2 + O_2 \rightarrow 2H_2O + \text{energy out}.$$

Unfortunately, in this case, the energy output is typically less than the energy input by a factor of two or so. This means that the ratio of the energy available from the hydrogen to the energy input into processing is about one to two. This is a characteristic of energy storage methods, as energy is always lost in the storage process. Since hydrogen may be viewed as a means of storing electricity, it is sometimes referred to as an electrofuel or e-fuel.

It is important to carefully analyze the process of using hydrogen as an energy storage method in order to fully understand its usefulness in future energy technologies. This chapter looks at some of the properties of hydrogen as a fuel, the methods of making hydrogen, its use as a transportation fuel, and its advantages and disadvantages over transportation technologies based on other fuels.

## 3.2 The properties of hydrogen

Hydrogen is the lightest of all elements and has a nucleus that contains just a single proton. Naturally occurring hydrogen consists of 99.985% hydrogen atoms with no neutrons in their nuclei (referred to as $^1H$) and 0.015% hydrogen atoms with one neutron in their nuclei (referred to as $^2H$ or deuterium). Since it is by far the most abundant component of natural hydrogen, it is the properties of $^1H$ that are relevant to its use as a fuel.

Some properties of hydrogen ($^1H$) are given in table 3.1. At standard temperature (293 K) and pressure (1 atm), hydrogen is a gas with a very low density. A comparison with the density of air (1.2 kg $m^{-3}$) shows that hydrogen is less than 8% as dense as air. This low density has a severe impact on the use of hydrogen as a fuel, particularly in portable applications. This is made obvious by a comparison between the energy density of hydrogen and that of gasoline. Gasoline has an energy density of about 34.8 MJ $L^{-1}$ or about 47 MJ $kg^{-1}$. Per unit weight, the table shows that the

**Table 3.1.** Properties of molecular hydrogen, $H_2$.

| Quantity | Value |
|---|---|
| Boiling point (1 atm) | 20.3 K |
| Gas density (293 K, 1 atm) | 0.09 kg $m^{-3}$ |
| Heat of combustion | 142 MJ $kg^{-1}$ |
| Heat of combustion (gas 293 K, 1 atm) | 12.7 MJ $m^{-3}$ |

energy density of hydrogen is about three times that of gasoline. However, because of its low density (at standard temperature and pressure), the volumetric energy density of hydrogen gas is only about 0.04% that of gasoline. It is clear that in order to make hydrogen a viable portable fuel it is necessary to increase its density considerably. The possible ways of approaching this problem are discussed in detail in the next section.

## 3.3 Hydrogen storage

As noted above, it is necessary to increase the density of hydrogen in order to utilize it as a transportation fuel. There are two obvious ways of doing this: by compressing hydrogen gas to a high pressure or by cooling it to a sufficiently low temperature, i.e. about 20 K, so that it becomes a liquid. It is also possible to cool the hydrogen gas to a temperature that is not sufficient to liquefy it and then to compress it, thereby increasing its density as a result of both pressure and temperature, as described by the ideal gas law. Other approaches to increasing the volumetric energy density of hydrogen include absorbing it inside a solid, adsorbing it onto the surface of a solid, and using it to produce a hydrogen-containing gaseous or liquid compound. In all cases except the last, the hydrogen can be recovered and used to create energy by reacting with oxygen. In this section, three possible hydrogen storage approaches, namely, compression, liquefaction, and storage in solids, are discussed in detail. The possibility of using hydrogen to form gaseous or liquid compounds, which can then be used as fuel, is discussed in sections 3.10–3.12.

### 3.3.1 Compressed hydrogen

At low pressures, hydrogen behaves like an ideal gas and its density, $\rho$, increases linearly with pressure, $P$, according to the ideal gas law:

$$\rho = \text{constant} \times \frac{P}{T},$$

where $T$ is the absolute temperature. For pressures above about 200 atmospheres, the density does not increase with pressure as fast as the ideal gas law indicates. The ideal gas law is based on the assumption that the molecules in a gas do not interact with each other. This is a good assumption as long as the molecules are far enough apart. As the pressure and density increase, the molecules get closer together and at some point (about 200 atmospheres for hydrogen) the molecules interact with one another sufficiently that the ideal gas law breaks down.

For road vehicle use, two pressures, i.e. 350 atmospheres and 700 atmospheres, have commonly been utilized for hydrogen storage in fuel tanks. Table 3.2 provides some properties of hydrogen gas at these two pressures. Included in the table is the relative volumetric energy density of hydrogen compared to that of gasoline. While the volumetric energy density of compressed hydrogen does not equal that of gasoline, reaching only about 16% of gasoline energy density at the highest pressure, it is, however, sufficient for hydrogen to be considered as a portable transportation fuel. Depending on the efficiency of energy conversion, as discussed in detail below,

**Table 3.2.** Properties of compressed molecular hydrogen gas.

| Quantity | Value |
|---|---|
| Gas density (293 K, 350 atm) | 24 kg $m^{-3}$ |
| Gas density (293 K, 700 atm) | 40 kg $m^{-3}$ |
| Heat of combustion (gas 293 K, 350 atm) | 3400 MJ $m^{-3}$ |
| Heat of combustion (gas 293 K, 700 atm) | 5700 MJ $m^{-3}$ |
| Volumetric energy density relative to gasoline (350 atm) | 0.098 |
| Volumetric energy density relative to gasoline (700 atm) | 0.164 |

several approaches may be taken in order to use hydrogen as a road vehicle fuel in place of gasoline:

- increase the fuel tank volume;
- decrease the vehicle range;
- add a second energy source (i.e. a hybrid vehicle).

Choosing an appropriate pressure for hydrogen storage for road vehicle use requires the consideration of a number of factors. It is obvious from the table that increasing the pressure increases the density and therefore increases the mass of hydrogen that can be stored in a tank of a specific size. However, as pressure increases, the strength of the tank also needs to be increased. This involves additional cost and the potential for reduced safety. In addition, compressing hydrogen requires energy. To achieve the pressures shown in the table, 10%–20% of the energy content of the hydrogen has to be expended. Finally, as the pressure is increased, the breakdown of the ideal gas law means that the benefits of additional compression diminish. For example, increasing the pressure from 700 atmospheres to 1400 atmospheres only increases the density from about 40 kg $m^{-3}$ to about 50 kg $m^{-3}$. As a result of these considerations, the pressures that have been used for hydrogen storage in previously manufactured vehicles fall into the range given in the table.

### 3.3.2 Liquefied hydrogen

Hydrogen can be liquefied at atmospheric pressure by cooling it to below about 20 K. Some of the relevant properties of liquid hydrogen are summarized in table 3.3.

The table indicates that there is some improvement in volumetric energy density compared to compressed hydrogen at 700 atmospheres. However, there are some potential drawbacks to the use of liquid hydrogen as a vehicle fuel. First, the liquefaction process utilizes about 30% of the energy content of the hydrogen. Second, storage tanks and vehicle fuel tanks must be well insulated to minimize liquid hydrogen loss by evaporation. Third, the transfer of liquid hydrogen requires special care to avoid liquid loss and/or contamination with condensates from the air. Finally, no matter how well the liquid hydrogen is insulated, some evaporation occurs. The rate of evaporative loss depends not only on the quality of the insulation but also on the surface-to-volume ratio of the container. The surface-to-volume ratio increases with decreasing container size, so that while a large central storage tank

**Table 3.3.** Properties of liquid hydrogen.

| Quantity | Value |
|---|---|
| Liquid density (20.3 K) | 71 kg $m^{-3}$ |
| Heat of combustion ($H_2$ liquid) | 10 100 MJ $m^{-3}$ |
| Volumetric energy density relative to gasoline | 0.29 |

may lose minimal hydrogen over a period of years, a small, vehicle-size tank can lose a substantial fraction of liquid hydrogen over a period of a couple of months.

### 3.3.3 Cryo-compressed hydrogen

Both compressed hydrogen storage and liquefied hydrogen storage have been used for automotive applications, as described below. However, both have advantages as well as disadvantages. Compressed hydrogen gas requires less sophisticated handling techniques than liquefied hydrogen but has the disadvantages of requiring high-pressure tanks and a somewhat lower energy density. Liquefied hydrogen has a higher energy density but suffers from the need for more complex handling techniques and losses due to boiloff. Recent work on hydrogen storage has included the possibility of cryo-compression methods which help to resolve some, although not all, of the disadvantages of other storage approaches. Cryo-compression stores hydrogen in the liquid phase under pressure. This approach increases the volumetric capacity above that of the liquid phase at standard pressure, and because the storage vessel is sealed, it reduces boiloff. It utilizes pressures that are significantly lower than the 700 atmospheres typically used for compressed gas storage and therefore requires less robust tank materials and construction. For example, storing liquid hydrogen at 240 atmospheres and 20 K provides a density of about 87 $kg{\cdot}m^{-3}$, compared to 70 $kg{\cdot}m^{-3}$ at ambient pressure.

Figure 3.1 shows a typical storage vessel for cryo-compressed hydrogen. The inner storage vessel is constructed from carbon fiber lined with aluminum. The inner vessel is surrounded by a vacuum jacket incorporating multiple layers of metalized plastic sheet to reduce radiative losses. The outer vacuum vessel is made of stainless steel.

Cryo-compressed hydrogen storage has not yet been used for hydrogen internal combustion engine or fuel cell vehicles, but if these vehicles become more prevalent in the future, cryo-compressed storage may offer some benefits.

### 3.3.4 Hydrogen storage in solids

While hydrogen vehicles constructed thus far have utilized either high-pressure hydrogen gas or liquid hydrogen as a fuel, another possibility for portable hydrogen storage may be developed for future transportation use. This is the use of solid materials for hydrogen storage. Hydrogen may be stored by chemical bonding with some of the atoms of the solid, typically in the form of metal hydrides, or it may be stored by being adsorbed onto the surface of the material. We look at some examples of these two approaches.

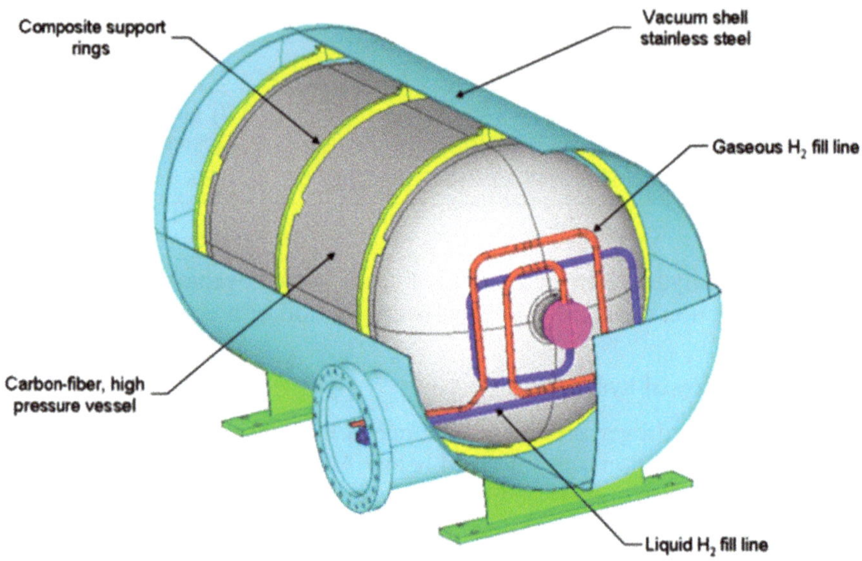


**Figure 3.1.** Vessel for storing cryo-compressed hydrogen. The outer shell is 58 cm in diameter and 129 cm in length. Reprinted from Aceves *et al* (2009), copyright (2009), with permission from Elsevier.

### *3.3.4.1 Metal hydrides*

Hydrogen readily reacts with many elemental metals and metallic compounds. These materials typically contain a metallic element from the left side of the periodic table. One example of the reaction between hydrogen and a metal that results in the formation of a metal hydride is the reaction between hydrogen and magnesium:

$$Mg + H_2 \rightarrow MgH_2.$$

This reaction is exothermic, meaning that it releases energy and occurs spontaneously. It is, therefore, a readily accessible means of storing hydrogen. Table 3.4 compares the hydrogen storage ability of magnesium with those of compressed gas storage and liquid storage. It can be seen that magnesium hydride has advantages in terms of hydrogen content per unit volume and per unit mass. In fact, $MgH_2$ has about 45% the energy per unit volume of gasoline (3.5).

Once hydrogen is stored in the form of a metal hydride, it is necessary to release the hydrogen so that it can be utilized as fuel in a vehicle. The simple approach would be to decompose the magnesium hydride according to the following reaction:

$$MgH_2 \rightarrow Mg + H_2$$

Since the formation of magnesium hydride is an exothermic process, it follows that the decomposition of magnesium hydride is an endothermic process. That means that energy must be supplied in order to release the hydrogen from the hydride. The energy is typically supplied by heating the hydride and this would need to be done on board the vehicle. There are several potential drawbacks to this approach. First, it considerably complicates the fuel system of the vehicle. Second, the rate at which hydrogen can be liberated from the hydride can limit the rate at which fuel is

**Table 3.4.** Hydrogen stored in a 100 L container in different forms. It is assumed that the tank for compressed or liquid hydrogen has a mass of 100 kg.

| Form | $H_2$ (kg/100 L) | Total weight (kg) | $H_2$ (weight %) |
|---|---|---|---|
| Gas at 350 atm | 2.0 | 102 | 2.0 |
| Gas at 700 atm | 3.5 | 104 | 3.4 |
| Liquid | 7.1 | 107 | 6.6 |
| $MgH_2$ | 11.0 | 145 | 7.6 |

**Table 3.5.** Comparison of the percentages by weight of hydrogen in various solid storage materials. Adapted from data in Dunlap (2020).

| Material | Weight percent hydrogen |
|---|---|
| $MgH_2$ | 7.6 |
| Buckminsterfullerene | 4.8 |
| Carbon nanotubes | 2.8 |
| Zeolite | 2.0 |

available for vehicle propulsion. Finally, the need to input energy in order to recover the hydrogen reduces the overall efficiency of the vehicle.

Another approach to the recovery of hydrogen from magnesium hydride is to utilize the following reaction:

$$MgH_2 + 2H_2O \rightarrow 2H_2 + Mg(OH)_2.$$

In other words, the hydride is reacted with water. Unfortunately, in this process, metallic magnesium is not recovered for further hydrogen storage. Instead, magnesium hydroxide is produced, which must be reprocessed in order to recover the elemental magnesium.

Research into the approach described above, as well as the development of new materials that react with hydrogen, has been ongoing for many years. To date, materials and procedures for hydrogen storage that would provide a viable alternative to gas or liquid storage for vehicle use have not been developed.

### *3.3.4.2 Hydrogen adsorption*

Hydrogen may be adsorbed onto the surface of a solid and bonded with the surface atoms. This type of adsorption is referred to as physisorption. The ability of hydrogen atoms to adhere to a surface may be increased by increasing the pressure of the hydrogen or by decreasing the temperature of the surface. In order to maximize the amount of hydrogen that can be stored in a specific volume or mass of storage material, it is important to maximize the surface area that is available. The best approach to maximizing surface area is through the use of materials that have a high degree of microporosity on the atomic scale. Two common examples of this type of material, zeolites and nanostructured carbon, are described here.

Zeolites make up a class of microporous aluminosilicates, which means that their compositions are combinations of aluminum oxides and silicon oxides. They typically contain an element from the left side of the periodic table such as calcium, sodium, potassium, or magnesium (or frequently a combination of these). Mordenite, whose structure is illustrated in figure 3.2, has the composition (Ca, $Na_2$,$K_2$)$Al_2Si_{10}O_{24}\cdot 7H_2O$. A common feature of all zeolites is their microstructure consisting of a framework with regularly spaced open volume pores, as seen in the figure. The pores in zeolites are the ideal size for adsorbing significant quantities of various atomic or molecular species, including hydrogen.

A wide variety of zeolites occur naturally as a result of the reaction between volcanic rocks and alkaline ground water. Zeolites can also be produced synthetically and those which are used for commercial purposes are the manufactured varieties, which allow for careful control of composition and pore size. Such zeolites can be designed with a pore size that is just compatible with the size of the hydrogen atom, so that the hydrogen readily sticks inside the zeolite structure.

Another class of materials with large microscopic surface areas is the nanostructured carbons. Pierson (1994) and Dunlap (2018) have provided overviews of the different structures that carbon can assume. Some naturally occurring crystallographic forms of carbon include diamond and graphite. Numerous other carbon structures can be synthesized in the laboratory. Many of these can be described by starting with the graphite structure shown in figure 3.3. This structure is made up of layers of carbon atoms arranged in a hexagonal array. The carbon atoms in the layers are tightly bonded together, while the bonding between the layers is much weaker. This feature is responsible for one of the important applications of graphite, that is, as a lubricant. Frictional forces can cause the planes of atoms to shear, allowing the hexagonal layers to slide past one another.

From the 1940s onward, it was speculated that it might be possible to separate a single hexagonal layer of carbon atoms from the graphite structure. This single

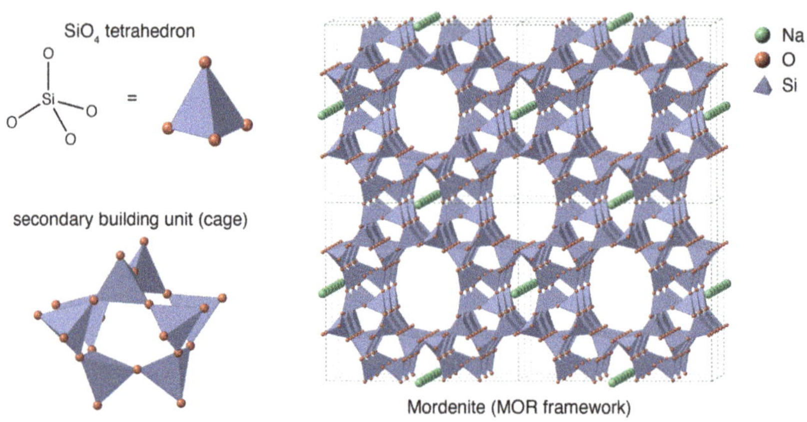


**Figure 3.2.** Atomic structure of a zeolite (mordenite). This image (Coudert F-X 2020) has been obtained by the author from the Wikimedia website where it was made available under a CC BY 4.0 licence, https://creativecommons.org/licenses/by/4.0/deed.en. It is included within this chapter on that basis. It is attributed to François-Xavier Coudert.

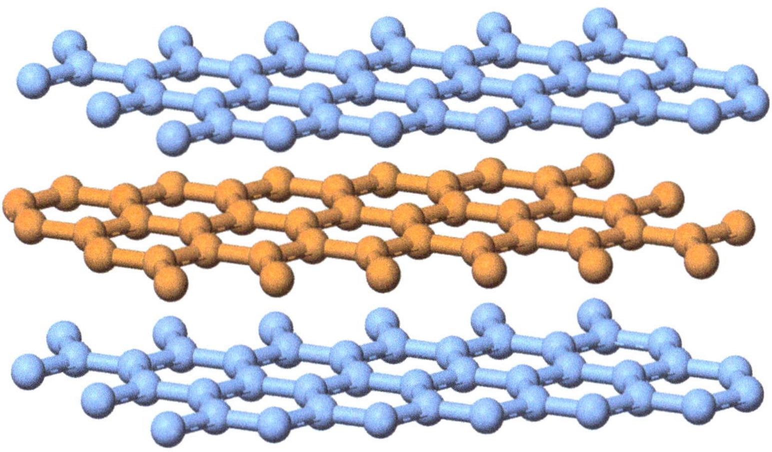

**Figure 3.3.** Layers of carbon atoms in graphite. This image (Benjah-bmm27 2007a) has been obtained by the author from the Wikimedia website, where it is stated to have been released into the public domain. It is included within this chapter on that basis.

hexagonal carbon sheet is referred to as graphene. In 2004 Andre Geim, Konstantin Novoselov, and co-workers first isolated and identified graphene in the laboratory (Novoselov *et al* 2004). For this work, Geim and Novoselov were awarded the 2010 Nobel Prize in Physics. A representation of the graphene structure is illustrated in figure 3.4.

In principle, the graphene structure can be extended indefinitely, leading to a macroscopic two-dimensional material. The mechanical, optical, electrical, magnetic, and other properties of such materials are of great interest and have a number of potential applications. Graphene is of interest for the present discussion, as it is the basis of another unusual carbon structure. A finite-width strip of graphene may be rolled up to create a nanoscopic tube. This structure is referred to as a carbon nanotube and it was discovered simultaneously in 1993 by Iijima and Ichihashi (1993) and Bethune *et al* (1993). Some examples of carbon nanotubes are shown in figures 3.5 and 3.6. It is interesting to note that a comparison of figures 3.5 and 3.6 shows that there are two ways of connecting the sides of the graphene sheet to form the tube. The resulting two types of nanotubes are referred to as zigzag nanotubes (as shown in figure 3.5, where a line connecting neighboring carbon atoms form a zigzag around the circumference of the tube) and armchair nanotubes (as shown in figure 3.6, where the line connecting neighboring carbon atoms forms an armchair-like shape).

By choosing the width of the graphene strip, i.e. the number of hexagonal units perpendicular to the tube axis, the diameter of the hole through the center of the tube can be controlled. The use of different sample preparation techniques provides the ability to control this structural property of the resulting nanotubes and allows for the preparation of nanotubes that are of an appropriate size for the adsorption of hydrogen atoms. It is interesting to note that carbon and hydrogen have a propensity for bonding, and this is the basis for the enormous number of different hydrocarbon molecules that are known to exist.

**Figure 3.4.** The structure of a graphene sheet. This image (AlexanderAlUS 2010) has been obtained by the author from the Wikimedia website where it was made available under a CC BY-SA 3.0 licence, https://creativecommons.org/licenses/by-sa/3.0/deed.en. It is included within this chapter on that basis. It is attributed to AlexanderAlUS.

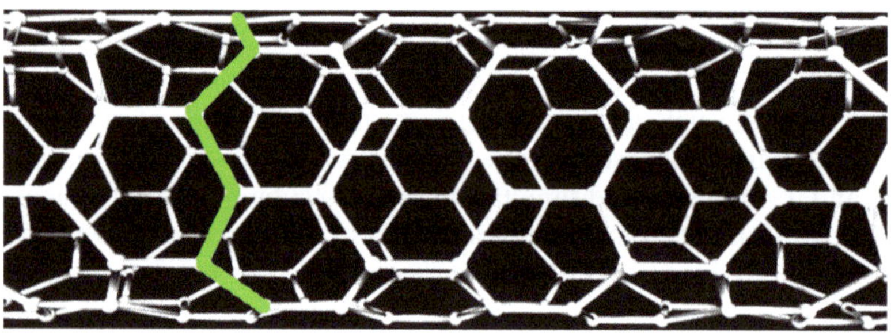

**Figure 3.5.** The structure of a zigzag carbon nanotube. This image (Materialscientist 2009a) has been obtained by the author from the Wikimedia website where it was made available under a CC BY-SA 3.0 licence, https://creativecommons.org/licenses/by-sa/3.0/deed.en. It is included within this chapter on that basis. It is attributed to Materialscientist.

Another interesting carbon nanostructure that is of potential relevance for the storage of hydrogen is the buckminsterfullerene molecule. This molecule consists of 60 carbon atoms arranged as illustrated in figure 3.7. This molecule may be described on the basis of the graphene sheet shown in figure 3.4. If one of the hexagonal cells of carbon atoms is replaced by a pentagonal cell, this causes the flat sheet to curve. If a suitable number of selected hexagons are replaced by pentagons, the flat sheet curls up and forms a sphere-like structure consisting of 60 carbon

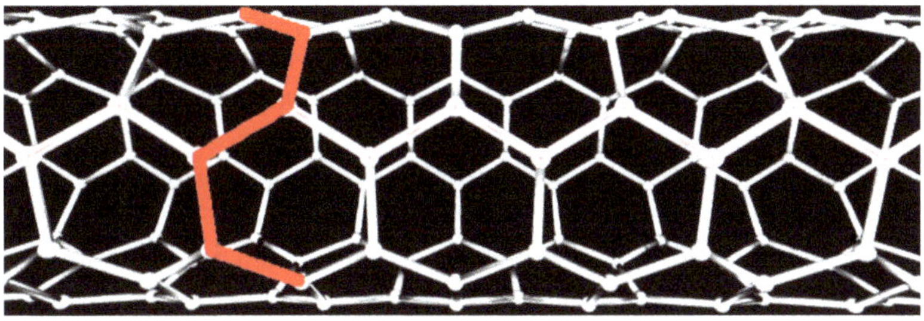

**Figure 3.6.** The structure of an armchair carbon nanotube. This image (Materialscientist 2009b) has been obtained by the author from the Wikimedia website where it was made available under a CC BY-SA 3.0 licence, https://creativecommons.org/licenses/by-sa/3.0/deed.en. It is included within this chapter on that basis. It is attributed to Materialscientist.

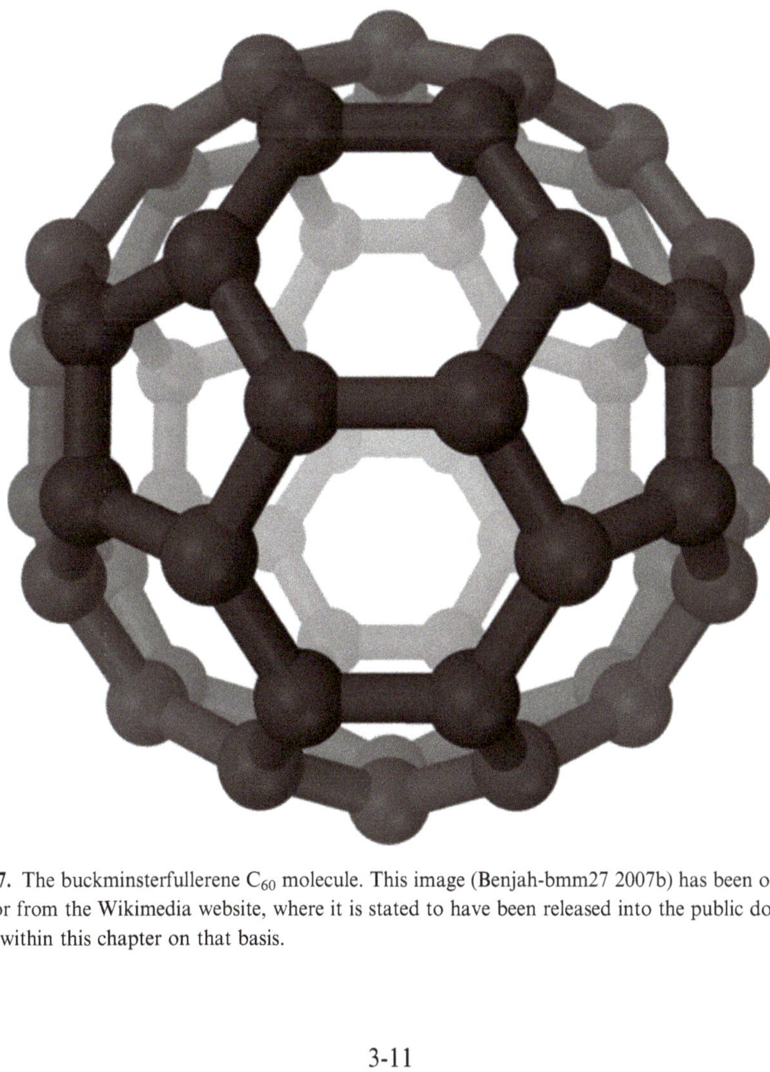

**Figure 3.7.** The buckminsterfullerene $C_{60}$ molecule. This image (Benjah-bmm27 2007b) has been obtained by the author from the Wikimedia website, where it is stated to have been released into the public domain. It is included within this chapter on that basis.

atoms as shown in the figure. From a geometric standpoint, this structure is referred to as a truncated icosahedron. It is the shape that is used for a standard soccer ball, in which hexagonal and pentagonal panels are sewn together to produce the truncated icosahedral shape.

The truncated icosahedron is also the shape of the geodesic dome popularized by the American architect Richard Buckminster Fuller (1895–1983). In his honor, $C_{60}$ is referred to as buckminsterfullerene and the truncated icosahedral molecule is called a 'bucky ball.' By selectively changing different hexagons in the graphene sheet to pentagons, closed structures with different numbers (more or less than 60) of carbon atoms may be constructed. These materials have been observed experimentally and are collectively referred to as fullerenes.

The experimental discovery of $C_{60}$ actually predates the experimental discovery of carbon nanotubes and graphene. The earliest theoretical work dates from the mid-20th century, and the first experimental observation of $C_{60}$ was made in 1985 by Robert Curl, Harold Kroto, and Richard Smalley (Kroto *et al* 1985). Curl, Kroto, and Smalley were awarded the 1996 Nobel Prize in Chemistry for their discovery.

The $C_{60}$ molecule is a suitable size for adsorbing hydrogen atoms on its inner surface. It has been theoretically determined that 58 hydrogen atoms can bond endohedrally (i.e. inside) in the $C_{60}$ molecule (Pupysheva *et al* 2008). The difficulty in actually utilizing this approach is that the carbon–carbon bonds in the $C_{60}$ molecule are very strong. It is therefore difficult to insert the hydrogen atoms into the interior of the molecule and likewise to remove them when the hydrogen is needed. This problem can be overcome by first bonding larger molecular complexes to the fullerene molecule and thereby breaking some of the carbon–carbon bonds and opening up a hole in the $C_{60}$ structure through which hydrogen can be inserted. This procedure has been described by Iwamatsu *et al* (2003, 2005).

While there has been much research on the use of solids to store hydrogen in recent years, there is as yet no practical technology that utilizes this approach in a portable hydrogen supply for road vehicle use. Future research may yield technology that will make such an approach to portable hydrogen storage a viable possibility.

## 3.4 The production of hydrogen

Elemental hydrogen does not naturally occur on Earth; therefore, hydrogen must be produced artificially. In fact, hydrogen is an important industrial material (aside from its possible use for energy storage), and there is a well-established industry for its production. At present, more than half of the hydrogen that is produced commercially is used for making ammonia (see below). As noted below in section 3.10, most ammonia is utilized in the production of fertilizers. Most of the remaining hydrogen that is produced commercially is used in a variety of industrial applications including petroleum refining and methanol production. Only a small fraction of hydrogen is used as a means of energy storage. Since the hydrogen industry that has been developed over the years has not been specifically designed with the goal of combating climate change, its production and utilization methods are not aimed at reducing greenhouse gas emissions. In fact, the current methods of hydrogen production, as described below, are particularly detrimental to atmospheric carbon dioxide levels. The frequent claim made by the proponents of a hydrogen economy, namely that hydrogen-powered vehicles run on water, must be viewed in the context of how the hydrogen is produced. There are a number of different technologies for hydrogen production and several of the most relevant are discussed in this section.

### 3.4.1 Steam reforming

At present, about 95% of commercially produced hydrogen is made by steam reforming. This process is based on a high-temperature reaction between methane and water and is basically a means of extracting the hydrogen from water and naturally occurring methane, i.e. natural gas. The reaction is:

$$CH_4 + H_2O \rightarrow 3H_2 + CO. \tag{3.1}$$

The product of this reaction is sometimes referred to as syngas (or synthesis gas). The carbon monoxide produced in this reaction can be further reacted with water according to the so-called water–gas shift reaction:

$$CO + H_2O \rightarrow CO_2 + H_2, \tag{3.2}$$

giving an overall reaction of

$$CH_4 + 2H_2O \rightarrow 4H_2 + CO_2. \tag{3.3}$$

The energy input, in the form of heat, that is required to drive this reaction accounts for about 20% of the total energy content of the methane, giving a net efficiency of about 80% for steam reforming. The hydrogen thus produced can be used to generate energy by combining it with oxygen either by combustion or in a fuel cell by the following reaction:

$$2H_2 + O_2 \rightarrow 2H_2O + \text{energy}.$$

So, although the combustion of hydrogen produces only water as a by-product, the production of hydrogen from methane by gas reforming produces carbon dioxide and is, from the standpoint of greenhouse gas emissions, equivalent to burning the methane directly:

$$CH_4 + 2O_2 \rightarrow CO_2 + 2H_2O. \tag{3.4}$$

### 3.4.2 Methane pyrolysis

Gas reforming is an attractive means of commercial hydrogen production for industry because it is relatively straightforward and generally more efficient than other methods. An alternative method of producing hydrogen from methane that is more suitable from an environmental standpoint is methane pyrolysis. In this technique, methane is heated and is directly broken up into its constituents by the reaction

$$CH_4 + \text{heat} \rightarrow 2H_2 + C. \tag{3.5}$$

In other words, hydrogen gas is produced along with solid carbon. The solid carbon can be used in various industrial process (such as steel production) and does not directly contribute to atmospheric greenhouse gas concentrations. In addition, it requires less energy input than the electrolysis methods described below.

Methane pyrolysis and gas reforming rely on the continued availability of methane as a starting material. Unless the methane is produced by biological means, as discussed in section 4.4, the production of hydrogen is linked to the utilization of fossil fuels (i.e. natural gas). In order to consider hydrogen as a future energy storage mechanism for transportation, it is essential to develop technologies for the sustainable production of hydrogen. Two possibilities are discussed below. One, electrolysis, is a well-established technology, while the other, biological hydrogen production, is a developing technology.

### 3.4.3 Electrolysis

When one thinks of hydrogen production, electrolysis often comes to mind (Chisholm and Cronin 2016). This is, perhaps, due to the simplicity of the process and the fact that this technique that can literally be implemented using household items. In fact, the principle of hydrogen production by the electrolysis of water can be demonstrated by placing an AA battery in a glass of salt water. Bubbles of hydrogen gas ($H_2$) form on the negative terminal of the battery, while oxygen ($O_2$) forms on the positive terminal. A slightly more sophisticated version of the home-made electrolysis cell is illustrated in figure 3.8. A 9 VDC battery (or a similar voltage AC adapter from an electronic device) is connected to two graphite rods (pencil leads). The graphite rods are placed in a glass of water with salt or an acid added (to increase the electrical conductivity). Hydrogen and oxygen gas are formed by the decomposition of the water according to the following reaction:

$$2H_2O \rightarrow 2H_2 + O_2.$$

The hydrogen gas is released on the negative electrode and the oxygen gas is released on the positive electrode, as shown in the figure.

Electrolysis cells that are used for industrial purposes are often referred to as electrolyzers. There are three commercially relevant approaches to electrolyzer design: the alkaline water electrolyzer, the polymer electrolyte membrane (PEM) electrolyzer, and the high-temperature electrolyzer (Rashid *et al* 2015), as described below.

#### *3.4.3.1 Alkaline water electrolyzers*

Figure 3.9 shows the basic design of an alkaline water electrolyzer (Santos *et al* 2013). Two electrodes in a liquid electrolyte, typically an alkaline solution of potassium hydroxide (KOH) or sodium hydroxide (NaOH), are separated by a non-conducting membrane that allows $OH^-$ ions to pass from the negative electrode (cathode) to the positive electrode (anode) while negatively charged electrons flow from the anode to the cathode through an external circuit. Note that the meanings of anode and cathode are opposite to those used for lithium-ion batteries, as the current

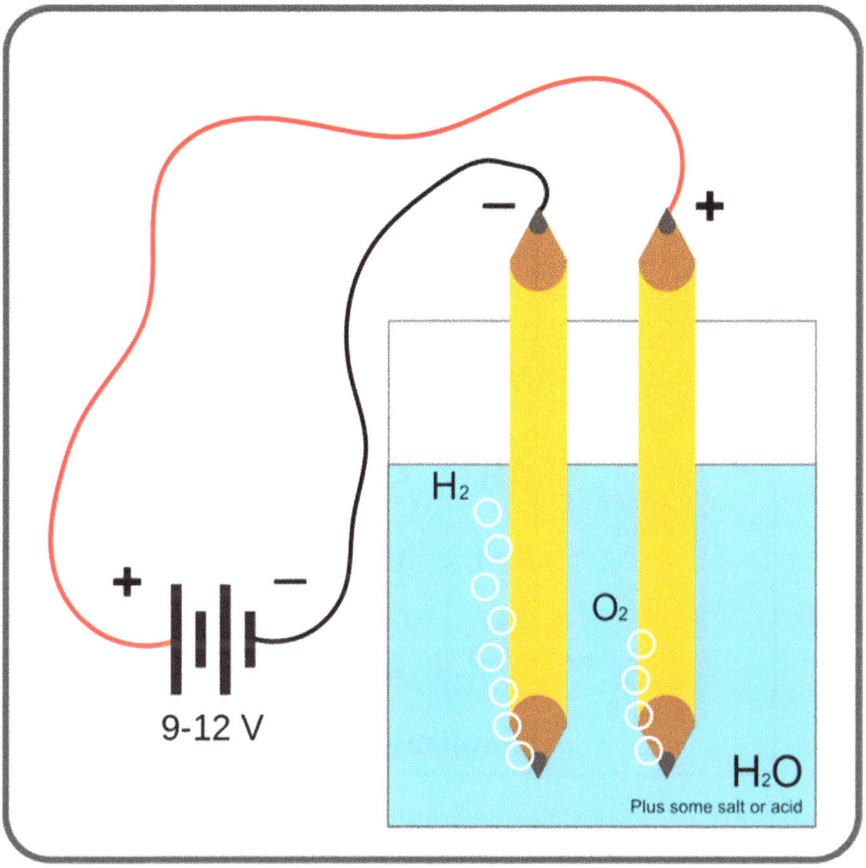


**Figure 3.8.** A simple home-made electrolysis cell for generating hydrogen gas. This image (Dilmen N 2010) has been obtained by the author from the Wikimedia website where it was made available under a CC BY-SA 3.0 licence, https://creativecommons.org/licenses/by-sa/3.0/deed.en. It is included within this chapter on that basis. It is attributed to Nevit Dilmen.

flow is defined by negative charges rather than positive lithium ions. The half reaction at the anode is

$$2OH^{-} \rightarrow \tfrac{1}{2}O_2 + H_2O + 2e^{-},$$

and the half reaction at the cathode is

$$2H_2O + 2e^{-} \rightarrow H_2 + 2OH^{-}.$$

This gives the following total reaction (by adding the reactions at the two electrodes together):

$$H_2O \rightarrow H_2 + \tfrac{1}{2}O_2. \tag{3.6}$$

In other words, each water molecule is broken up into one $H_2$ molecule and half an $O_2$ molecule.

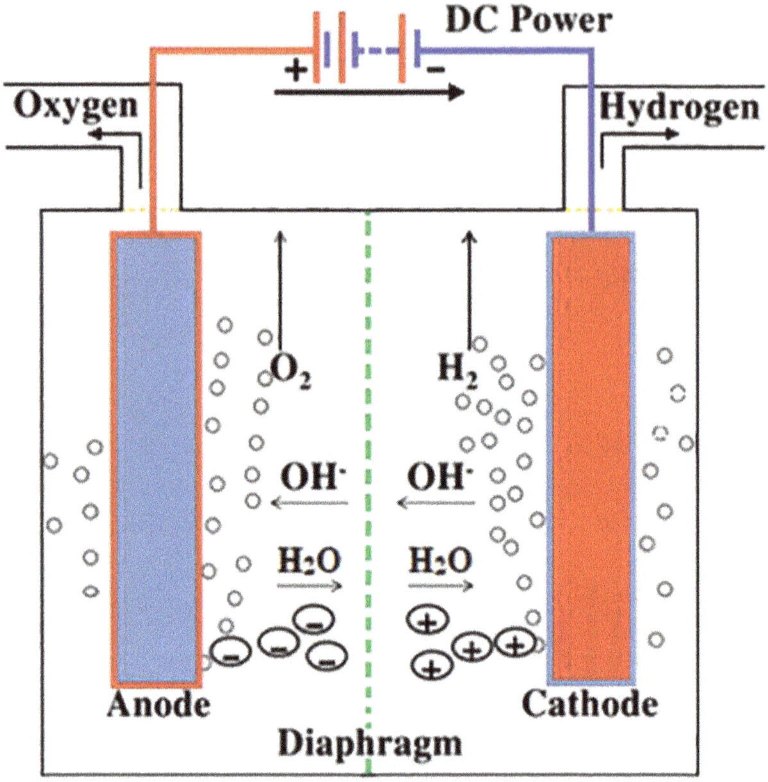


**Figure 3.9.** Basic design of an alkaline water electrolyzer. Reproduced from Naimi and Antar (2018). CC BY 3.0.

*3.4.3.2 Polymer electrolyte membrane electrolyzers*

Figure 3.10 shows the basic design of a PEM electrolyzer (sometimes called a proton exchange membrane electrolyzer). In this device, the anode and cathode (typically made from iridium and platinum, respectively) are separated by a solid electrolyte made from a plastic (polymer) which allows specific ions to pass between the electrodes. Water reacts at the anode, producing oxygen gas along with a hydrogen ion (a proton) and an electron. The hydrogen ion passes through the membrane to the cathode, while the electron travels through an external circuit. At the cathode, the electrons recombine with the hydrogen ions to produce hydrogen gas.

For the PEM electrolyzer, the half reaction at the anode is

$$H_2O \rightarrow \tfrac{1}{2}O_2 + 2H^+ + 2e^-,$$

and the half reaction at the cathode is

$$2H^+ + 2e^- \rightarrow H_2.$$

This gives the same total reaction as that shown for the alkaline water electrolyzer in equation (3.6).

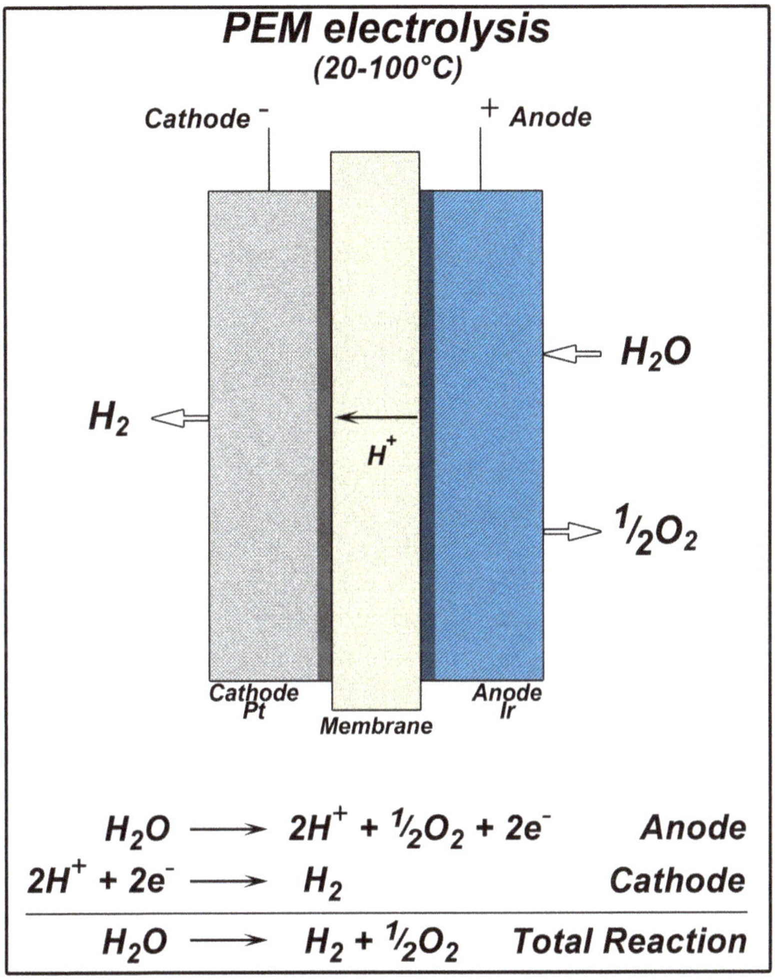


**Figure 3.10.** Basic design of a PEM electrolyzer. This image (Davidlfritz 2013) has been obtained by the author from the Wikimedia website where it was made available under the GNU Free Documentation License, https://en.wikipedia.org/wiki/en:GNU_Free_Documentation_License. It is included within this chapter on that basis. It is attributed to Davidlfritz.

### *3.4.3.3 High-temperature electrolyzers*

At around 2500 °C, water undergoes thermolysis or thermal decomposition and breaks down into oxygen and hydrogen. Thus, thermal energy aids the electrolysis process, making it more efficient (Hauch *et al* 2008). This is the basic principle of the high-temperature electrolyzer, which typically operates in the range of 500 °C. This high-temperature electrolyzer utilizes an electrolyte that consists of solid zirconium dioxide ($ZrO_2$) doped with yttrium (III) oxide ($Y_2O_3$). This material is commonly known as YSZ. Common electrode materials are Ni-doped YSZ for the cathode and lanthanum strontium manganite, $La_{1-x}Sr_xMnO_3$ (commonly known as LSM), for the anode.

High-temperature electrolyzers are economically advantageous in situations in which an inexpensive source of heat is available, thereby reducing the necessary electrical energy per unit hydrogen generated. One potential situation in which this occurs is the utilization of a nuclear fission reactor for electricity generation. About 40% of the energy produced by a nuclear reactor is converted into electricity, while the remaining 60% or so is expelled as waste heat. Thus, the reactor is able to provide both electricity and heat to power a high-temperature electrolyzer and to make use of at least a portion of the waste heat that would otherwise just be expelled to the environment. The possible implementation of very-high-temperature helium-cooled fission reactors in the future makes the option of combining electricity and hydrogen production at the same facility potentially viable. Figure 3.11 shows a diagram of such a facility.

### 3.4.4 Biological hydrogen production techniques

Hydrogen produced by biological reactions is generally referred to as biohydrogen. A number of fairly distinct processes can be used for the production of biohydrogen. One straightforward (although not necessarily efficient) method of biohydrogen production is the use of traditional gas reforming or methane pyrolysis techniques with methane produced by the anaerobic fermentation of biological waste, as described in section 4.4 (see the review by Poggi-Varaldo *et al* (2014)). Other

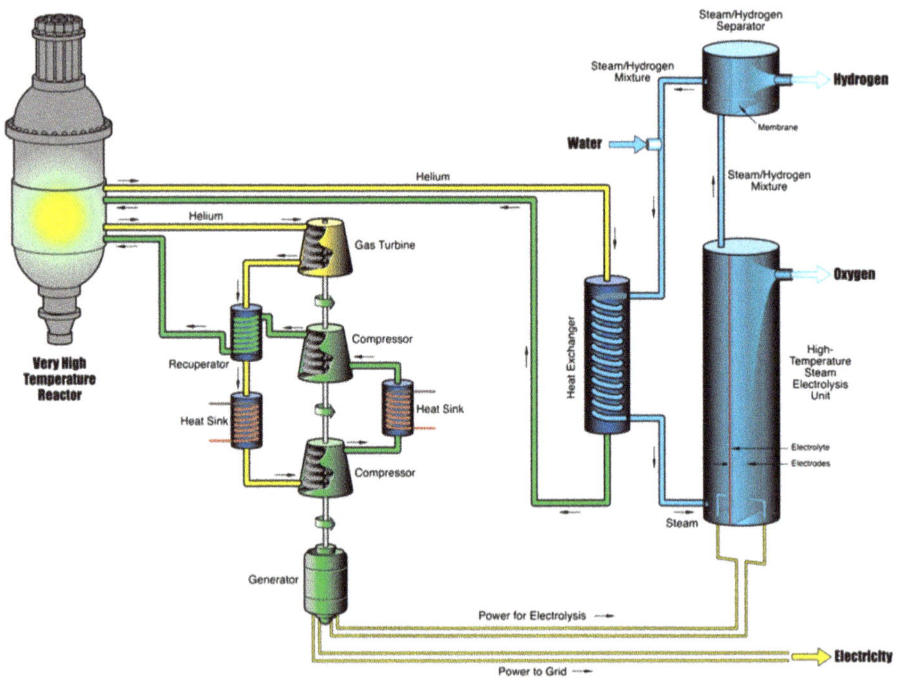


**Figure 3.11.** High-temperature electrolyzer coupled to an advanced very-high-temperature nuclear fission reactor. Reprinted from O'Brien *et al* (2009), copyright (2009), with permission from Elsevier.

approaches, as described in this section, involve the direct release of hydrogen in biological processes.

Under normal circumstances, plants produce glucose through the process of photosynthesis. In this process, carbon dioxide and water are combined using an energy input obtained from sunlight, and glucose ($C_6H_{12}O_6$) and oxygen are released:

$$6CO_2 + 6H_2O + \text{Sunlight} \rightarrow C_6H_{12}O_6 + 6O_2.$$

It has been found that this reaction can be modified by certain species of algae that adjust the sulfur concentration in the environment (see e.g. Benemann (1996), Melis *et al* (2000), and Kosourov *et al* (2018)). Specifically, certain algae, when deprived of sulfur, produce hydrogen rather than oxygen during photosynthesis. This process is referred to as biophotolysis and represents the decomposition of water into hydrogen and oxygen. The first step of the process is

$$2H_2O \rightarrow 4H^+ + O_2 + 4e^-,$$

and this is followed by

$$4H^+ + 4e^- \rightarrow 2H_2.$$

The results of experiments on the effects of sulfur deprivation on the alga *Chlamydomonas reinhardtii* are shown in figure 3.12. Hydrogen produced in this manner using energy from sunlight to drive biological processes is sometimes referred to as a solar fuel. The potential of this approach for the production of

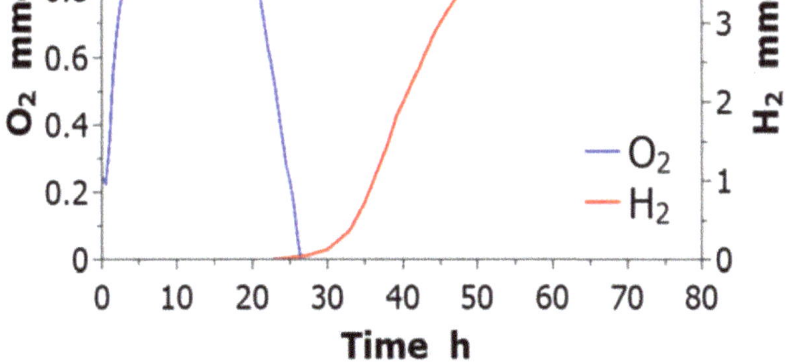


**Figure 3.12.** Oxygen and hydrogen production from the alga *Chlamydomonas reinhardtii* as a function of the duration of sulfur deprivation. Reproduced from Kosourov *et al* (2018) with permission of The Royal Society of Chemistry.

hydrogen is clear, but the development of this technique as a viable carbon-free approach to hydrogen production is in the early stages.

The development of efficient carbon-free hydrogen production methods along with their widespread implementation are essential components of an effective hydrogen economy. However, the development of viable hydrogen-powered vehicles and the implementation of an effective hydrogen distribution network (as discussed in the latter part of this chapter) can proceed in parallel with the implementation of efficient carbon-free hydrogen production facilities, as greenhouse-gas-emitting hydrogen production techniques, which are in common use, can be used in the interim. However, an effective, and environmentally sustainable hydrogen economy requires a transition to alternative hydrogen production technologies.

### 3.4.5 Colors of hydrogen

In recent years it has become common to refer to hydrogen by color. Since hydrogen is a colorless gas, the term color is not used in the literal sense. Rather, it is used to classify hydrogen produced by different methods in terms of its relative environmental effects. In this sense the term 'green hydrogen' is used to designate carbon-free hydrogen in the same way that the term 'green energy' is used to designate carbon-free energy. The classification of hydrogen by color is not standardized. While the term 'green hydrogen' is universally used, other classifications are not as standardized and are sometimes not as well defined. However, a consideration of the color of hydrogen that is used as a fuel makes it clear that the choice of hydrogen source is crucial in determining the degree to which hydrogen energy is beneficial to the environment. The hydrogen designations that are presented below reflect the most commonly used definitions of hydrogen color.

*Green hydrogen*—Green hydrogen is produced by the electrolysis of water using electricity that has been generated by renewable, carbon-free sources, such as solar or wind. It is carbon neutral, as no $CO_2$ is used in either the production of the electricity or in the electrolysis process. This is, from an environment standpoint, the most desirable approach to hydrogen production.

*Blue hydrogen*—Blue hydrogen is produced from fossil fuels, typically by steam reforming of methane as shown in equation (3.3). The carbon dioxide that is produced by the combustion of blue hydrogen is captured. The captured carbon dioxide is either sequestered as described in section 1.7 or used in industrial processes or fuel production, as described in sections 3.10 and 5.4. If the carbon is sequestered, the process is carbon neutral, assuming there is no leakage. If the carbon dioxide is used to produce fuel such as methanol, as discussed in section 3.10, the carbon is released during the combustion of the fuel.

*Turquoise hydrogen*—Turquoise hydrogen is produced by methane pyrolysis. This process consists of the breaking down of the methane molecule at high temperatures, as shown in equation (3.5), thereby producing hydrogen and solid carbon. The solid carbon can be stored much more easily than carbon dioxide gas. This process is carbon neutral subject to a couple of conditions: first, the solid carbon that is produced cannot be used as fuel, or otherwise converted into carbon dioxide and released to the atmosphere; second, the energy used to produce the high temperatures needed to drive the reaction in equation (3.5) must be carbon free. At present, the commercial viability of methane pyrolysis as a means of production has not been demonstrated.

*Gray hydrogen*—Gray hydrogen is produced from fossil fuels, typically by the steam reforming of methane. In this case, the carbon dioxide that is produced is not captured. This results in a net carbon contribution to the atmosphere. This is the method by which the largest percentage of currently utilized hydrogen is produced.

*Black hydrogen and brown hydrogen*—Black hydrogen and brown hydrogen are produced from coal, typically as an intermediate step in the production of methane by coal gasification, as described in section 3.10. Black hydrogen is produced from bituminous coal, while brown hydrogen is produced from lignite. Large quantities of carbon monoxide and carbon dioxide are produced in this process and, unless they are captured, these contribute to atmospheric carbon.

*Pink, purple, and red hydrogen*—Pink, purple, and red hydrogen are produced by energy provided by a nuclear reactor. Since the nuclear reactor is a low-carbon energy source, this process is a low-carbon approach to hydrogen production. The terms pink, purple, and red are sometimes used interchangeably for nuclear-produced hydrogen, but they are sometimes used to designate specific processes. The term pink hydrogen is usually used to denote hydrogen that is prepared using traditional electrolysis techniques from electricity produced by a nuclear reactor. Purple hydrogen refers to hydrogen produced using combined electricity and thermal energy from a conventional nuclear reactor, as described above in the section on high-temperature electrolysis. Red hydrogen is produced by the high-temperature decomposition of water using heat from a high-temperature nuclear reactor.

*White hydrogen*—The term white hydrogen has been used for naturally occurring hydrogen that is released during the fracking process. White hydrogen has no commercial significance.

## 3.5 Hydrogen internal combustion engine vehicles

One of the earliest powered road vehicles was constructed in 1807 by François Isaac de Rivaz and utilized a hydrogen-fueled internal combustion engine. In recent years, the two major automobile manufacturers that have been most active in pursuing the development of hydrogen-powered vehicles with internal combustion engines are Mazda and BMW. A brief history of their exploration of hydrogen internal combustion vehicles follows.

### 3.5.1 BMW

BMW's first experimental hydrogen internal combustion vehicle was a modified 1979 BMW 520 (designated the 520h). More recently, BMW has produced the Hydrogen 7, as shown in figure 3.13. This vehicle was produced from 2007–2009 and was based on the gasoline-powered BMW 760Li. It shared the 6.0 L V-12 engine used in the 760Li, except that it was modified to run on both hydrogen and gasoline.

**Figure 3.13.** 2007 BMW Hydrogen 7. Reproduced from More Cars (2016) Flickr (CC BY-2.0).

It was, therefore, a type of hybrid, as it used two fuel sources, although it was distinct from the battery/gasoline hybrids that are discussed in section 2.10. Vehicles that can utilize two different chemical fuel sources are sometimes referred to as bi-fuel vehicles (see section 4.5). The Hydrogen 7 is somewhat unique, compared to other recent hydrogen-powered road vehicles, as it utilizes liquid hydrogen stored in an insulated fuel tank rather than compressed hydrogen. While the availability of pressurized hydrogen for vehicle use is rather limited (see section 3.9), the availability of liquid hydrogen is even more limited.

Some of the specifications of the BMW Hydrogen 7 are given in table 3.6 for operation in the gasoline and hydrogen modes. Two notable pieces of information shown in the table are the engine power and the fuel consumption. The engine is designed to produce a maximum power of 191 kW operating on either fuel. The fuel consumption in the hydrogen mode is substantially greater than it is in the gasoline mode. This difference in fuel consumption is to be expected on the basis of the difference between the energy content per unit volume of liquid hydrogen (see table 3.3) compared to that of gasoline. Because of the differences in fuel consumption and the properties of the two fuels, each fuel is provided to the engine by a different injector system. Gasoline is supplied to the cylinders using a conventional direct fuel injection approach. In the hydrogen mode, external mixing of the fuel and air is accomplished in the intake manifold.

BMW promoted the Hydrogen 7 as a possible hydrogen-powered production vehicle. Although the Hydrogen 7 was never mass-produced, BMW made about 100 of these vehicles, which were leased to high-profile individuals intended to act as ambassadors for hydrogen vehicle technology. There are at least three concerns related to the possible mass production of a road vehicle such as the BMW Hydrogen 7. First, the potential price of a production Hydrogen 7 is not known. However, based on the fact that the base price of the gasoline 760Li is around 120 000 USD, the

Hydrogen 7 would undoubtedly be aimed at a fairly limited market. Second, the unreasonably low overall efficiency of internal combustion hydrogen vehicles (see section 3.7) would make the vehicle very expensive to operate and would not be compatible with the future need for an efficient energy infrastructure. In addition, the overall environmental impact of a transportation technology that requires such a significant number of energy conversions from the primary energy source to energy at the vehicle's wheels is not consistent with a sustainable approach. Finally, the concerns discussed in section 3.9 for the development of an infrastructure for hydrogen road vehicles are only exacerbated by the use of liquid, rather than pressurized, fuel. More recently, BMW has followed the more common route for hydrogen vehicles and has researched the development of fuel cell vehicles.

### 3.5.2 Mazda

Mazda's early interest in internal combustion engine hydrogen vehicles results largely from their efforts to develop the Wankel (rotary) gasoline engine. Mazda started to develop and promote rotary engine vehicles in the 1960s and began substantial production in the 1970s in order to carve out a niche in the Japanese automobile market to compete with Toyota, Nissan, and Honda. Their research suggested that the rotary engine would be the ideal platform for the development of hydrogen internal combustion technology.

Mazda produced a prototype hydrogen internal combustion engine vehicle in the early 1990s that utilized a two-rotor Wankel engine. This vehicle, the HR-X, is shown in figure 3.14. More recent Mazda hydrogen vehicles have been based on production gasoline engine vehicles. Figure 3.15 shows the Mazda RX-8 Hydrogen RE introduced in 2003. This vehicle uses the 1.3 L two-rotor Wankel engine from the production RX-8, which has been adapted for use with both gasoline and hydrogen. Mazda's hydrogen vehicles differ from those produced by BMW in that the fuel is stored on board as a gas in a pressurized tank. Some relevant specifications of the Mazda RX-8 Hydrogen RE are given in table 3.7. Mazda made a limited number of these vehicles available for lease in Japan in 2006 at a cost of 420 000 JPY per month (about 3600 USD per month at the time).

Another significant Mazda hydrogen internal combustion vehicle was the Mazda Premacy Hydrogen RE Hybrid introduced in 2005 (see figure 3.16). This vehicle was based on the gasoline Premacy, which was a minivan powered by a conventional four-cylinder internal combustion engine. In the Premacy Hydrogen RE Hybrid, the drivetrain was replaced with that from the RX-8 Hydrogen RE. The Premacy

**Table 3.6.** Specifications of the BMW Hydrogen 7 operating in the gasoline and hydrogen modes.

| Specification | Gasoline mode | Hydrogen mode |
|---|---|---|
| Fuel volume (L) | 73.8 | 110 |
| Engine power (kW) | 191 | 191 |
| Fuel consumption (L/100 km) | 13.9 | 50.0 |
| Range (km) | 500 | 201 |

**Figure 3.14.** 1991 Mazda HR-X. This image (160SX 2007) has been obtained by the author from the Wikimedia website where it was made available under the GNU Free Documentation License, https://en.wikipedia.org/wiki/en:GNU_Free_Documentation_License. It is included within this chapter on that basis. It is attributed to 160SX.

**Figure 3.15.** 2009 Mazda RX-8 Hydrogen RE. Reproduced from Tabercil (2010) Flickr (CC BY-SA 2.0)

**Table 3.7.** Specifications of the Mazda RX-8 Hydrogen RE operating in the gasoline and hydrogen modes.

| Specification | Gasoline mode | Hydrogen mode |
|---|---|---|
| Fuel volume (L) | 61 | 110* |
| Engine power (kW) | 154 | 80 |
| Fuel consumption (L/100 km) | 11.5 | 34** |
| Range (km) | 530 | 100 |

*Compressed hydrogen gas at 350 atmospheres.
**Equivalent in liters of liquid hydrogen per 100 km.

**Figure 3.16.** The 2007 Mazda Premacy Hydrogen RE Hybrid. This image (Hatsukari715 2007) has been obtained by the author from the Wikimedia website, where it is stated to have been released into the public domain. It is included within this chapter on that basis.

Hydrogen RE Hybrid took the concept of the hybrid vehicle one step further than the RX-8 Hydrogen RE. The Premacy was capable of using either hydrogen or gasoline in its rotary internal combustion engine. It also had batteries that were charged during regenerative braking (see section 2.10) and could provide power to the wheels through an electric motor. The Premacy underwent a major redesign in 2007, and the newer version was sometimes referred to as the Mazda Premacy Hydrogen RE Plug-In Hybrid. The basic layout was similar to that of a plug-in series hybrid in which the internal combustion engine (running on either hydrogen or gasoline) operates a generator to charge batteries (in this case, lithium-ion) to operate an electric motor that drives the wheels. The batteries can also be charged (as in the case of a conventional plug-in hybrid) from an external power source. In 2009, Mazda began leasing a very limited number of Premacy hydrogen vehicles in Japan.

### 3.5.3 Other hydrogen internal combustion engine vehicles

In recent years, the interest in hydrogen internal combustion engine vehicles has shifted away from light-duty road vehicles towards heavy-duty vehicles. Many of the major concerns for the viability of this technology for light-duty vehicles are less significant deterrents to the development of heavy-duty vehicles, such as trucks or buses. Larger vehicles are less adversely affected by larger fuel volumes and the widespread availability of fueling stations, as discussed in section 3.8. In addition, economic factors may be viewed differently for commercial vehicles. An experimental heavy-duty vehicle developed in Japan is illustrated in figure 3.17. The large fuel tanks for liquid hydrogen are clearly visible in the image.

## 3.6 Fuel cells

An alternative to using hydrogen in an internal combustion engine to drive a road vehicle is to use the hydrogen in a fuel cell. A fuel cell is basically an electrolyzer operating in reverse. In other words, it takes hydrogen and oxygen and combines them to produce water and electricity, as follows:

$$2H_2 + O_2 \rightarrow 2H_2O + \text{electricity}.$$

**Figure 3.17.** Experimental hydrogen internal combustion engine truck utilizing liquid hydrogen as a fuel, 2011. This vehicle was developed by the Musashi Institute of Technology. This image (Comyu 2011) has been obtained by the author from the Wikimedia website where it was made available under a CC BY-SA 3.0 licence, https://creativecommons.org/licenses/by-sa/3.0/deed.en. It is included within this chapter on that basis. It is attributed to Comyu.

The electricity can then be used to operate an electric motor to drive the vehicle's wheels (as well as charging its batteries). Hydrogen fuel cell vehicles are typically hybrids that use hydrogen and batteries to provide energy to propel the vehicle.

Fuel cells are by no means modern devices. The first hydrogen fuel cell was developed in the late 1830s. At about the same time, around 1838 and 1839, reports describing hydrogen fuel cells were published by the Welsh physicist Sir William Grove (1811–96) and by the German–Swiss chemist Christian Friedrich Schönbein (1799–1868). There are a number of different types of fuel cell and the choice of one that is suitable for use in a commercial road vehicle is not necessarily straightforward. We begin here with an overview of the properties of some different fuel cell designs.

### 3.6.1 Alkaline fuel cells

Alkaline fuel cells are analogous to the alkaline water electrolyzers described above, and the half reactions at their electrodes are the reverse of those given above for the alkaline water electrolyzers. Alkaline fuel cells were the first practical hydrogen fuel cells and were developed by the British engineer Francis Thomas Bacon (1904–92) in 1932. Like the alkaline water electrolyzers, alkaline fuel cells use an alkaline electrolyte, typically potassium hydroxide, and transport $OH^-$ between the electrodes. A diagram showing the design and general operation of an alkaline fuel cell is shown in figure 3.18.

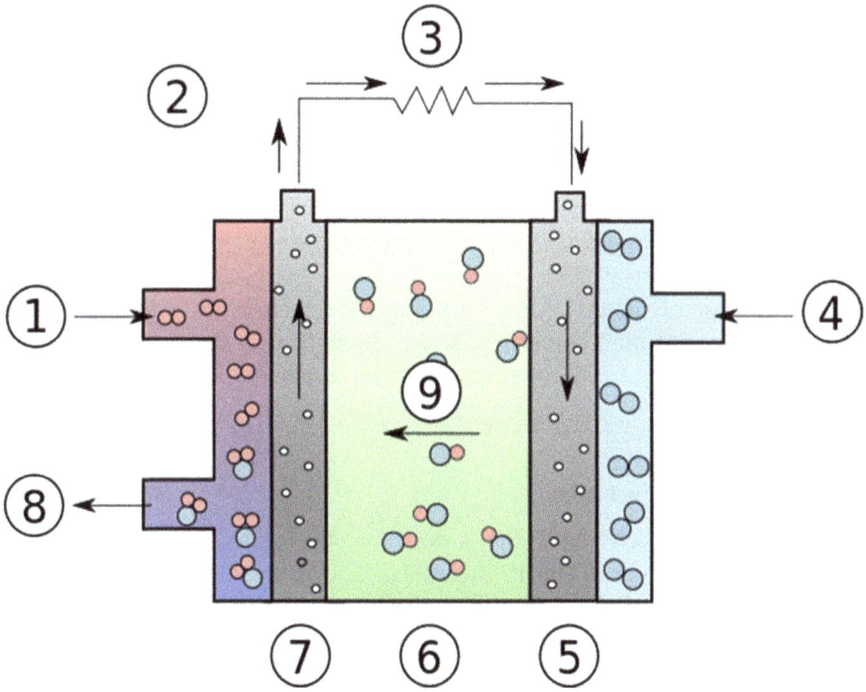


**Figure 3.18.** Diagram of an alkaline fuel cell. (1) $H_2$ input, (2) electron flow through the external circuit, (3) load, (4) $O_2$ input, (5) cathode, (6) electrolyte, (7) anode, (8) $H_2O$ output, (9) $OH^-$ ions. This image (Ring 2007) has been obtained by the author from the Wikimedia website where it was made available under the GNU Free Documentation licence, https://en.wikipedia.org/wiki/en:GNU_Free_Documentation_License. It is included within this chapter on that basis. It is attributed to Darryl Ring.

Alkaline fuel cells are typically fairly low-power devices with electrical outputs of about 10 kW or less. Like the other fuel cell designs discussed below, these fuel cells operate at elevated temperatures. However, the temperature requirements for alkaline fuel cells are typically around 80 °C and are less demanding than most other fuel cell designs. Alkaline fuel cells have been used by NASA to provide electrical power on spacecraft where high power is generally not required. They also have the added feature that the water produced as as output by this device is of high purity and can serve as a source of drinking water for astronauts.

Alkaline fuel cells typically use a platinum catalyst in order to optimize performance. However, alkaline fuel cells utilizing more economical components are under development and may be viable for more widespread applications in the future.

### 3.6.2 Polymer electrolyte membrane fuel cells

PEM fuel cells, also known as proton exchange membrane fuel cells, are analogous to PEM electrolyzers and the half reactions at their electrodes are the reverse of those given above for PEM electrolyzers. Like PEM electrolyzers, these fuel cells use a polymer membrane to separate the anode and cathode and transport charge through the electrolyte in the form of $H^+$ ions. The general design and operation of a PEM fuel cell is illustrated in figure 3.19.

Overall, PEM fuel cells are, at least at present, the most suitable hydrogen fuel cells for use in vehicles. This is due to their characteristics, such as their operating temperature and power output. PEM fuel cells are among the hydrogen fuel cells that operate at the lowest temperatures; they have typical operating temperatures between 50 °C and 100 °C. They are also capable of outputs in the range of 100 kW and higher. This is substantially greater than the output available from alkaline fuel cells and is compatible with the power required for light-duty road vehicles.

Among the disadvantages of PEM fuel cells for vehicle use is the fact that they typically utilize a platinum catalyst. Current research into PEM fuel cells includes improving efficiency and finding economical replacements for expensive platinum catalysts.

### 3.6.3 Phosphoric acid fuel cells

Phosphoric acid fuel cells have the same basic design as the original fuel cells developed in the first half of the 19th century. They are generally considered to be inappropriate for use in road vehicles because they require higher operating temperatures than alkaline or PEM fuel cells, typically 150 °C–200 °C, and they tend to be quite large. They have, however, found applications in stationary facilities for backup emergency power in, for example, police stations, hospitals, etc. As they operate at elevated temperatures, the water that is produced is in the form of steam and can be used for heating applications, thereby providing both electricity and heat during emergency situations.

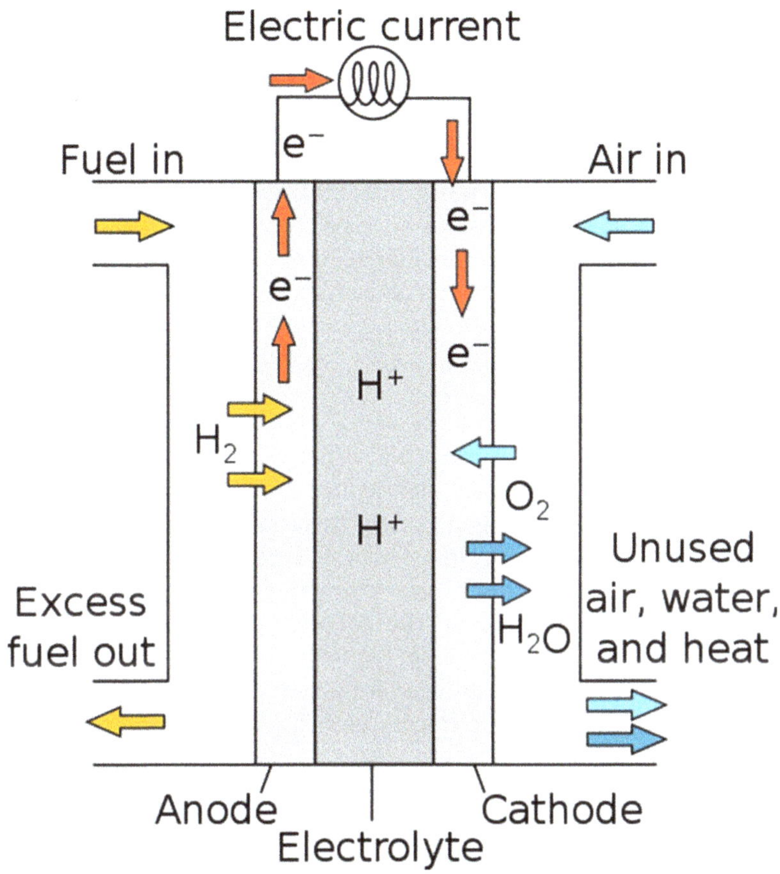


**Figure 3.19.** Design and operation of a PEM fuel cell. This image (Mattuci 2015) has been obtained by the author from the Wikimedia website, where it is stated to have been released into the public domain. It is included within this chapter on that basis.

### 3.6.4 Solid oxide fuel cells and molten carbonate fuel cells

Solid oxide fuel cells and molten carbonate fuel cells utilize different chemical reactions at their electrode but are similar in their characteristics and applications. Solid oxide fuel cells (SOFCs) are similar in design to the solid oxide electrolyzer described above but operate in reverse. Molten carbonate fuel cells use a liquid electrolyte that typically consists of a mixture of molten sodium carbonate and potassium carbonate. Both operate at high temperatures, typically over 500 °C, and can produce substantial power, in excess of 1 MW. Because of the high operating temperatures, these fuel cells do not necessarily require the use of expensive metals as catalysts. Their high power output makes them suitable for applications similar to those described above for phosphoric acid fuel cells. They also have the advantage that they can run on hydrogen that contains significant amounts of hydrocarbon impurities. In fact, solid oxide fuel cells and molten carbonate fuel cells can be designed to operate exclusively on light hydrocarbons, such as methane ($CH_4$), butane ($C_4H_{10}$), and methanol ($CH_3OH$). This ability opens up the possibility of utilizing different fuels, including some which are in liquid form and are more convenient to distribute than hydrogen. Unfortunately, the disadvantage of utilizing light hydrocarbons in a fuel cell is that the output is not only water but water plus carbon dioxide. For example, methane used in a fuel cell produces the same by-products as burned methane, as shown in equation (3.4). Thus, the impact of this approach on greenhouse gas emissions must be considered carefully. The potential applications of these fuel cells are similar to those of phosphoric acid fuel cells, although their larger power outputs make them more suitable for backup power for large facilities or remote communities.

### 3.6.5 Hybrid fuel cell systems

Hybrid fuel cell systems are, in some ways, the reverse of the high-temperature electrolyzer described above. The high-temperature electrolyzer uses electrical energy in conjunction with thermal energy to produce hydrogen. A hybrid fuel cell system produces electricity from an intermediate-temperature or high-temperature fuel cell and also utilizes waste thermal energy to produce additional electricity. A typical example of a hybrid fuel cell system is the SOFC–gas turbine system shown in figure 3.20. In this case, biomass gas (see chapter 4) is heated by a heat exchanger that recycles waste heat from the system. The biogas heats and provides fuel for a SOFC which produces electricity. Incompletely reacted biogas is

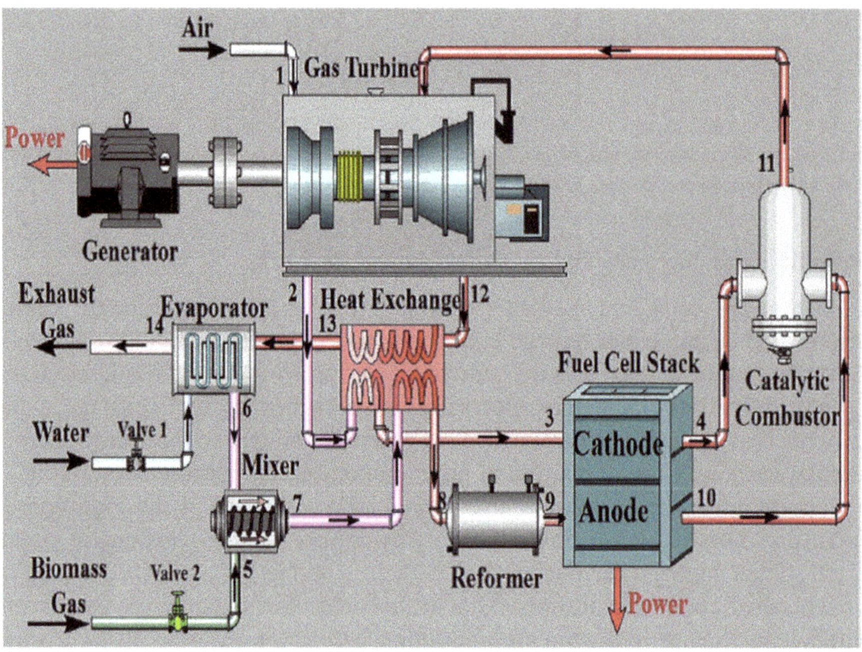


**Figure 3.20.** A hybrid SOFC–gas turbine system. Reprinted from Lv *et al* (2015), copyright (2015), with permission from Elsevier.

burned in a catalytic combustor and is used to drive a gas turbine. The gas turbine drives a generator to produce additional electricity. The overall efficiencies of SOFC–gas turbine systems can be over 80% (Azizi and Brouwer 2018, Tarroja *et al* 2010). The systems described here are large and are not suitable for direct road vehicle use. However, they can provide grid power which can then be used to provide energy for vehicles, such as battery electric vehicles.

The overall efficiency of fuel cells used for the direct generation of electricity is typically around 60%, or about three times the efficiency of a hydrogen internal combustion engine. Combined with the high efficiency of an electric motor, the overall efficiency of converting chemical energy stored in hydrogen fuel to mechanical energy at the wheels of a vehicle can be in excess of 50%. It is also convenient to incorporate battery or supercapacitor storage in conjunction with a fuel cell to increase total energy capacity. The major drawback of fuel cells at present is the requirement for the most suitable designs to incorporate platinum as a catalyst, which substantially increases the cost of their production. Much research has been undertaken in recent years to find viable platinum-free materials for use in fuel cells in order to reduce their cost.

## 3.7 Hydrogen fuel cell vehicles

In 1959, Allis-Chalmers constructed a hydrogen-fuel-cell-powered tractor that was exhibited at state fairs across the United States. This tractor utilized 1008 individual alkaline fuel cells configured to provide 15 kW of output at 750 VDC. The first fuel-cell-powered road vehicle was the Chevrolet Electrovan (also called the GM Electrovan), produced in 1966 by General Motors. This vehicle was manufactured by converting a GMC Handivan to operate on PEM fuel cells. The fuel cells produced an average continuous output of 32 kW and a peak output of 160 kW. The Electrovan had a range of about 240 km and a top speed of about 110 km $h^{-1}$. Fuel for the fuel cell was supplied by two tanks, one for liquid hydrogen and one for liquid oxygen. The fuel tanks and fuel cell stack were both heavy and occupied considerable volume. This resulted in a total vehicle weight of around 3200 kg and the vehicle only had room for the driver and two passengers. While the Chevrolet Electrovan demonstrated the possibility of a fuel-cell-powered road vehicle, it also illustrated some of the potential technical difficulties and the need for more compact fuel cell technology. It also showed that fuel cell vehicles were, at least at that time, prohibitively expensive to produce.

From the late 1960s to the 1990s, a number of experimental fuel cell vehicles were constructed at university research laboratories and by some automobile manufacturers around the world. By the late 1990s, several automobile manufacturers took the possibility of commercializing fuel cell vehicles seriously. The development in 2001 of a suitable high-pressure tank that could store hydrogen at 700 atmospheres increased the potential range of fuel cell vehicles and made them more attractive as an alternative to battery electric vehicles. The first fuel cell vehicles approved for road use were the Honda FCX (see figure 3.21) and the Toyota FCHV-4 (see figure 3.22). These both became available for lease in limited numbers towards the

**Figure 3.21.** Honda FCX fuel cell vehicle. This image (Ida Tengu 2007a) has been obtained by the author from the Wikimedia website where it was made available under the GNU Free Documentation license, https://en.wikipedia.org/wiki/en:GNU_Free_Documentation_License. It is included within this chapter on that basis. It is attributed to Ida Tengu.

**Figure 3.22.** Toyota FCHV-4 fuel cell vehicle. This image (Ida Tengu 2007b) has been obtained by the author from the Wikimedia website where it was made available under the GNU Free Documentation license, https://en.wikipedia.org/wiki/en:GNU_Free_Documentation_License. It is included within this chapter on that basis. It is attributed to Ida Tengu.

end of 2002. The lease prices were approximately 11 500 USD per month for the Honda and 17 200 USD per month for the Toyota. Over the next decade or so, several automobile manufacturers released fuel cell vehicles for lease. Notable among these manufacturers were Chevrolet, Ford, Honda, Hyundai, Mercedes-Benz, Nissan, and Toyota. Over the years, lease prices dropped considerably, reaching the range of 600–800 USD per month by the mid-2010s. Leases were available in selected locations where hydrogen filling stations existed (see section 3.9). The most common lease locations were California, Japan, and some parts of Europe.

As of 2021, three hydrogen fuel cell vehicles are available for lease or purchase by the general public. These are the Honda Clarity Fuel Cell, the Toyota Mirai, and the Hyundai Nexo, as shown in figures 3.23, 3.24, and 3.25, respectively. Table 3.8 provides some of the specifications of these vehicles. All of these vehicles are hydrogen fuel cell/battery hybrids and utilize hydrogen stored as a gas at high pressure. They are sold or leased at specific locations where hydrogen fueling stations are available and hydrogen vehicles are being promoted. As can be seen in the following images, the currently available hydrogen fuel cell vehicles are compact to mid-size passenger vehicles and SUVs. They fit into a category that is analogous to the mid-priced battery electric vehicles discussed in the previous chapter.

Using information presented here for a current production fuel cell vehicle (Toyota Mirai) and information from chapter 2 for a similar size battery electric

**Figure 3.23.** Honda Clarity Fuel Cell. This image (Comyu 2015) has been obtained by the author from the Wikimedia website where it was made available under a CC BY-SA 4.0 licence, https://creativecommons.org/licenses/by-sa/4.0/deed.en. It is included within this chapter on that basis. It is attributed to Comyu.

**Figure 3.24.** Toyota Mirai fuel cell vehicle. This image (Tokumeigakarinoaoshima 2020) has been obtained by the author from the Wikimedia website where it was made available under a CC BY-SA 4.0 licence, https://creativecommons.org/licenses/by-sa/4.0/deed.en. It is included within this chapter on that basis. It is attributed to Tokumeigakarinoaoshima.

**Figure 3.25.** Hyundai Nexo fuel cell vehicle. This image (M 93 2019) has been obtained by the author from the Wikimedia website where it was made available under a CC BY-SA 3.0 DE licence, https://creativecommons.org/licenses/by-sa/3.0/de/deed.en. It is included within this chapter on that basis. It is attributed to M 93.

**Table 3.8.** Specifications of commercially available light-duty fuel cell vehicles as of 2021.

| Vehicle | Year introduced | Weight (kg) | Hydrogen storage (kg) | Power (kW) | Equivalent fuel consumption (L/100km) | Range (km) | Price (USD) | Availability |
|---|---|---|---|---|---|---|---|---|
| Toyota Mirai | 2015 | 1950 | 5.0 | 136 | 3.6 | 502 | 58 500 | California<br>Europe<br>Japan<br>Québec<br>UAE |
| Honda Clarity | 2016 | 1875 | 5.4 | 103 | 3.5 | 589 | 369/mo | California<br>Europe<br>Japan |
| Hyundai Nexo | 2018 | 1873 | 6.3 | 120 | 4.1 | 600 | 58 300 | California<br>Europe<br>South Korea |

**Table 3.9.** Comparison of typical current models of battery electric and fuel cell passenger vehicles.

| Vehicle | Power (kW) | Range (km) | Equivalent fuel consumption (L/100 km) | Price (USD) |
|---|---|---|---|---|
| Toyota Mirai | 136 | 502 | 3.6 | 58 300 |
| Tesla Model 3 | 211 | 423 | 1.7 | 37 990 |

vehicle (Tesla Model 3), we can compare these two transportation technologies. Such a comparison is summarized in table 3.9 and shows that, in general, the fuel cell vehicle has the following characteristics (relative to the battery electric vehicle):

- less available power;
- greater range;
- greater equivalent fuel consumption;
- higher purchase price.

The reasons for these differences are discussed below.

For fuel cell vehicles, the available power is determined by the operating characteristics of the fuel cell. The greater the desired power output, the greater the fuel cell cost (as well as its size). The sizes of the fuel cells that are used in commercially available vehicles are, to a large extent, chosen for financial reasons. The power available to battery electric vehicles is dependent on the capacity of the batteries. The battery capacity is generally chosen to provide an acceptable range, and the benefit of this approach is that it provides a generous amount of power.

The greater range of the fuel cell vehicle is largely the result of the greater energy storage capacity of hydrogen, compared to batteries. While more expensive battery electric vehicles often have greater ranges than the base model of (say) the Tesla Model 3, this comes at the expense of a significant increase in weight. As discussed in the previous chapter, it is important to note that in addition to their potentially greater range, hydrogen fuel cell vehicles can be refueled in a few minutes (as is the case for gasoline-powered vehicles) compared to the much longer recharging times of battery electric vehicles.

The greater fuel consumption of the fuel cell vehicle (by about a factor of two) is a direct result of the relative efficiencies of the two technologies that are used for vehicle propulsion. It is clear from the discussion in chapter 2 that the efficiency of a battery electric vehicle in converting the energy stored in the battery into energy supplied to the wheels is about 90%. In contrast, as has been discussed earlier in this chapter, the overall efficiency of the fuel cell vehicle is basically limited by the efficiency of the fuel cell and can be in the order of about 50%. It is important to realize, however, that a complete analysis of vehicle efficiency is somewhat more complex and is discussed in further detail in chapter 6.

While battery electric vehicles are typically more expensive at present than gasoline internal combustion vehicles with similar design characteristics (as seen in table 2.2), fuel cell vehicles are even more expensive than comparable battery electric vehicles (as seen in table 3.9). The main reason for the high cost of fuel cell vehicles is the high cost of the fuel cells themselves. As discussed previously, current fuel cells typically utilize expensive platinum catalysts. Any significant reduction in the price of fuel cell vehicles would likely depend on the substantial reduction or elimination of platinum (or other expensive elements) from the fuel cell design. In fact, estimates of the production costs of fuel cell vehicles such as the Toyota Mirai have suggested that automobile manufacturers may be losing as much as $100 000 per fuel cell vehicle that they sell (Ayre 2014). As was the case for the battery electric General Motors EV1 in the 1990s, it is also the case that lease prices do not accurately reflect the manufacturing costs. From an economic standpoint, the long-term sustainability of this technology is uncertain.

While the fuel cell vehicles that are currently available to the public are rather mainstream highway-capable passenger vehicles, a British company, Riversimple, is taking a somewhat different approach to this technology. The company is developing the Riversimple Rasa, a two-seat vehicle shown in figure 3.26. Its anticipated specifications are summarized in table 3.10.

The Riversimple Rasa is more or less analogous to a neighborhood electric vehicle, such as a Citroën Ami or Renault Twizy, as described in the previous chapter. The Rasa is a fuel cell–supercapacitor hybrid in which the supercapacitor is charged by regenerative braking and is used primarily as a power source to supplement the output from the fuel cell. The Rasa is intended to be available in the near future as a lease-only vehicle. The anticipated lease terms are shown in table 3.10 and include a basic monthly fee plus a per-kilometer (or per-mile) charge.

**Figure 3.26.** Riversimple Rasa fuel cell vehicle. Reproduced from (Lamb 2016) Flickr (CC BY-SA 2.0).

**Table 3.10.** Specifications of the Riversimple Rasa fuel cell vehicle.

| Characteristic | Value |
|---|---|
| Weight | 580 kg |
| Top speed | 97 km $h^{-1}$ |
| Fuel cell output | 8.5 kW |
| Maximum continuous motor output | 16 kW |
| Maximum peak motor output | 55 kW |
| Supercapacitor capacity | 0.53 kWh |
| Equivalent fuel consumption | 1.2 L/100 km |
| Range | 483 km |
| Lease price | 480 USD/mo + 0.14 USD $km^{-1}$ |

The lease covers all fuel costs as well as maintenance. The future of fuel cell road vehicles in general may be questionable, due to both infrastructure considerations, as discussed below, and the economic sustainability of manufacturing these vehicles. Whether any fuel cell vehicles can be competitive with vehicles implementing other carbon-free technologies remains to be seen. A comparison of the specifications along with the cost of leasing for the Riverside Rasa with those of (say) the Toyota Mirai suggests that marketing 'neighborhood' style fuel cell vehicles may be particularly challenging. In fact, the drawbacks of battery electric vehicles, such as range and recharge time, which are alleviated by fuel cell technology are less of an issue for neighborhood vehicles than they are for highway-capable vehicles.

**Figure 3.27.** A hydrogen fuel cell Hyundai ElecCity bus on the Jamsu Bridge, Seoul, North Korea. This image (*Youngjin 2019) has been obtained by the author from the Wikimedia website where it was made available under a CC BY-SA 3.0 licence, https://creativecommons.org/licenses/by-sa/3.0/deed.en. It is included within this chapter on that basis. It is attributed to *Youngjin.

As discussed in the previous chapter for battery electric trucks and buses, the use of hydrogen fuel cells for commercial heavy-duty vehicles benefits from the possibility of using larger fuel storage tanks compared to those of light-duty passenger vehicles. Fuel cell trucks have only recently become commercially available. Concept and demonstration trucks have been produced by Nikola Motor Company and by Kenworth. The first commercially available fuel cell truck was the Hyundai Xcient Fuel Cell, which was introduced in 2020. Concept and demonstration fuel cell city buses have been presented by a number of manufacturers, including Mercedes-Benz, Ford, Škoda, and Toyota, beginning in the early 2000s. The first commercially available fuel cell bus was the Hyundai ElecCity introduced in 2020 (see figure 3.27).

## 3.8 Hydrogen infrastructure

The successful implementation of any transportation technology includes the establishment of suitable fueling infrastructure. This was discussed in terms of battery electric vehicles in the previous chapter. However, the availability of infrastructure is closely linked to the availability of vehicles. Vehicles cannot operate without suitable infrastructure and infrastructure is not developed if there are no users. This is a point of great concern for hydrogen-powered vehicles, for which the situation is significantly different than for battery electric vehicles, as we will see in this chapter. Here, we look at the development of a hydrogen vehicle infrastructure.

The commercial sale (and lease) of fuel cell vehicles began in earnest in 2015 with the public availability of the Toyota Mirai. The number of fuel cell vehicles registered as a function of time in the United States is illustrated in figure 3.28.

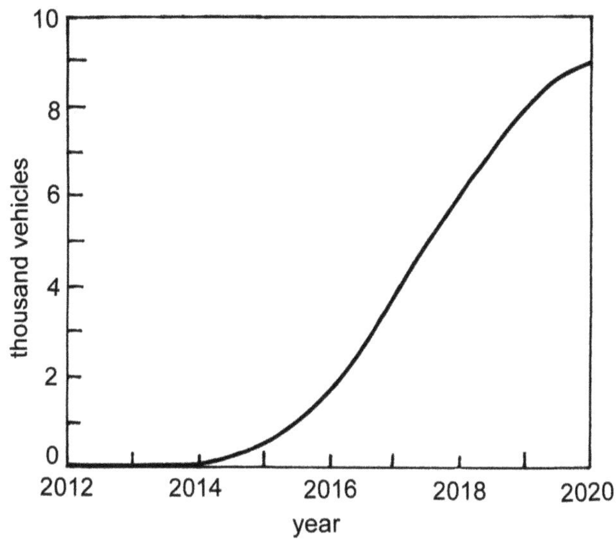


**Figure 3.28.** Number of fuel cell vehicles on the road in the United States as a function of year from 2012 to 2020. Data adapted from INSIDEEVs (2021).

**Table 3.11.** Percentages of fuel cell vehicles in different regions of the world. Data adapted from Samsun *et al* (2020).

| Region | Percent of total |
|---|---|
| North America | 32 |
| China | 25 |
| South Korea | 20 |
| Japan | 14 |
| Europe | 9 |

The graph shows that the number of fuel cell vehicles on the road grew roughly linearly from 2015 to 2019 but began to level off in 2020. In fact, the number of new fuel cell vehicles sold (or leased) per year from 2017 to 2019 was between about 2000 and 2500, but this number dropped to less than 1000 in 2020. In recent years about a third of all fuel cell vehicles have been sold in North America, as shown in table 3.11. It can also be noted that the majority of fuel cell vehicles worldwide are light-duty passenger vehicles, as shown in table 3.12.

As can be deduced from the above data, the total number of passenger vehicles worldwide that utilize hydrogen as a fuel is around 20 000. This is in comparison to about seven million battery electric vehicles and about three million plug-in hybrids. It is quite obvious that the that the total number of vehicles, as well as growth rate of these vehicles, is substantially less for hydrogen-powered vehicles than for battery electric vehicles. In fact, the growth of battery vehicles, as seen in figure 2.28, is much

**Table 3.12.** Percentages of the different types of fuel cell road vehicle worldwide in 2019. Data adapted from Samsun *et al* (2020).

| Type | Percent of total |
|---|---|
| Light-duty passenger vehicle | 75 |
| Bus | 18 |
| Truck | 7 |

faster than linear, while the growth of hydrogen vehicles, as seen in figure 3.28, has become slower than linear in recent years. The actual growth of hydrogen-powered vehicles has been substantially slower than most estimates that were made during the early development of commercial fuel cell vehicles. One notable example of an overly optimistic prediction is a Pike Research study completed in 2010 (see Industry Week Staff (2010)), which predicted that there would be 2.8 million fuel cell vehicles on the road by 2020. A more recent Navigant Research study (see Green Car Congress (2015)) predicted that annual fuel cell vehicle sales would reach 228 000 by 2024. Clearly, the development of hydrogen-powered vehicles has not lived up to the expectations of five to ten years ago. Before we look at possible reasons for this situation, we need to consider the supportive infrastructure that has developed for fuel cell vehicles.

As discussed previously, hydrogen for vehicle use can take the form of a liquid (at low temperature), as used in the BMW Hydrogen 7 or a gas (at high pressure), as used for Mazda internal combustion engine vehicles and all recent fuel cell vehicles. Liquid hydrogen is not likely practical for widespread adoption for vehicle use and is currently only available in an extremely limited number of locations. This discussion, therefore, deals primarily with infrastructure utilizing high-pressure hydrogen gas.

There are two possible types of fueling station: those which produce the hydrogen on-site and those which utilize hydrogen produced off-site. In the first case, hydrogen is produced at the fueling station by the steam reforming of natural gas or (for sustainable carbon-free stations) by the electrolysis of water. In the latter case, hydrogen is produced at a centralized facility and delivered to the fueling station by pipeline or by transport truck, which fills a storage tank at the station. At the fueling station, high-pressure hydrogen gas is dispensed through a hose as illustrated in figure 3.29. Public hydrogen fueling stations may include hoses for dispensing hydrogen gas at 350 atmospheres and at 700 atmospheres (see figure 3.30); in this respect, they are similar to pumps at conventional gasoline stations which often have two hoses, one to dispense gasoline and one to dispense diesel fuel. Some locations have phased out 350 atmosphere hydrogen in favor of 700 atmosphere hydrogen, which is the standard pressure used in current hydrogen vehicles. Light-duty vehicles typically use gas at 700 atmospheres, while some larger vehicles such as trucks and buses which have larger tanks use hydrogen at 350 atmospheres. There has been some consideration of home hydrogen fueling

**Figure 3.29.** Hydrogen fueling station. This image (Braun 2017) has been obtained by the author from the Wikimedia website where it was made available under a CC BY-SA 4.0 licence, https://creativecommons.org/licenses/by-sa/4.0/deed.en. It is included within this chapter on that basis. It is attributed to Dr. Artur Braun (Arturbraun).

**Figure 3.30.** High-pressure hydrogen dispensing pump showing hoses for 350 atmosphere gas (H35) and 700 atmosphere gas (H70). Reproduced from U.S. Department of Energy (2015) Flickr (public domain).

**Table 3.13.** Numbers of public hydrogen fueling stations in different locations as of 2020. Data are shown for U.S. states, Canadian provinces, and countries outside North America.

| Location | Number of stations |
|---|---|
| Japan | 114 |
| Germany | 84 |
| California | 43 |
| South Korea | 33 |
| China | 27 |
| United Kingdom | 11 |
| Denmark | 6 |
| Norway | 6 |
| France | 5 |
| Netherlands | 4 |
| Sweden | 4 |
| British Columbia | 3 |
| Iceland | 3 |
| Switzerland | 3 |
| Belgium | 2 |
| Connecticut | 1 |
| Hawaii | 1 |
| Italy | 1 |
| Québec | 1 |
| **World Total** | **352** |

capabilities (see Naratek (2021)), which are analogous to home charging stations for battery electric vehicles. While a home charging station for a battery electric vehicle is relatively inexpensive and can typically be installed by a local electrician, a home hydrogen fueling station is most likely beyond current or foreseeable viability from a technical and economic standpoint.

The numbers of hydrogen fueling stations in different locations worldwide are summarized in table 3.13. The availability of hydrogen as a fuel is consistent with the availability of current-model fuel cell vehicles, as seen in table 3.8, and the fraction of vehicles on the road in these locations (see table 3.11). It is clear that the infrastructure for hydrogen vehicles and the sales of such vehicles are confined to fairly localized regions. The total number of hydrogen fueling stations worldwide, 352, can be compared to the number of public charging stations for battery electric vehicles, which is about 850 000. This lack of hydrogen fueling stations is in contrast to the predictions made about a decade ago. A Pike Research study in 2010 (see USDOE (2012)) predicted that the number of hydrogen fueling stations would grow from the 200 available at that time to 5200 by the year 2020.

It is interesting to consider the possible reasons for the lack of expected growth in hydrogen transportation technologies. The lack of a significant number of fueling stations is certainly an important factor for the small number of hydrogen-fueled vehicles on the road; in turn, the small number of hydrogen-fueled vehicles on the road provides little incentive for the development of more fueling stations. Some factors that contribute to this situation are discussed below.

*Vehicle cost*—Fuel cell vehicles are more expensive than their battery or gasoline counterparts. Even at their current price, they may not be economically sustainable for automobile manufacturers (Ayre 2014). Although there are three commercially available fuel cell-based passenger vehicles available, many manufacturers who have been involved in fuel cell vehicle development have scaled back or eliminated their development programs. These manufacturers include Daimler (Quartier 2017), Hyundai (Williams 2017), and Honda (Allen 2019). Factors that favor the development of battery electric vehicles that have contributed to this situation include declining battery costs, decreased charging times, and increased driving range.

*Fueling station cost*—Hydrogen fueling stations are expensive, costing $1 million to $4 million each (Kurtz *et al* 2019). Consequently, there is little motivation for companies to invest in this infrastructure unless there are sufficient numbers of fuel cell vehicles on the road. The development of battery electric vehicle infrastructure since the 1990s has had the advantage that users could charge their vehicles at home overnight, even when an extensive public charging infrastructure was not available. This approach is still applicable, particularly for urban vehicle use and plug-in hybrids.

*Efficiency*—Hydrogen transportation technology is much less efficient than battery electric vehicle technology (Dunlap 2013). This is apparent from a simple comparison of the equivalent fuel consumption of battery electric vehicles and fuel cell vehicles, as shown in table 3.9. This point is discussed in more quantitative detail in chapter 6.

*Availability of carbon-free hydrogen*—At present, about 95% of hydrogen is produced by the steam reforming of natural gas (methane) and produces carbon dioxide emissions. This hydrogen is used primarily for industrial processes, such as fertilizer manufacturing and oil refining. Although environmental issues may influence hydrogen production at some point in the future, the current industrial infrastructure, which has been established over many years, is geared to the production of large quantities of hydrogen at minimal cost, regardless of the environmental consequences. Therefore, the development of an environmentally sustainable fuel cell vehicle industry would include not only the production of vehicles and the creation of a distribution network, but also the establishment of suitable hydrogen production facilities.

From the above discussion, it can be seen that there are a number of serious obstacles to overcome in order to make hydrogen-powered vehicles, and fuel cell vehicles in particular, a viable transportation alternative. Both the automobile industry and the fuel distribution industry would have to take the financial risk of developing technology and infrastructure, respectively, without a long-term guarantee that their efforts would be financially productive. Unfortunately, competition from simpler, less expensive, and already better-established alternative technologies, particularly battery electric vehicles, makes the future prospects of hydrogen vehicles uncertain at best. Progress in the development of a hydrogen economy for transportation has been too little, too late.

There are, however, other options for the use of hydrogen as an energy carrier related to the transportation industry. Hydrogen can be converted into liquid fuels, and some of these may be readily integrated into the existing transportation infrastructure. The remaining sections of this chapter discuss two of these possibilities. In addition, options for the use of hydrogen as a fuel for other forms of transportation are considered in chapter 5.

## 3.9 Ammonia

### 3.9.1 The properties of ammonia

Ammonia, $NH_3$, is a colorless gas at room temperature and atmospheric pressure. Some of its relevant properties are summarized in table 3.14. The household cleaner that is commonly referred to as 'ammonia' is actually a solution of 5%–10% ammonia gas dissolved in water. Today, ammonia is an important industrial chemical. About 80% of the ammonia that is produced commercially is used for the manufacture of fertilizers. The remaining 20% or so is used for the manufacture of a variety of products including cleaners, fibers, plastics, dyes, and explosives. It also has potential as a refrigerant and had been used fairly commonly for this purpose in the past.

At present, nearly all commercially produced ammonia is made by the reaction of hydrogen with nitrogen from the atmosphere by the so-called Haber–Bosch process:

$$3H_2 + N_2 \rightarrow 2NH_3.$$

This process is most efficient at a temperature of around 450 °C and at a pressure of about 100 atmospheres. So, in a sense, as discussed further below, ammonia may be thought of as a medium for storing hydrogen. Most commercial hydrogen used in such processes is produced from methane by steam reforming, as described above. Thus, at present, ammonia is not produced in an environmentally friendly, carbon-free manner. However, as with the potential use of hydrogen as a carbon-free fuel, carbon-free ammonia can be produced from green hydrogen. Using this approach, the utilization of ammonia is not only carbon free, but also is not contingent on the long-term availability of fossil fuels.

The potential use of ammonia as a fuel is attractive, since it is the only potentially viable fuel for road vehicle use (aside from pure hydrogen) that does not contain carbon (see Thomas and Parks (2006)). As such, only hydrogen and ammonia can

**Table 3.14.** Some important properties of ammonia.

| Property | Value |
|---|---|
| Gas density (STP) | 0.77 kg $m^{-3}$ |
| Liquid density | 771 kg $m^{-3}$ |
| Boiling point (1 atm) | 239.8 K |
| Heat of combustion | 22.5 MJ $kg^{-1}$ |
| Volumetric energy density relative to gasoline (liquid) | 0.50 |

undergo combustion without releasing carbon dioxide. The combustion process for ammonia is as follows:

$$4NH_3 + 3O_2 \rightarrow 2N_2 + 6H_2O,$$

i.e. only nitrogen and water are produced as combustion by-products.

As with other fuels, ammonia has some advantages and some disadvantages. The advantages of ammonia as a fuel include the following:

- It is easier to liquefy (because of its higher boiling point) than hydrogen; its liquefaction process therefore requires less energy.
- As noted above, it is the only fuel (other than pure hydrogen) that does not contain carbon and thus does not produce $CO_2$.
- Its volumetric energy density in its liquid state is about half that of gasoline.

Some of the possible disadvantages of ammonia as a fuel include the following:

- Ammonia gas is very toxic. In concentrations of a few thousand ppm in air, it can be fatal. An essential aspect of its use is to avoid leaks that could release ammonia gas into the atmosphere.
- Ammonia is highly corrosive, so storage and utilization systems must be designed appropriately.
- Ammonia is very hygroscopic, so water contamination through atmospheric exposure must be avoided.

In order to assess the viability of ammonia as a transportation fuel, it is important to begin by looking at the ways in which it might actually be utilized. Possible approaches to an ammonia-powered road vehicle include the following:

- Ammonia-fueled internal combustion engine vehicles.
- Onboard cracking of the ammonia to produce hydrogen for use in a hydrogen internal combustion engine vehicle or in a hydrogen-fuel-cell-powered vehicle.
- Direct ammonia-fuel-cell-powered vehicle.

We next consider these three possibilities.

### 3.9.2 Ammonia-fueled internal combustion engine vehicles

Ammonia may undergo combustion, as noted above, to produce nitrogen and water along with excess heat. If this combustion process occurs in an internal combustion engine, the heat is converted to mechanical energy and can be used to transport a vehicle. The efficiency of this process is limited by the thermodynamic efficiency of the heat engine and is typically around 20%.

An internal combustion engine that utilizes ammonia as a fuel can use either spark ignition (similarly to gasoline engines) or pressure ignition (similarly to diesel engines). However, ammonia is less combustible than gasoline or hydrogen, and the design of the engine must take this into account. Specifically, it is much more difficult to optimize the fuel/air mixture for all operating speeds and loads in an ammonia engine than it is in a gasoline engine. In order to ensure efficient fuel burning, a combustion promoter is typically added to the ammonia. This promoter is a more combustible fuel, such as gasoline. In order to avoid fossil fuel carbon emissions, biodiesel (see chapter 4) may be used instead of gasoline. It is also possible to crack a fraction of the ammonia to produce hydrogen (see next section) and use this as a combustion promoter. In any case, because the optimal ratio of ammonia to combustion promoter is a function of engine speed and load, it is most convenient to provide these two components with separate fuel distribution systems. Overall, the utilization of ammonia as a fuel in an internal combustion engine is more complex than for most of the possible alternatives, such as pure hydrogen (see above), methanol (see below), or ethanol (see chapter 4).

### 3.9.3 Onboard cracking

Ammonia may be converted to hydrogen by a process referred to as cracking:

$$2NH_3 \rightarrow N_2 + 3H_2.$$

Thus, if the ammonia is produced from hydrogen according to a process such as the Haber–Bosch process, then the ammonia is merely a storage medium for the hydrogen. The potential advantage of this approach compared to the direct use of hydrogen is the greater volumetric energy density of ammonia and the greater ease of liquefying, storing, and/or transporting ammonia because of its higher boiling temperature.

A typical system for the onboard cracking of ammonia is illustrated in figure 3.31. This diagram shows that the ammonia is heated in a furnace to facilitate cracking. Temperatures in excess of 600 °C are typically necessary, and the cracking is performed in the reactor in the presence of an appropriate metal catalyst. The purity of the resulting hydrogen (as evidenced by the lack of residual ammonia) increases with cracking temperature and ammonia pressure. The hydrogen produced by a system such as that illustrated in figure 3.31 may be used to power a vehicle as discussed above for the direct use of hydrogen, i.e. by combustion in an internal combustion engine or by use in a hydrogen fuel cell.

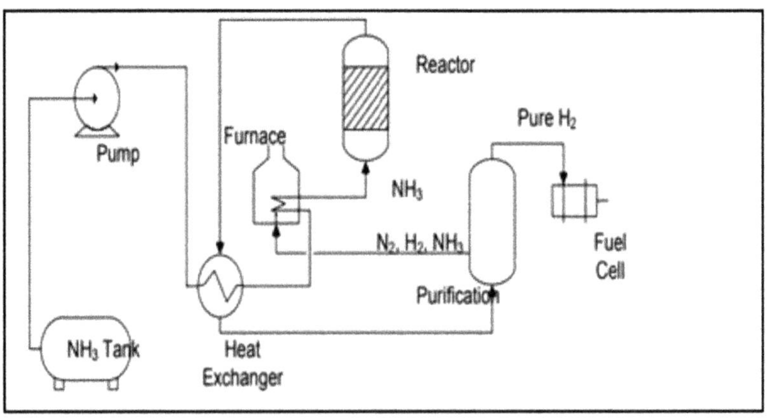


**Figure 3.31.** Ammonia cracking system for the onboard production of hydrogen. Reproduced from Abd Ali *et al* (2018), copyright. © IOP Publishing Ltd. CC BY 3.0.

The already low efficiency associated with the use of hydrogen in an internal combustion engine is further exacerbated by the inefficiency of the conversion of hydrogen to ammonia and back again. This suggests that the use of ammonia in a fuel cell would be a preferable approach. Unfortunately, hydrogen proton exchange membrane fuel cells are sensitive to ammonia impurities. Typical impurity levels of less than about 0.1 ppm are required. Typical cracking systems operating at 900 °C and pressures of around 10 atmospheres have residual ammonia concentrations of 100 ppm. Thus, a substantial improvement in cracking technologies will be necessary to make this approach feasible.

### 3.9.4 Direct ammonia-fuel-cell-powered vehicles

Perhaps the best option for the use of ammonia as a transportation fuel is the use of a fuel cell that is specifically designed for the direct use of ammonia. The comparison between direct ammonia fuel cells and ammonia internal combustion engines is analogous to that of hydrogen. The greater efficiency of the fuel cell approach at least partly compensates for the loss of efficiency that results from fuel production and storage. The development of direct ammonia fuel cells has recently been reviewed by Afif *et al* (2016).

Overall, the use of ammonia as a fuel or as a hydrogen storage medium has some advantages in terms of ease of storage and transport. However, compared to hydrogen, ammonia has its own concerns, such as toxicity, corrosivity, and complex vehicle design, as well as the potential reduction in overall efficiency that results from its additional conversion processes. All of these factors make its use somewhat questionable, at least for road transportation, unless significant technological advances can be made that alleviate these concerns. The use of ammonia for other modes of transportation is considered in chapter 5.

## 3.10 Methanol and dimethyl ether

### 3.10.1 The properties of methanol

Methanol ($CH_3OH$), sometimes referred to as methyl alcohol or methyl hydrate, is the lightest alcohol in the alkanol series, $C_nH_{2n+1}OH$. Some other members of this series of organic compounds are shown in table 3.15. As can be seen from the table, at

**Table 3.15.** Some alcohols in the alkanol series $C_nH_{2n+1}OH$ and some of their basic properties.

| Alcohol | $n$ | $C_nH_{2n+1}$ OH | Molecular mass (g mol$^{-1}$) | Boiling point (°C) | Density (kg m$^{-3}$) | Heat of combustion (MJ L$^{-1}$) | Volumetric energy density relative to that of gasoline |
|---|---|---|---|---|---|---|---|
| Methanol | 1 | $CH_3OH$ | 32.04 | 64.7 | 792 | 17.9 | 0.52 |
| Ethanol | 2 | $C_2H_5OH$ | 46.07 | 78.4 | 789 | 23.5 | 0.68 |
| Propanol | 3 | $C_3H_7OH$ | 60.10 | 82.3 | 785 | 26.3 | 0.76 |
| Butanol | 4 | $C_4H_9OH$ | 74.12 | 117.7 | 810 | 29.7 | 0.85 |
| Pentanol | 5 | $C_5H_{11}OH$ | 88.15 | 138 | 811 | 30.6 | 0.90 |

standard temperature and pressure, methanol is a liquid with a density similar to that of other alcohols and somewhat greater than that of gasoline (about 720 kg $m^{-3}$). Methanol is a commonly produced industrial chemical and about 20 million tonnes are produced annually worldwide. It is primarily used as a feedstock for the production of other chemical compounds. One of the principal chemicals produced from methanol, namely formaldehyde ($CH_2O$), is produced by the following process:

$$2CH_3OH + O_2 \rightarrow 2CH_2O + 2H_2O,$$

which in turn is used to produce various materials such as polymers, paints, and explosives. Methanol may also be used as an energy carrier, as described below, and may be used directly as a fuel. Energy is liberated by the combustion (or oxidation in a fuel cell) of methanol by the reaction

$$2CH_3OH + 3O_2 \rightarrow 2CO_2 + 4H_2O. \tag{3.7}$$

It is important to note that this reaction produces carbon dioxide; specifically, one mole of methanol produces one mole of carbon dioxide. The volumetric energy density of methanol is slightly over half of that of gasoline and is substantially greater than that of either pressurized hydrogen gas (at 700 atmospheres) or liquid hydrogen. The use of methanol as a fuel has attracted considerable attention in recent years, and methanol has been promoted as an energy carrier, particularly by Chemistry Nobel Laureate George Olah (1927–2017) (see Olah *et al* (2018)).

### 3.10.2 The production of methanol

Because methanol is an important industrial chemical that is produced in substantial quantities, there is a well-established infrastructure for its manufacture. In recent years, the demand for and production of methanol has increased substantially

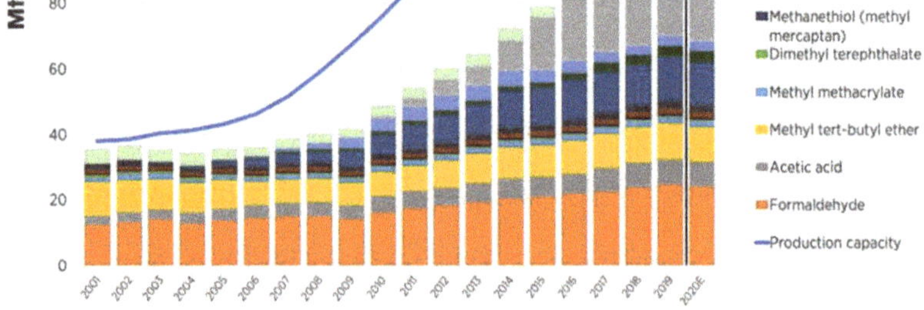


**Figure 3.32.** Global methanol use and production capacity from 2001 to 2020. Reproduced with permission from IRENA and Methanol Institute (2021).

worldwide, as shown in figure 3.32. Virtually all methanol that is produced today is made from syngas, which is a mixture of hydrogen, carbon monoxide, and sometimes carbon dioxide. Syngas is heated to a temperature of around 250 °C at a pressure of 50–100 atmospheres in the presence of a catalyst, such as copper oxide mixed with zinc oxide. Hydrogen and carbon monoxide react to produce methanol by the reaction

$$2H_2 + CO \rightarrow CH_3OH.$$

Hydrogen in the syngas can also react with carbon dioxide to produce methanol by the reaction

$$3H_2 + CO_2 \rightarrow CH_3OH + H_2O.$$

The water produced in this second reaction is recycled by means of the water–gas shift reaction (equation (3.2)). The efficacy of producing methanol from syngas depends on the relative composition of the three components of the syngas. This can be described quantitatively using the stoichiometric number, *S*. This number is defined as the number of moles of $H_2$ minus the number of moles of $CO_2$ divided by the number of moles of CO plus the number of moles of $CO_2$. A value of *S* of less than two means that the syngas is deficient in hydrogen, while a value of *S* greater than two means that the syngas has an excess of hydrogen. The exact composition of syngas depends on the methods by which the syngas is created, and this will be discussed further below.

A common method of producing syngas is by the steam reforming of natural gas (i.e. methane) by the reaction given in equation (3.1) in section 3.4. The carbon monoxide produced by this reaction can be further reacted with water by the water–gas shift reaction (equation (3.2)) to produce carbon dioxide and hydrogen. The percentages of carbon monoxide, hydrogen, and carbon dioxide and hence the stoichiometric number, *S*, are determined by the details of the process and can be controlled to yield syngas that is appropriate for specific applications.

Syngas can also be produced by the gasification of coal, which proceeds via the following steps:

*Pyrolysis*—Pyrolysis is the heating of a material in a low-oxygen atmosphere. In this case, volatile compounds in the coal are driven off, leaving material with a higher carbon content.

*Incomplete combustion*—Incomplete oxidation of the carbon produces carbon monoxide by the reaction

$$2C + O_2 \rightarrow 2CO.$$

*Gasification*—The carbon monoxide that is produced by combustion is reacted with steam at high temperature, as in equation (3.2). The remaining carbon is reacted with steam by the reaction

$$C + H_2O \rightarrow CO + H_2.$$

These reactions lead to a syngas that is a mixture of hydrogen, carbon monoxide, and carbon dioxide.

The production of methanol from fossil-fuel-derived syngas is not environmentally sustainable. Not only does it depend on the continued availability of fossil fuel resources, but it also contributes to greenhouse gas emissions. The carbon that is present in the fossil fuel becomes part of the syngas and subsequently represents the carbon content of the methanol. When the methanol is burned (or otherwise oxidized), this carbon is released as carbon dioxide.

We can also reconsider the production of methanol by the reaction of hydrogen and carbon dioxide. It is important to realize that it takes one mole of carbon dioxide to make one mole of methanol, and, as noted above, the combustion of one mole of methanol produces one mole of carbon dioxide. This means that there is no net production of carbon dioxide if the methanol is produced from hydrogen that is produced without carbon emissions. This approach has been discussed in detail recently by Bozzano and Manenti (2016). In a simple sustainable scheme, methanol could be produced using hydrogen that is made by the electrolysis of water using electricity produced by renewable technologies, such as wind power or solar power. The carbon dioxide can be obtained from the atmosphere, and this would balance the carbon dioxide that is produced by the use of the methanol. Alternatively, carbon dioxide can be obtained from industries that produce excess carbon dioxide. Methanol that is produced using this approach is sometimes referred to as *e-methanol* (IRENA and Methanol Institute 2021) and is distinct from biomethanol that is produced from biomass, as discussed in the next chapter. This approach has been taken by Carbon Recycling International, which produces e-methanol at the George Olah Renewable Methanol Plant in Iceland (see figure 3.33). This facility uses electrolytically produced hydrogen and carbon dioxide (a by-product of the country's extensive geothermal power generation system) to produce methanol. This

**Figure 3.33.** The George Olah Renewable Methanol Plant in Grindavik, Iceland. This facility has been operational since 2012. The Svartsengi Geothermal Power Station is seen in the background. Reproduced from Marlin *et al* (2018). CC BY 4.0.

plant is located in Grindavik, Iceland at the Svartsengi Geothermal Power Station and its operation has been described in detail by Marlin *et al* (2018). The use of the methanol that is produced at this plant is discussed further below.

### 3.10.3 Methanol as a fuel

The properties of methanol make it attractive as a transportation fuel. Current technology suggests two possible approaches to powering a vehicle with methanol: methanol internal combustion engines and methanol fuel cells. We consider each of these options below.

#### *3.10.3.1 Methanol internal combustion engine vehicles*

Methanol is an attractive substitute for gasoline in internal combustion engines. Table 3.16 gives some relevant properties of methanol in comparison with those of compressed hydrogen (described above), ethanol (described in the next chapter), and gasoline.

Compared to pressurized hydrogen gas, as discussed above, methanol has the clear advantage that at standard temperature and pressure it is a liquid. This greatly simplifies the necessary infrastructure needed to implement methanol transportation technology. In addition, methanol has about three times the energy density per unit volume of compressed hydrogen.

Compared to ethanol, which is discussed in the next chapter, methanol has a somewhat lower energy density per unit volume but is somewhat safer from a flammability standpoint. The autoignition temperature, as given in table 3.16, is the temperature at which the fuel would ignite without the need of a source of ignition (such as a spark). In a spark ignition engine, a higher autoignition temperature allows the fuel to resist ignition before being ignited by the spark. Thus, the high autoignition temperature of methanol allows for more efficient engine operation, as noted below. Generally, studies suggest that the production of methanol by environmentally sustainable techniques is less expensive than the production of ethanol.

Methanol has about half the heat of combustion per unit volume of gasoline. Therefore, a larger fuel tank is needed in order to obtain the same vehicle range. However, this is partly outweighed by the fact that methanol internal combustion engines are inherently more efficient than gasoline engines (Vancoillie *et al* 2013) and

**Table 3.16.** Properties of some transportation fuels.

| Fuel | Density ($kg\ m^{-3}$) | Heat of combustion ($MJ\ L^{-1}$) | Autoignition temperature (°C) |
|---|---|---|---|
| $H_2$ (700 atm) | 40 | 5.7 | 585 |
| Methanol | 792 | 17.9 | 585 |
| Ethanol | 789 | 23.5 | 365 |
| Gasoline | ~720 | 34.8 | 222 |

can have thermodynamic efficiencies of up to around 42%, about twice that of gasoline internal combustion engines (see Dunlap (2020)).

The potentially greater efficiency of a methanol internal combustion engine compared to one that utilizes gasoline can be explained on the basis of octane rating (sometimes called octane number). Gasoline is made up of a wide variety of hydrocarbons, which typically have between five and twelve carbon atoms per molecule. The hydrocarbon octane ($C_8H_{18}$) is a common component of gasoline. Gasoline that is sold for vehicle use is typically labeled with an octane rating. This does not directly reflect the amount of octane that is present in the gasoline; instead, it is a measure of the properties of the gasoline relative to the properties of octane. Specifically, it is a measure of the resistance of a fuel to detonation or self-ignition under pressure compared to that of pure octane. The scale is a percentage defined as the ratio of octane and heptane ($C_7H_{16}$) that would have the same self-ignition pressure. An octane rating of 100 would not indicate that the fuel was pure octane (or even that it contained octane) but rather that its self-ignition properties under pressure were the same as those of octane. It is therefore possible for some fuels to have octane ratings of greater than 100. The higher the octane rating, the more efficiently the fuel can be burned after spark ignition. Thus, for spark ignition engines, higher octane ratings are in principle more desirable. Typical octane ratings for gasoline in North America range from 87 for regular gasoline to 91 for premium gasoline. It is interesting to note that the same fuel may have different octane ratings in different locations. For example, gasoline with an octane rating of 91 in Boston might have an octane rating of 89 in Boulder, Colorado. This reason for this difference is related to atmospheric density. At higher elevations, air is less dense, meaning that the same volume of air–fuel mixture in the engine's cylinder has less mass, leading to reduced engine performance. Methanol, in comparison to gasoline, has an octane rating of 99, leading to improved engine performance.

The much higher autoignition temperature of methanol compared to that of gasoline makes it inherently safer from a combustion standpoint, although this same feature is a disadvantage when starting engines at low temperatures. One notable advantage of methanol (compared to petroleum-based fuels, such as gasoline) is its relatively low toxicity. It is also miscible with water and is biodegradable on a timescale of a few days (rather than a few months, as is the case for many components of gasoline).

The implementation of methanol transportation technology would be much simpler and much more straightforward than the implementation of a hydrogen transportation infrastructure. Since methanol is a liquid, the modifications to the existing gasoline distribution system that would be required to accommodate methanol are fairly minor. In fact, existing gasoline stations could be partially converted to methanol distribution during a transition period. In addition, traditional spark ignition internal combustion engines can run on methanol with relatively straightforward modifications. In addition to differences in volatility and burn temperature compared to those of gasoline, engine designs (as well fuel system designs) must account for the fact that methanol is more hygroscopic than gasoline and that common impurities in methanol can make it more corrosive.

In the early 1980s, Ford built a number of passenger vehicles that ran on pure methanol. These were mostly tested in California, but some were also used in New Zealand, Sweden, Norway, the United Kingdom, and Canada. While this program provided important design information for methanol internal combustion engines, the methanol production was not carbon free and the goal was more to reduce dependence on imported petroleum than to minimize greenhouse gas emissions. Flex fuel vehicles (as described in chapter 4) operating on methanol–gasoline mixtures up to M85 (i.e. 85% methanol, 15% gasoline) were tested in the 1990s, mostly in California and Sweden. At that time, ethanol–gasoline flex fuel vehicles seemed to be more successful, and efforts towards future flex fuel vehicle development shifted in that direction.

The utilization of methanol as a transportation fuel has been studied most extensively in China (Zhao 2019). Initial studies began around 2012, and by 2019 there were more than 10 000 methanol-powered road vehicles (mostly taxis produced by the Chinese automobile manufacturer Geely) in China. Studies have shown that these vehicles have performed very well; the typical methanol consumption for a passenger vehicle was 13.5 L/100 km, compared to 8 L/100 km for a comparable gasoline-powered vehicle. In order to support the growing number of methanol-powered vehicles in China, a network of methanol fueling stations has also been developed.

Although studies of methanol internal combustion engine vehicles in China are quite advanced, the program is not directly aimed (at present, anyway) towards a sustainable transportation system, as virtually all of the methanol used in China is produced by coal gasification. The extensive coal deposits in China are a motivation for the development of transportation technologies that can make use of these resources.

Since 2016, a methanol program in Iceland has taken a more environmentally sustainable approach. As noted above, the George Olah Renewable Methanol Plant in Grindavik produces methanol with a minimal carbon footprint using electrolytically derived green hydrogen and waste carbon dioxide from a geothermal generating station. Geely, based on their extensive methanol vehicle experience in China, has provided a fleet of six methanol internal combustion vehicles to evaluate system performance. These vehicles are versions of their gasoline-powered Emgrand EC7 (figure 3.34) that can run on either methanol or gasoline. A small tank of gasoline is used to start the vehicle, because of the low vapor pressure of methanol and the often-cold Icelandic climate. Once the engine is running, the fuel supply automatically switches over to the larger methanol tank. Methanol fueling stations have been established to support these vehicles, and the results of this test program have been very positive.

#### *3.10.3.2 Methanol fuel cell vehicles*

A somewhat different approach to the use of methanol as a transportation fuel is to adapt a fuel cell vehicle to use methanol. There are two types of methanol fuel cell system: direct methanol fuel cells and reformed methanol fuel cells. Direct methanol fuel cells are by far the simpler approach to using methanol in a fuel cell. They are a type of PEM or proton exchange membrane fuel cell that utilizes methanol according to the reaction in equation (3.7). As in the case of the direct oxidation

**Figure 3.34.** Geely Emgrand EC7. This image (KotVadim 2012) has been obtained by the author from the Wikimedia website where it was made available under a CC BY-SA 4.0 licence, https://creativecommons.org/licenses/by-sa/4.0/deed.en. It is included within this chapter on that basis. It is attributed to KotVadim.

of methanol, these fuel cells produce carbon dioxide as one of their by-products and the comments above concerning the carbon footprint of methanol as a fuel are relevant. Unfortunately, despite their relative simplicity, for now at least, direct methanol fuel cells are only capable of producing limited power and are not suitable for road vehicle use.

Reformed methanol fuel cells utilize a system for converting methanol into hydrogen, which is then used in a conventional hydrogen PEM fuel cell. This is much like the cracking system described above for possible ammonia-powered fuel cell vehicles. Figure 3.35 shows a basic diagram of a reformed methanol fuel cell system. The overall reforming reaction combines methanol with water to produce hydrogen and carbon dioxide:

$$CH_3OH + H_2O \rightarrow 3H_2 + CO_2.$$

The reaction in the PEM fuel cell is

$$2H_2 + O_2 \rightarrow 2H_2O.$$

Carbon dioxide is emitted when the methanol is reformed, but this balances the carbon dioxide that is used to produce the methanol using technology such as that implemented in Iceland. So, if the hydrogen is produced in a carbon-free manner, then the overall energy production process does not contribute to atmospheric greenhouse gas.

Gumpert Aiways Automobile, a German–Chinese collaboration, is the first company to announce the sale of a methanol fuel cell vehicle. It plans to produce

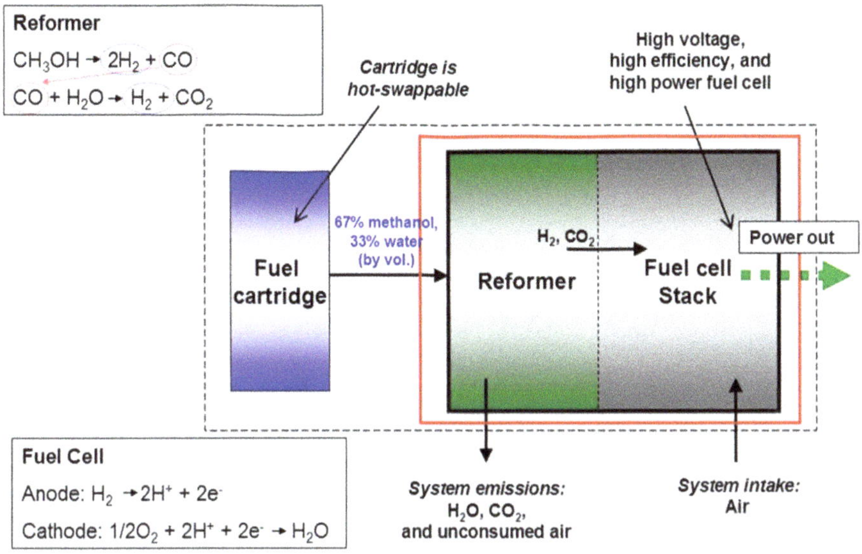


**Figure 3.35.** Reformed methanol fuel cell system showing the reformer system for the onboard generation of hydrogen for a fuel cell vehicle. This image (Prescop 2007) has been obtained by the author from the Wikipedia website, where it is stated to have been released into the public domain. It is included within this chapter on that basis.

a limited number (500) of vehicles and is taking advanced orders as of 2022 (Roland Gumpert 2022). The vehicle, the Gumpert Nathalie, is a hybrid methanol fuel cell/battery hybrid and is shown in figure 3.36. The Nathalie uses a reformed methanol fuel cell provided by Denmark-based Blue World Technologies and has a range of up to 1200 km.

It is interesting to compare the hybrid approach of the Nathalie with that of production hydrogen fuel cell vehicles, for example, the Toyota Mirai. Table 3.17 shows a comparison between some of the specifications of the Nathalie and those of the Mirai. The Mirai has a fuel cell that can supply virtually all of the power that the electric motors can accommodate. The battery has a small capacity and is primarily a means of storing energy recovered by regenerative braking. Thus, the Mirai is analogous to a mild gasoline–battery hybrid. The Nathalie has a fuel cell with a small output power and a battery with a quite significant capacity. In this case, the fuel cell maintains the battery charge, while the battery provides the power to the electric motors. Thus, the Nathalie is more analogous to an extended-range battery electric vehicle, such as the BMW i3.

While the Gumpert Nathalie is the first vehicle to utilize methanol fuel cell technology, it is planned to be produced in very limited numbers and is aimed at a quite different customer base than (say) the Toyota Mirai or the BMW i3. Both the Mirai and the i3 have price tags that are compatible with the automobile budgets of a large fraction of the public (although the Mirai may not be entirely economically sustainable at its current selling price). The Nathalie, on the other hand, is a high-performance sports car and is aimed at a very small and specialized segment of the

**Figure 3.36.** Gumpert Nathalie methanol fuel cell sports car. This image (Migl 2019) has been obtained by the author from the Wikimedia website where it was made available under a CC BY-SA 4.0 licence, https://creativecommons.org/licenses/by-sa/4.0/deed.en. It is included within this chapter on that basis. It is attributed to Alexander Migl.

**Table 3.17.** Some basic specifications of the Gumpert Nathalie and a comparison with the Toyota Mirai.

| Vehicle | Maximum fuel cell output (kW) | Fuel storage | Battery capacity (kWh) | Maximum motor power (kW) | Price (USD) |
|---|---|---|---|---|---|
| Gumpert Nathalie | 15 | 65 L methanol | 70 | 400 | 455 000 |
| Toyota Mirai | 128 | 5.0 kg $H_2$ | 1.24 | 136 | 58 500 |

automobile market. Although it is an interesting and important proof-of-concept achievement that demonstrates the use of methanol as a fuel, it does not in itself provide evidence that this technology could, at least in the short term, provide a carbon-free transportation option for any significant portion of the general public.

Karma Automotive, which took over production of the Fisker Karma (now renamed the Karma Revero) has announced that it is developing a methanol fuel cell vehicle. As is the case for the Nathalie, the methanol fuel cell will be supplied by Blue World Technologies.

Technical advances have been made for both internal combustion engine vehicles and fuel cell vehicles that are powered by methanol. From an economic standpoint, the internal combustion engine approach seems likely to be more popular with the automobile-buying public. However, the current minimal availability of methanol fuel outside of China and Iceland may hinder the rapid acceptance of these vehicles. On the other hand, the development of an appropriate methanol fueling infrastructure is fairly inexpensive and straightforward, compared to that required for hydrogen, and can be accomplished by the slow conversion of many existing facilities.

### 3.10.4 Dimethyl ether

Methanol can be used as a more or less direct substitute for gasoline in spark ignition internal combustion engines. Dimethyl ether or $CH_3OCH_3$, often simplified to $C_2H_6O$, can be used as a more or less direct substitute for diesel fuel in compression ignition internal combustion engines. As in the case of methanol internal combustion engines, diesel engines require only fairly minor modifications to utilize dimethyl ether as a fuel. Table 3.18 gives some of the relevant properties of dimethyl ether.

Dimethyl ether has some current laboratory uses and is also sometimes used as an aerosol propellant and welding torch fuel. At present, about three million tonnes of dimethyl ether are produced annually. It is produced almost exclusively by the dehydration of methanol:

$$2CH_3OH \rightarrow C_2H_6O + H_2O.$$

In the commercial process, the methanol is produced from fossil fuel-derived syngas, meaning that, at present, dimethyl ether is not a carbon-neutral product. A direct one-step process of synthesis from syngas is also possible. Carbon-neutral dimethyl ether can be prepared using methanol produced with electrolytically generated hydrogen and carbon dioxide, as described above. It may also be produced by biological processes, as discussed further in the next chapter.

Energy is obtained from dimethyl ether by the combustion process

$$C_2H_6O + 3O_2 \rightarrow 2CO_2 + 3H_2O.$$

One mole of carbon dioxide produces one mole of methanol, and two moles of methanol produce one mole of dimethyl ether. The combustion of one mole of dimethyl ether produces two moles of carbon dioxide. So, a simple analysis shows that the overall process is carbon neutral if the hydrogen used to produce the methanol is carbon free.

The possible use of dimethyl ether as a transportation fuel is greatly facilitated by the similarities of its properties to those of propane. Propane, which is derived from fossil fuel sources, is a common fuel for a variety of applications, such as portable stoves, backup generators, and torches. Propane is a major component of liquified petroleum gas (LPG), and in many countries, it is a component of so-called autogas, which is used as a fuel for internal combustion vehicles. At ambient temperature and pressure, propane is a gas. It has a boiling temperature of –42.2 °C but can be liquefied at room temperature under a pressure of around 10 atmospheres. There is, therefore, an extensive infrastructure for the storage and distribution of propane in liquid form.

**Table 3.18.** Properties of dimethyl ether.

| Property | Value |
|---|---|
| Molecular mass | 46.069 g $mol^{-1}$ |
| Boiling temperature | –24 °C |
| Density (liquid) | 735 kg $m^{-3}$ |
| Heat of combustion (liquid) | 21.2 MJ $L^{-1}$ |
| Volumetric energy density relative to diesel | 0.67 |

Dimethyl ether can be liquified at room temperature under a pressure of about five atmospheres, so the existing storage and distribution infrastructure used for propane (or LPG) can readily be adapted for use with dimethyl ether. Liquefied dimethyl ether has an energy density per unit volume of about two thirds that of fossil fuel diesel. As for other alternative liquid fuels, such as ethanol or methanol, this lower energy density must be taken into consideration in the design of vehicles and fueling infrastructure.

Although dimethyl ether is a potential substitute for diesel fuel in light-duty or heavy-duty road vehicles or other modes of transportation, its use for this purpose has not attracted a lot of attention. One study which has been reported was a collaborative project by Oak Ridge National Laboratory, Volvo North America, and the Pennsylvania State University (Szybist *et al* 2014) that had the objective of constructing a dimethyl-ether-powered truck. The results suggested that dimethyl ether could provide an efficient and environmentally advantageous alternative to conventional diesel fuel.

## References


160SX 2007 MAZDA HR-X('91) https://commons.wikimedia.org/wiki/File:MAZDA_HR-X(%2791)_00.jpg

Aceves S M, Espinosa-Loza F, Ledesma-Orozco E, Ross T O, Weisberg A H, Brunner T C and Kircher O 2010 High-density automotive hydrogen storage with cryogenic capable pressure vessels *Int. J. Hydrog. Energy* **35** 1219–26

Abd Ali Z D, Allami S and Jwad B H 2018 Ammonia as hydrogen storage media, sustainable method to hydrogen evolution *J. Phys. Conf. Ser.* **1032** 012062

Afif A, Radenahmad N, Cheok Q, Shams S, Kim J H and Azad A K 2016 Ammonia-fed fuel cells: a comprehensive review *Renew. Sustain. Energy Rev.* **60** 822–35

AlexanderAlUS 2010 Graphen https://commons.wikimedia.org/wiki/File:Graphen.jpg

Allen J 2019 Honda: now is the right time to embrace electric cars *The Sunday Times,* 4 November 2019

Ayre J 2014 *CleanTechnia* (19 November 2014) https://cleantechnica.com/2014/11/19/toyota-lose-100000-every-hydrogen-fcv-sold/

Azizi M A and Brouwer J 2018 'Progress in solid oxide fuel cell-gas turbine hybrid power systems: system design and analysis, transient operation, controls and optimization' *Appl. Energy* **215** 237–89

Benemann J 1996 Hydrogen biotechnology: progress and prospects *Nat. Biotechnol.* **14** 1101–3

Benjah-bmm27 2007a Graphite-layers-side-3D-balls https://commons.wikimedia.org/wiki/File:Graphite-layers-side-3D-balls.png

Benjah-bmm27 2007b Buckminsterfullerene https://commons.wikimedia.org/wiki/File:Buckminsterfullerene-perspective-3D-balls.png

Bethune D S, Kiang C H, De Vries M S, Gorman G, Savoy R, Vazquez J and Beyers R 1993 Cobalt-catalyzed growth of carbon nanotubes with single-atomic-layer walls *Nature* **363** 605–7

Bozzano G and Manenti F 2016 Efficient methanol synthesis: perspectives, technologies and optimization strategies *Prog. Energy Combust. Sci.* **56** 71–105

Braun A 2017 Air Liquide 700 bar Wasserstoff-Tankstelle in Düsseldorf https://commons.wikimedia.org/wiki/File:Air_Liquide_700_bar_Wasserstoff-Tankstelle_in_D%C3%BCsseldorf_c.jpg

Chisholm G and Cronin L 2016 Hydrogen from water electrolysis ed T M Letcher *Storing Energy—With Special Reference to Renewable Energy Sources* (Amsterdam: Elsevier), ch 16 pp 315–43

Comyu 2011 Musashi 9, Liquid hydrogen truck by Musashi Institute of Technology (Tokyo City University) https://commons.wikimedia.org/wiki/File:Musashi_Institute_of_Technology_Musashi_9_Liquid_hydrogen_truck.jpg

Comyu 2015 Honda CLARITY FUEL CELL, displayed at Tokyo Motor Show 2015 https://commons.wikimedia.org/wiki/File:Honda_CLARITY_FUEL_CELL_(Garnet)_at_Tokyo_Motor_Show_2015.jpg

Davidlfritz 2013 Schematic of the basic operating principle of a polymer electrolyte membrane electrolysis cell https://commons.wikimedia.org/wiki/File:PEMelectrolysis.jpg

Coudert F-X 2020 Zeolite structure as an assembly of tetrahedra https://commons.wikimedia.org/wiki/File:Zeolite_structure_as_an_assembly_of_tetrahedra.png

Dilmen N 2010 Electrolysis https://commons.wikimedia.org/wiki/File:Electrolysis.svg

Dunlap R A 2013 A simple and objective carbon footprint analysis for alternative transportation technologies *Energy Environ. Res.* **3** 33–9

Dunlap R A 2018 *Novel Microstructures for Solids* (San Rafael, CA: Morgan & Claypool)

Dunlap R A 2020 *Renewable Energy—Volume 3: Electrical, Magnetic and Chemical Energy Storage Methods* (San Rafael, CA: Morgan & Claypool)

Green Car Congress 2015 Navigant forecasts annual fuel cell vehicles sales to exceed 228,000 units by 2024. Available at www.greencarcongress.com/2015/12/20151202-navigant.html

Hatsukari715 2007 Mazda Premacy Hydrogen RE Hybrid in Tokyo Motor Show 2007 https://commons.wikimedia.org/wiki/File:Mazda_Premacy_HRE_Hybrid.JPG

Hauch A, Ebbesen S D, Jensen S H and Mogensen M 2008 Highly efficient high temperature electrolysis *J. Mater. Chem.* **18** 2331–40

Ida Tengu 2007a Honda FCX first model https://commons.wikimedia.org/wiki/File:FCX-Honda.JPG

Ida Tengu 2007b Toyota FCHV https://commons.wikimedia.org/wiki/File:Toyota_FCHV_MegaWEB.JPG

Iijima S and Ichihashi T 1993 Single-shell carbon nanotubes of 1-nm diameter *Nature* **363** 603–5

Industry Week Staff 2010 Fuel cell vehicle sales to total 2.8 million by 2020 – new study says fuel cell cars and trucks will surpass 2.8 million vehicles globally by 2020 *Industry Week* 26 February 2010. Available at www.industryweek.com/the-economy/environment/article/21940504/fuel-cell-vehicle-sales-to-total-28-million-by-2020

INSIDEEVs 2021 US: hydrogen fuel cell car sales collapsed in 2020 *InsideEVs* 27 January 2021. Available at https://insideevs.com/news/482386/us-hydrogen-fuel-cell-car-sales-2020/

IRENA and Methanol Institute 2021 *Innovation Outlook: Renewable Methanol* (Abu Dhabi: International Renewable Energy Agency) Available at www.methanol.org/wp-content/uploads/2020/04/IRENA_Innovation_Renewable_Methanol_2021.pdf

Iwamatsu S, Murata S, Andoh Y, Minoura M, Kobayashi K, Mizorogi N and Nagase S 2005 Open-cage fullerene derivatives suitable for the encapsulation of a hydrogen molecule *J. Org. Chem.* **70** 4820–5

Iwamatsu S C, Ono F and Murata S 2003 A novel migrative addition reaction of hydrazines to the diketone derivative of $C_{60}$ *Chem. Commun.* **2003** 1268–9

Kosourov S, Jokel M, Aro E M and Allahverdiyeva Y 2018 A new approach for sustained and efficient $H_2$ photoproduction by *Chlamydomonas reinhardtii Energy Environ. Sci.* **11** 1431–6

KotVadim 2012 Geely Emgrand EC7 https://commons.wikimedia.org/wiki/File:Geely_Emgrand_EC7.jpg

Kroto H W, Health J R, O'Brien S C, Curl R F and Smalley R E 1985 $C_{60}$: buckminsterfullerene *Nature* **318** 162–3

Kurtz J, Sprik S and Bradley T H 2019 Review of transportation hydrogen infrastructure performance and reliability *Int. J. Hydrog. Energy* **44** 12010–23

Lamb M 2016 FoS20162016_0625_104829AA https://flickr.com/photos/94169616@N03/27290289084

Lv X, Lu C, Wang Y and Weng Y 2015 Effect of operating parameters on a hybrid system of intermediate-temperature solid oxide fuel cell and gas turbine *Energy* **91** 10–19

M 93 2019 Hyundai Nexo https://commons.wikimedia.org/wiki/File:Hyundai_Nexo_%E2%80%93_h_16032019.jpg

Marlin D S, Sarron E and Sigurbjörnsson Ó 2018 Process advantages of direct $CO_2$ to methanol synthesis *Front. Chem.* **6** 446

Materialscientist 2009a ZigzagCNT https://commons.wikimedia.org/wiki/File:ZigzagCNT.png

Materialscientist 2009b ArmchairCNT https://commons.wikimedia.org/wiki/File:ArmchairCNT.png

Mattuci 2015 Proton Exchange Fuel Cell Diagram https://commons.wikimedia.org/wiki/File:Proton_Exchange_Fuel_Cell_Diagram.svg

Melis A, Zhang L, Forestier M, Ghirardi M L and Seibert M 2000 Sustained photobiological hydrogen gas production upon reversible inactivation of oxygen evolution in the green alga *Chlamydomonas reinhardtii Plant Physiol.* **122** 127–36

Migl A 2019 Gumpert Nathalie at Geneva Motor Show 2019 https://commons.wikimedia.org/wiki/File:Gumpert_Nathalie_Genf_2019_1Y7A5052.jpg

MMSA (Methanol Market Services Asia) 2020 *Methanol Price and Supply/Demand* (Singapore: Methanol Institute) Available at www.methanol.org/methanol-pricesupply-demand/

More Cars 2016 BMW Hydrogen 7 https://flickr.com/photos/146538874@N06/32402740931

Naimi Y and Antar A 2018 Hydrogen generation by water electrolysis *Advances in Hydrogen Generation Technologies* ed M Eyvaz (London: IntechOpen)) Available at www.intechopen.com/books/advances-in-hydrogen-generation-technologies

Naratek 2021 www.naratek.com/en/technology-at-hand/2019/home-hydrogen-fueling-stations.php

Novoselov K S, Geim A K, Morozov S V, Jiang D, Zhang Y, Dubonos S V, Grigorieva I V and Firsov A A 2004 Electric field effect in atomically thin carbon films *Science* **306** 666–9

O'Brien J E, McKellar M G, Harvego E A and Stoots C M 2009 High-temperature electrolysis for large-scale hydrogen and syngas production from nuclear energy—summary of system simulation and economic analyses *Int. J. Hydrog. Energy* **35** 4808–19

Olah G A, Goeppert A and Surya Prakash S K 2018 *Beyond Oil and Gas: The Methanol Economy* 3rd edn (Weinheim: Wiley-VHC)

Pierson H O 1994 *Handbook of Carbon, Graphite, Diamond and Fullerenes: Properties, Processing and Applications* (Norwich, NY: William Andrew) https://www.sciencedirect.com/book/9780815513391/handbook-of-carbon-graphite-diamonds-and-fullerenes

Poggi-Varaldo H M *et al* 2014 Biohydrogen, biomethane and bioelectricity as crucial components of biorefinery of organic wastes: a review *Waste Manage. Res.* **32** 353–65

Prescop T 2007 Reformed methanol fuel cell (block diagram) https://en.wikipedia.org/wiki/File:Reformed_methanol_fuel_cell_(block_diagram).jpg

Pupysheva O V, Farajian A A and Yakobson B I 2008 Fullerene nanocage capacity for hydrogen storage *Nano Lett.* **8** 767–74

Quartier D 2017 Hydrogen: BMW yes, Daimler not anymore. Available at www.fleeteurope.com/en/connected/global/article/hydrogen-bmw-yes-daimler-not-anymore

Rashid M M, Al Mesfer M K, Naseem H and Danish M 2015 Hydrogen production by water electrolysis: a review of alkaline water electrolysis, PEM water electrolysis and high temperature water electrolysis *Int. J. Eng. Adv. Technol.* **4** 80–93 https://www.ijeat.org/portfolio-item/c3749024315/

Ring D 2007 Alkaline cell https://commons.wikimedia.org/wiki/File:Alkalinecell.svg

Roland Gumpert 2022 www.rolandgumpert.com/en/get-your-nathalie/

Samsun R C, Antoni L and Rex M 2020 Mobile fuel cell application: tracking market trends *IEA Technology Collaboration Programme—Advance Fuel Cells.* Available at www.ieafuelcell.com/fileadmin/publications/2020_AFCTCP_Mobile_FC_Application_Tracking_Market_Trends_2020.pdf

Santos D M F, Sequeira C A C and Figeuiredo J L 2013 Hydrogen production by alkaline water electrolysis *Quím. Nova* **36** 1176–93

Szybist J P, McLaughlin S and Iyer S 2014 Emissions and performance benchmarking of a prototype dimethyl ether-fueled heavy-duty truck *Oak Ridge National Laboratory Report* ORNL/TM-2014/59 February 2014. Available at https://afdc.energy.gov/files/u/publication/ornl_dme_tm-2014-59.pdf

Tabercil 2010 Car Show186, 2009 Mazda RX-8 Hydrogen RE https://www.flickr.com/photos/44775455@N00/4371210171

Tarroja B, Mueller F, Maclay J and Brouwer J 2010 Parametric thermodynamic analysis of a solid oxide fuel cell gas turbine system design space *J. Eng. Gas Turbines Power* **132** 072301

Thomas G and Parks G 2006 *Potential Roles of Ammonia in a Hydrogen Economy: A Study of Issues Related to the Use Ammonia for On-Board Vehicular Hydrogen Storage* (Washington, DC: U.S. Department of Energy) available at www.energy.gov/sites/prod/files/2015/01/f19/fcto_nh3_h2_storage_white_paper_2006.pdf

Tokumeigakarinoaoshima 2020 Toyota MIRAI Z (JPD20) with the factory-installed option https://commons.wikimedia.org/wiki/File:Toyota_MIRAI_Z_(JPD20)_with_the_factory-installed_option.jpg

U.S. Department of Energy 2012 *2011 Fuel Cell Technologies—Market Report* available at www1.eere.energy.gov/hydrogenandfuelcells/pdfs/2011_market_report.pdf

U.S. Department of Energy 2015 IMG_0204 https://flickr.com/photos/37916456@N02/15723751263

Vancoillie J, Demuynck J, Sileghem L, Van De Ginste M, Verhelst S, Brabant L and Van Hoorebeke L 2013 The potential of methanol as a fuel for flex-fuel and dedicated spark-ignition engines *Appl. Energy* **102** 140–9

Williams K 2017 The switch from hydrogen to electric vehicles continues, now Hyundai makes the move *Seeking Alpha* (1 September 2017) available at https://seekingalpha.com/article/4103682-switch-from-hydrogen-to-electric-vehicles-continues-now-hyundai-makes-move

*Youngjin 2019 A hydrogen fuel Hyundai ElecCity is seen on route 405 at Jamsu Bridge https://commons.wikimedia.org/wiki/File:Seoul_405_Hydrogen_fuel_ElecCity.jpg

Zhao K 2019 *A Brief Review of China's Methanol Vehicle Pilot and Policy* (Washington, DC: Methanol Institute) (20 March 2019) available at www.methanol.org/wp-content/uploads/2019/03/A-Brief-Review-of-Chinas-Methanol-Vehicle-Pilot-and-Policy-20-March-2019.pdf

**IOP** Publishing

Transportation Technologies for a Sustainable Future
Renewable energy options for road, rail, marine and air transportation
**Richard A Dunlap**

# Chapter 4

# Biofuels

## 4.1 Introduction

Battery electric vehicles and vehicles that use hydrogen (or one of its derivatives) for which the hydrogen is produced in a carbon-free manner using electrolysis both rely on electricity generated using renewable technologies. An alternate approach is the use of biological processes to produce fuels that are suitable for vehicle propulsion. Biofuels are defined as fuels that are produced from biomass by contemporary processes. This distinguishes them from fossil fuels that have been produced from biomass through ancient geological processes. In a sense, biomass itself may be considered a biofuel. Wood or other solid biological material may be used to produce energy by direct combustion. In fact, wood was humanity's earliest source of energy for heating. Today, wood remains an important source of energy in many developing countries and is also being used as a replacement for fossil fuels, particularly coal, in power stations, as noted in chapter 1. From the standpoint of transportation, the term biofuels generally refers to liquid or gaseous forms of fuel that have been derived from solid biomass.

There are several ways of categorizing biofuels. One approach is based on the feedstock and the method of production. In this respect, biofuels may be designated as first-generation, second-generation, third-generation, or fourth-generation biofuels, as described below.

### 4.1.1 First-generation biofuels

First-generation biofuels are those that are produced directly from crops that are traditionally grown on land for food. These include, for example, ethanol made by the fermentation of the sugar or starch content of plant material and biodiesel made by the transesterification of the oil in plant material.

doi:10.1088/978-0-7503-5306-9ch4

### 4.1.2 Second-generation biofuels

Second-generation biofuels are produced from plant material that is either waste material from agriculture or material that is not grown for food purposes. These materials include wood, grasses, bagasse, and waste vegetable oil.

### 4.1.3 Third-generation biofuels

Third-generation biofuels are those which are produced from algae that is grown in coastal waters or in tanks or ponds on land.

### 4.1.4 Fourth-generation biofuels

Fourth-generation biofuels are produced by advanced technologies. Some of these approaches include solar fuels, as noted in section 3.4.

In this chapter we consider some of the more commonly used biofuels and some biofuels that show promise as future transportation fuels. These are arranged in terms of the characteristics of the final fuel and its potential applications in road transportation.

## 4.2 Bioethanol

### 4.2.1 Properties of ethanol

Some relevant properties of ethanol, $CH_3CH_2OH$, or more concisely, $C_2H_5OH$, are summarized in table 4.1. Compared to methanol, as discussed in the last chapter, ethanol has a somewhat greater heat of combustion per unit volume (see table 3.15), but its heat of combustion per unit volume is still only about two thirds that of gasoline. Its autoignition temperature is higher than that of gasoline but lower than those of methanol and hydrogen. Thus, it is reasonably safe from a flammability standpoint and allows for improved engine efficiency over gasoline. Ethanol has low toxicity and is readily biodegradable. Therefore, from an environmental standpoint it is much safer than gasoline, or even methanol.

Energy is obtained from ethanol by combustion or other methods of oxidation (such as fuel cells). The oxidation process for ethanol is

$$C_2H_5OH + 3O_2 \rightarrow 2CO_2 + 3H_2O.$$

**Table 4.1.** Some properties of ethanol.

| Property | Value |
|---|---|
| Molecular mass | 46.07 g mol$^{-1}$ |
| Boiling point | 78.4 °C |
| Density | 789 kg m$^{-3}$ |
| Heat of combustion | 23.5 MJ L$^{-1}$ |
| Volumetric energy density relative to gasoline | 0.68 |
| Octane rating | 99 |
| Autoignition temperature | 365 °C |

Carbon dioxide is released during the oxidation of ethanol (at the ratio of two moles of carbon dioxide for one mole of ethanol), and this fact must be included in an overall assessment of ethanol's carbon footprint.

### 4.2.2 Uses of ethanol

Ethanol has a significant number of applications. These include the following:

#### *4.2.2.1 Fuel applications*

Ethanol can be used as a fuel in a conventional gasoline internal combustion engine with fairly minor modifications, as described below. Ethanol may be used either in its pure form or blended with gasoline in different proportions for so-called flex-fuel vehicles. Residential space heating accounts for a minor part of ethanol fuel use.

#### *4.2.2.2 Recreational applications*

Ethanol is a psychoactive drug that acts as a central nervous system depressant. Historical evidence suggests that it was produced for recreational human consumption as early as 7000 BCE in Northern China. Evidence of wine production dates to around 6000 BCE in the South Caucasus region and beer was probably first produced around 600 BCE in Egypt.

#### *4.2.2.3 Industrial applications*

Ethanol is a very versatile solvent, and this property makes it of considerable use for industrial processes. It is an important component of cleaning chemicals and its properties as a solvent make it useful for the preparation of a wide variety of industrial products. Many chemicals that are insoluble in water can be dissolved in ethanol. The low boiling point of ethanol also makes it easy to remove from solutions. Ethanol is a common component of many widely used products, such as paints. In addition to being an important solvent and component of final products, ethanol is also an important precursor or feedstock for other important chemicals, such as acetic acid and ethyl ethers.

#### *4.2.2.4 Medical applications*

Ethanol is commonly used as a medical antiseptic because of its antibacterial and antifungal properties. It is a typical component in hand sanitizers and medical wipes. It is interesting that a mixture of 70% ethanol in water is more effective at eliminating microbes than pure ethanol. This is because the presence of water makes it easier for the ethanol to penetrate the cell membrane. Ethanol is also a common solvent used in the preparation of medications that are insoluble in water. Liquid preparations of products such as cough medications, pain relief medication, and mouth wash often use ethanol as a major component. Ethanol is also used in liquid medications as an antimicrobial preservative to extend shelf life.

Current annual world ethanol production is around $8 \times 10^{10}$ L. A breakdown of its use shows that about 82% is used for fuel, 12% is used for beverages, and 6% is used for medical and industrial applications.

### 4.2.3 Methods of ethanol production

Typically, ethanol that is used for industrial applications is produced from fossil fuels, while ethanol that is used for fuel or human consumption is produced by biological methods. These are described below.

#### *4.2.3.1 Fossil fuel ethanol production*

Fossil-fuel-based ethanol is most commonly produced by the hydration of ethylene ($C_2H_4$):

$$C_2H_4 + H_2O \rightarrow C_2H_5OH.$$

Ethylene, in turn, is made by the steam cracking of heavier hydrocarbons that are present in petroleum. This process breaks up the heavier hydrocarbons into lighter ones. For example, the steam cracking of ethane ($C_2H_6$) produces ethylene by the following process:

$$C_2H_6 \rightarrow C_2H_4 + H_2.$$

Ethylene is frequently produced from natural gas. Natural gas is mostly methane, $CH_4$, but also contains quantities of heavier hydrocarbons. These compounds are primarily from the alkane series of hydrocarbons, which have the composition $C_nH_{2n+2}$. The first few hydrocarbons in this series are methane ($n = 1$), ethane ($n = 2$), propane ($n = 3$), butane ($n = 4$), and pentane ($n = 5$). Typical natural gas deposits contain up to about 6% of ethane. The cracking of natural gas produces a mixture of hydrocarbons which can be separated into its components (including ethane) by distillation.

If ethanol is produced from fossil fuel sources, then the carbon dioxide that is released during its combustion is a net carbon contribution to the atmosphere.

#### *4.2.3.2 Biological ethanol production*

The typical route for the production of biological ethanol, commonly called bioethanol, is by the fermentation of sugars or starches. The simplest situation may be described by beginning with photosynthesis. Photosynthesis is the process by which plants use sunlight to convert carbon dioxide and water into glucose by the reaction

$$6CO_2 + 6H_2O \rightarrow C_6H_{12}O_6 + 6O_2.$$

This is the reaction by which plants absorb carbon dioxide from the atmosphere and emit oxygen. Glucose may be converted into ethanol by fermentation according to the process

$$C_6H_{12}O_6 \rightarrow 2C_2H_5OH + 2CO_2.$$

Thus, six moles of carbon dioxide are absorbed from the atmosphere to produce one mole of glucose. One mole of glucose in turn produces two moles of ethanol and two moles of carbon dioxide during fermentation. The combustion of two moles of ethanol produces four additional moles of carbon dioxide. This means that overall, the process absorbs six moles of carbon dioxide from the atmosphere and releases six moles of carbon dioxide to the atmosphere. Therefore, the utilization of energy stored in bioethanol is carbon neutral.

Fermentation processes, as described above, are the result of the action of certain yeasts, such as *Saccharomyces cerevisiae*, on sugar. The ideal temperatures for these processes are in the range of 35 °C to 40 °C. Since ethanol is toxic to the yeasts that produce it, there is a maximum ethanol concentration (about 18% by volume) that can be produced by fermentation. Higher concentrations of ethanol in fuels and alcoholic beverages are the result of distillation.

Plants can also modify glucose to make fructose, which has the same chemical formula but a different molecular structure. Plants can also bind a glucose molecule to a fructose molecule to make sucrose, which is common sugar. Ethanol can be produced from fructose or sucrose by fermentation processes, as described above. In addition, plants can bond multiple glucose molecules together into chains. This is somewhat more complex, and we have to look in detail at the structure of glucose to understand why this is the case. Two isomers of glucose exist, designated the L-form (the left-handed form) and its mirror image the D-form (the right-handed form). Only D-glucose is biologically active and can be utilized by cells. D-glucose can exist in two forms, referred to as α-glucose and β-glucose, that differ in the orientation of the bond connected to one of the OH radicals. This difference is very important when the glucose molecules are bonded together to form chains. When α-glucose molecules are bonded together, they form a starch, which is a polysaccharide that the plant uses to store energy. When β-glucose molecules are bonded together, they form polymers such as cellulose, a polysaccharide that is found in the fibrous parts of plants. Both starches and cellulose have the basic composition, $(C_6H_{10}O_5)_n$, however, because of their structure, starches are easy to break down, while cellulose is not. Starches, such as grains, are commonly used for the production of ethanol; one example is the production of beer. This is done by first soaking the grain in water and allowing it to germinate. The germination process is stopped by drying the grain with hot air. This process is referred to as malting and produces the enzyme amylase, which converts the starch to sugar, which can then be fermented. The process of converting cellulose to ethanol is more complex and is discussed below.

The overall process of producing ethanol from sugar proceeds via the following three steps.

*Fermentation*—Sugars are fermented by the action of an appropriate yeast at a suitable temperature. Starches are fermented after they have been converted to sugar, as described above.

*Distillation*—As noted, fermentation processes are limited to producing an aqueous solution with up to about 18% ethanol. Water is removed by distillation, producing an azeotropic solution of about 96% ethanol and 4% water. This is sometimes referred to as hydrous ethanol. This amount of water is sufficiently low that the ethanol can be used in this form in an internal combustion engine. However, the presence of 4% water makes the ethanol immiscible with gasoline. Thus, for applications in which the ethanol is mixed with gasoline (see flex-fuel vehicles in section 4.2.4.1), the water must be removed by dehydration.

*Dehydration*—To remove the water from an azeotropic ethanol solution, it is mixed with a solvent such as benzene. The benzene mixes with the water, allowing the ethanol to be separated. The resulting ethanol is referred to as anhydrous ethanol. However, the benzene which is used is highly carcinogenic. To minimize health risks associated with ethanol dehydration, new techniques based on molecular sieves allow water and ethanol to be separated and, in principle, can be used for both distillation and dehydration.

Most current bioethanol production uses either corn or sugarcane as feedstock. The details of these two approaches are discussed further below in the sections on ethanol infrastructure and the environmental aspects of ethanol fuel.

#### *4.2.3.3 Cellulosic ethanol production*

There are fairly straightforward methods of producing ethanol by the fermentation of the sugar-containing components of plant matter and the starch-containing components of plant matter that have been converted into sugar. These techniques are not sufficient to produce ethanol from lignocellulosic biomass. Lignocellulose is comprised of the carbohydrate polymers cellulose and hemicellulose as well as the aromatic polymer lignin. Before these compounds can undergo fermentation to produce ethanol, their polymer chains need to be broken down. The basic steps used to prepare lignocellulosic material for fermentation are as follows:

*Pretreatment*—Pretreatment generally involves physical processes, such as grinding to reduce particle size, and chemical treatment to improve the efficacy of the subsequent treatments.

*Cellulolysis*—Cellulolysis is the treatment of the lignocellulosic material with cellulase. Cellulases are a group of enzymes that are produced mainly by fungi, bacteria, and protozoans that catalyze cellulolytic reactions and break down cellulosic material into simple sugars.

*Separation of sugars from lignin*—The sugars that are produced during cellulolysis are tightly bound to lignin, which is not broken down by the cellulase. The sugars must therefore be separated from the lignin. This is typically done by chemical treatment with, for example, sodium hydroxide.

Following the three steps above, the resulting sugars that have been extracted from the lignocellulose can be fermented, distilled, and dehydrated as described above for the production of ethanol from glucose.

The lignocelluosic feedstock used for ethanol production can be either wood or agricultural biomass. Agricultural biomass includes plant (non-wood) material that is grown for ethanol production (Huang *et al* 2009) and waste material from agricultural activity. Switchgrass (*Panicum virgatum*) and Miscanthus×giganteus are grasses that have been considered as possible feedstocks for cellulosic ethanol production. Both are fast-growing grasses which have good tolerance for adverse environmental conditions. Waste agricultural material includes (for example) woodchips from sawmills, bagasse, and waste paper. Bagasse is the solid by-product that is left when liquid is extracted from plant matter. Although this term is sometimes used to refer to the solid residue that is obtained from any plant matter, it is also used specifically to refer to the dry matter associated with sugarcane or sorghum.

Scientifically, the principle of producing ethanol from cellulosic material is well established. However, from a practical standpoint, it has not been developed to the point where its economic viability or large-scale practicality have been demonstrated. A major difficulty in the commercial implementation of cellulosic ethanol production is the high cost of preparing the enzyme that is required for cellulolysis. Further research is needed to establish a practical and cost-effective route to producing ethanol by this approach.

### 4.2.4 Ethanol as a fuel

#### *4.2.4.1 Ethanol internal combustion engine vehicles*

Ethanol has been used as a fuel for internal combustion engine vehicles for more than a century. In fact, during the early part of the 20th century, it was a popular fuel for powering passenger vehicles. The Ford Model T (figure 4.1), produced from 1908 to 1927 was designed to use gasoline, ethanol, or any mixture of the two fuels, as well as pure kerosene. Part of the reason for the interest in ethanol as an automobile fuel during this era was due to the fact that the petroleum industry was still in its early stages of development. By the 1920s or so, the petroleum industry had expanded to the point that interest in ethanol as an automobile fuel declined. In the early 1970s, interest in ethanol and other potential alternative fuels was revived by the oil crisis and subsequent gasoline shortage (Farrell *et al* 2006). In order to mitigate the adverse effects of the limited petroleum supply, ethanol was mixed with

**Figure 4.1.** Ford Model T. Reproduced from (Dave7 2005) Flickr (CC BY-SA 2.0, https://creativecommons.org/licenses/by-sa/2.0/).

gasoline. In the United States, for example, the common blend of gasoline and ethanol in the 1970s was 10% ethanol and 90% gasoline. At the time, this was often referred to as gasohol. The current terminology for gasoline–ethanol mixtures is, for example, E10 for 10% ethanol and 90% gasoline and E85 for 85% ethanol and 15% gasoline.

Today, ethanol-containing fuel is available in different parts of the world with compositions ranging from around E5 up to E100. Standard internal combustion engines used in current passenger vehicles can typically operate on gasoline–ethanol mixtures of up to E10 without modification. For concentrations of ethanol greater than 10%, minor modifications to standard gasoline engines are required. The main problem results from the reactivity of some polymers to ethanol. This requires that certain components in the fuel delivery system are changed to materials that do not degrade in the presence of ethanol. This is a small, ~$100, addition to the cost of manufacturing a new vehicle. In principle, older vehicles can be modified to allow for the use of ethanol. However, this approach can cause problems, as organic deposits in the fuel system that are relatively insoluble in gasoline can be dissolved by ethanol-containing fuels, resulting in clogging in some engine components (Dunlap 2019). At present, up to 10% ethanol (E10) is added to gasoline in many countries. This is compatible with the fuel systems and engine components of virtually all gasoline-powered automobiles, and this ethanol content is frequently indicated on the gasoline pump (see figure 4.2). As noted in table 4.1, the heat of combustion of ethanol is 68% of that of gasoline. This means that a given volume of an ethanol–gasoline mixture produces less energy than the same volume of

**Figure 4.2.** Gasoline pump indicating that the fuel contains 10% ethanol, commonly referred to as E10. Image: Richard A Dunlap.

pure gasoline. This effect is partly offset by improved engine efficiency due to the higher octane rating of ethanol, around 99, compared to that of gasoline.

In recent years, a number of manufacturers have marketed so-called flex-fuel vehicles (short for 'flexible-fuel' vehicles). In general, a flex-fuel vehicle is an internal combustion engine vehicle that can run on either of two different fuels or any mixture of the two fuels. The two fuels are combined in the same fuel tank. Flex-fuel vehicles frequently utilize a mixture of gasoline with either methanol or ethanol. In North America, the term 'flex-fuel vehicle' generally refers to a vehicle that can run on any mixture of ethanol and gasoline with up to 85% ethanol (E85). Such vehicles are typically labeled with the logo shown in figure 4.3. Currently, worldwide, there are more than 60 million flex-fuel vehicles on the road. In the United States there are more than 20 million. It is, however, interesting to note that flex fuel, i.e. E85, is not as widely available as might be expected. Figure 4.4 shows the distribution of fueling stations in the United States that sell E85. The figure shows that there are slightly more than 3000 locations in the United States where E85 is available. This represents about 2% of all fueling locations, in comparison to more than 150 000 retail gasoline stations. For this reason, most owners of flex-fuel vehicles in the United States operate their vehicles on gasoline or E10 and do not make use of their vehicle's capability to utilize higher ethanol-content fuel.

**Figure 4.3.** 'Flex Fuel' logo on a Ford Escape. Image: Richard A Dunlap.

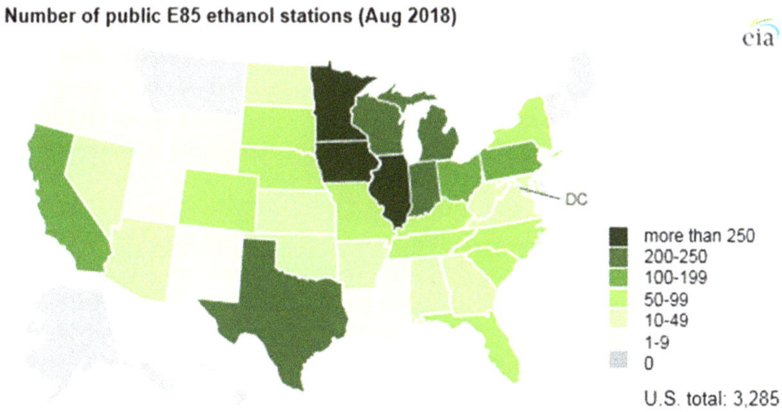


**Figure 4.4.** Number of public-access fueling stations selling E85 by state in the United States in 2018. Reproduced from EIA (2018) Image stated to be in the public domain.

The use of ethanol for vehicle fuel in other countries is somewhat different than in the United States. Two countries that have been very active in promoting flex-fuel vehicles and the use of ethanol as a general gasoline substitute have been Brazil and Sweden. While Sweden imports nearly all of the ethanol used for transportation, Brazil produces ethanol domestically from sugarcane. The use of ethanol as a transportation fuel in Brazil is, therefore, an excellent example of a complete infrastructure designed around this alternative fuel and is considered in some detail below.

In the early 1970s, Brazil implemented an alternative fuel strategy that mandated the use of gasoline–ethanol blends or pure ethanol (referred to as neat ethanol) as fuel for passenger automobiles. Gasoline internal combustion engines were modified to accommodate the more reactive nature of ethanol, as described above, and compression ratios were increased to optimize performance for ethanol fuel. Ethanol is less flammable than gasoline, as noted in the previous chapter. For this reason, some ethanol vehicles have a small gasoline tank to facilitate starting, similar to those of methanol vehicles in Iceland. The market share of vehicles operating on different fuels in Brazil since 1979 is shown in the graph in figure 4.5. Since the mid-2000s, flex-fuel vehicles have dominated the market. At present, Brazil has about half of the world's flex-fuel or ethanol-fueled road vehicles. This significant and rapid increase in ethanol use as a vehicle fuel in Brazil came about largely because of the development of the supporting infrastructure. At present, virtually all fueling stations in Brazil sell flex fuel (e.g. E85) or pure ethanol in comparison to about 2%, as noted above, in the United States.

The environmental consequences of the use of ethanol as a transportation fuel in Brazil and other countries are discussed in detail later in the chapter.

#### *4.2.4.2 Ethanol hybrid vehicles*

The possibility of using ethanol or ethanol–gasoline blends for fuel in internal combustion engine vehicles also opens up the possibility of using ethanol-containing

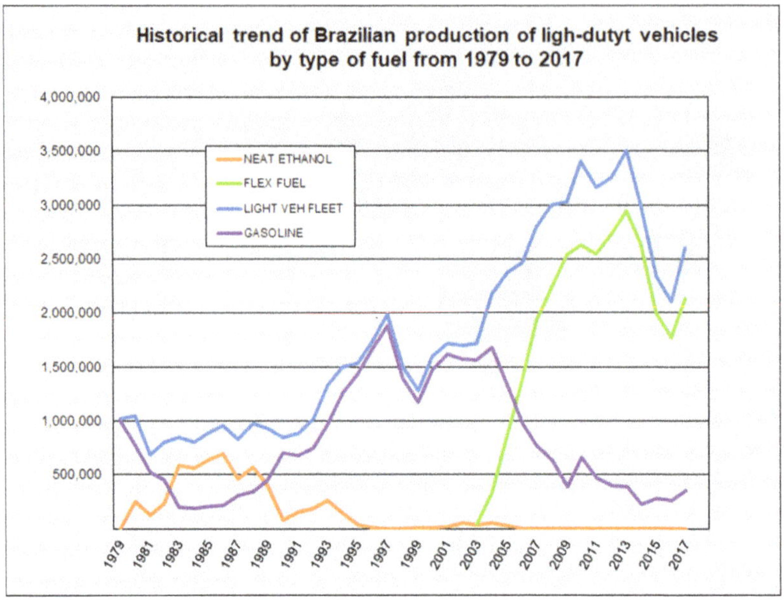


**Figure 4.5.** Annual number of light-duty vehicles produced in Brazil from 1979 to 2017 based on fuel. This image (Ortiz 2018) has been obtained by the author from the Wikimedia website where it was made available under a CC BY-SA 3.0 licence, https://creativecommons.org/licenses/by-sa/3.0/deed.en. It is included within this chapter on that basis. It is attributed to Mariordo.

fuels in hybrid vehicles. This approach was first put into practice in 2007 when Ford produced a fleet of 20 flex-fuel (E85)–battery hybrid Escape SUVs for testing. This was followed the next year with a prototype E85 plug-in hybrid, shown in figure 4.6. More recently, Toyota do Brasil has announced plans to market a Corolla-based E85 hybrid, as illustrated in figure 4.7 (Fagundes 2018).

#### *4.2.4.3 Reformed ethanol fuel cell vehicles*

Ethanol may be reformed to produce hydrogen for use in a fuel cell. This approach is analogous to that of the reformed methanol fuel cell shown in figure 3.35. The reformer produces hydrogen from ethanol by the reaction (Hou *et al* 2015)

$$C_2H_5OH + 3H_2O \rightarrow 6H_2 + 2CO_2,$$

and this is used to power a hydrogen fuel cell. In 2016, Nissan announced plans to develop an electric vehicle utilizing a reformed ethanol fuel cell. This is analogous to the approach used by Gumpert's reformed methanol fuel cell vehicle, the Nathalie, as described in the previous chapter. Nissan's vehicle was intended to use a 55:45 mixture of ethanol–water as fuel and produce hydrogen by the reaction above. The hydrogen was to be used to produce electricity by means of a solid oxide fuel cell. This is in contrast to the polymer electrolyte membrane fuel cells used in current production fuel cell vehicles. The solid oxide fuel cell has several very attractive properties. It can produce significant output power, it has good efficiency, can tolerate low-purity hydrogen and does not require an expensive metal as a catalyst.

**Figure 4.6.** Ford Escape E85 flex-fuel plug-in hybrid. At right from the top: charging socket, fuel filling door open, fuel filling door closed. This image (Ortiz 2010) has been obtained by the author from the Wikimedia website where it was made available under a CC BY-SA 3.0 licence, https://creativecommons.org/licenses/by-sa/3.0/deed.en. It is included within this chapter on that basis. It is attributed to Mariordo.

**Figure 4.7.** Toyota prototype flex-fuel hybrid for the Brazilian automobile market. Reproduced from Duran-Ortiz (2018) Flickr (CC BY-SA 2.0, https://creativecommons.org/licenses/by-sa/2.0/).

It is therefore much more cost effective than fuel cells such as those used in current fuel cell vehicles. The properties of solid oxide fuels have made them appropriate for applications such as stationary backup power for emergency facilities and remote locations. The main disadvantage of solid oxide fuel cells is that they operate at high temperatures and Nissan specifies an operating temperature of around 700 °C for its solid oxide fuel cell. The application of reformed ethanol solid oxide fuel cells for vehicle use has recently been reviewed by Dogdibegovic *et al* (2020).

### 4.2.5 Fuel ethanol and its environmental consequences

The use of ethanol as a fuel has many attractive features. The infrastructure for distributing ethanol to consumers can readily be adapted from existing gasoline infrastructure and represents a very small fraction of the investment required to implement a hydrogen infrastructure. In addition, vehicles that operate on ethanol can be produced by making simple and inexpensive modifications to existing gasoline internal combustion engine vehicle designs. In fact, a gradual transition from gasoline-powered vehicles to ethanol-powered vehicles is possible by means of increased utilization of flex-fuel vehicles. Compared to battery electric vehicles, ethanol vehicles can have ranges and fueling times that are similar to gasoline vehicles. Finally, the use of ethanol promotes energy independence, and cost analyses suggest that ethanol can be produced at a competitive (or lower) price than gasoline.

The proponents of ethanol fuel use point out that it is readily renewable by the growth of sugar- or starch-containing crops and that the growth and combustion process is carbon free. However, an actual detailed analysis of the environmental consequences of ethanol fuel use is quite complex. Many factors influence the overall impact of ethanol fuel including the type of crop, the methods of processing the plant matter, the approach to producing ethanol, and the details of the distribution network. Perhaps most important in this analysis is the extent to which fossil fuels are used in this process. We can begin with an assessment of fuel ethanol production and use in the United States.

Figure 4.8 shows world fuel ethanol production for the leading countries from 2007 to 2019. From these data, it can be seen that in recent years about 53% of ethanol was produced in the United States, 31% in Brazil, and the remaining 16% in

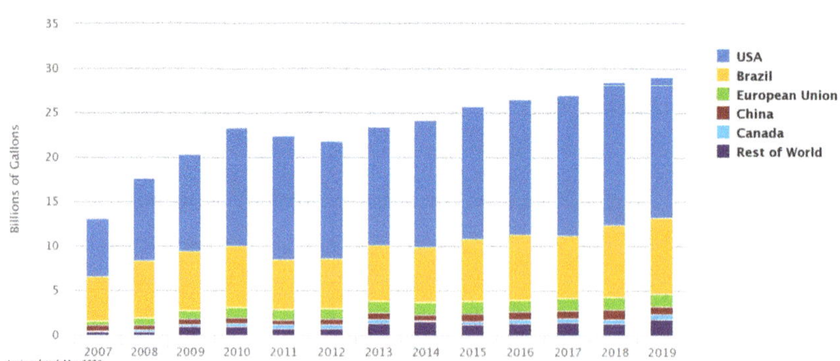


**Figure 4.8.** World fuel ethanol production from 2007 to 2019 by country. Reproduced from U.S. Department of Energy (2022). Image stated to be in the public domain.

the rest of the world. In 2019, 537 billion liters of gasoline were used in the United States compared to 59 billion liters of fuel ethanol. As noted above, most fuel ethanol in the United States is mixed with gasoline to produce E10 and is sold for use in traditional gasoline internal combustion vehicles. In terms of energy content, fuel ethanol represents about 7% of the energy content of gasoline in the United States.

Bioethanol produced in the United States is made almost entirely from corn. Table 4.2 gives some properties of corn that are relevant to its use in ethanol production. Commercial processes utilize only the corn kernels. Possible future cellulosic ethanol production technologies could also make use of the corn stover (i.e. stalks and leaves), thereby increasing productivity per unit land area by about 25%–50% over the values given below.

The hardiness zone provides information concerning the geographic range where a given plant can readily be grown. Plant hardiness zones in the United States are shown in figure 4.9. It can be seen in the figure that the hardiness zones that are compatible with the growth of corn, i.e. zones 4–8, cover most of the United States except for some very southerly regions where the climate is too warm for corn growth. While it would seem that the compatible climate and the large land area of the United States would make ethanol production from corn for transportation applications a viable and environmentally conscious option, there are several points that need to be considered in detail. These factors include the loss of agricultural land for food production, the ratio of energy gained to energy expended, and the net environmental impact of ethanol production from corn.

We can make a rough estimate of the effects that extensive bioethanol use in the United States would have on food production. According to the United States Energy Information Administration (EIA 2021a), $4.67 \times 10^{11}$ L of gasoline was consumed in the country in 2020. This was somewhat less than usual due to the COVID-19 pandemic, but is a suitable value to use for this exercise. Although, according to table 4.1, ethanol has only 68% of the energy content of gasoline per unit volume, this is offset to some extent by the greater efficiency of its use in internal combustion engines or reformed fuel cells. Assuming that we would require the same volume of ethanol to replace gasoline as a transportation fuel and using the mean ethanol yield per land area as given in table 4.2, means that we would require $(4.67 \times 10^{11}\ \mathrm{L})/(3.9 \times 10^{5}\ \mathrm{L\ km^{-2}}) = 1.2 \times 10^{6}\ \mathrm{km^2}$ of land area. The total arable land area of the continental United States is estimated to be $2.7 \times 10^{6}\ \mathrm{km^2}$ (Dunlap 2019), and this means that 45% of all farmlands in the United States would have to be devoted to the growth of corn for fuel ethanol production to replace the gasoline that is used (primarily) for transportation. This is consistent with the current use of about 3.7% of arable land in the United States for ethanol production (as of 2010), in

**Table 4.2.** Properties of corn and sugarcane as feedstocks for bioethanol production. Data adapted from Dunlap (2019) and Sanderson (2006) and references therein.

| Feedstock | Hardiness zones | Annual yield ($10^5$ L km$^{-2}$) |
|---|---|---|
| Corn | 4–8 | 3.8–4.0 |
| Sugarcane | 9–13 | 6.8–8.0 |

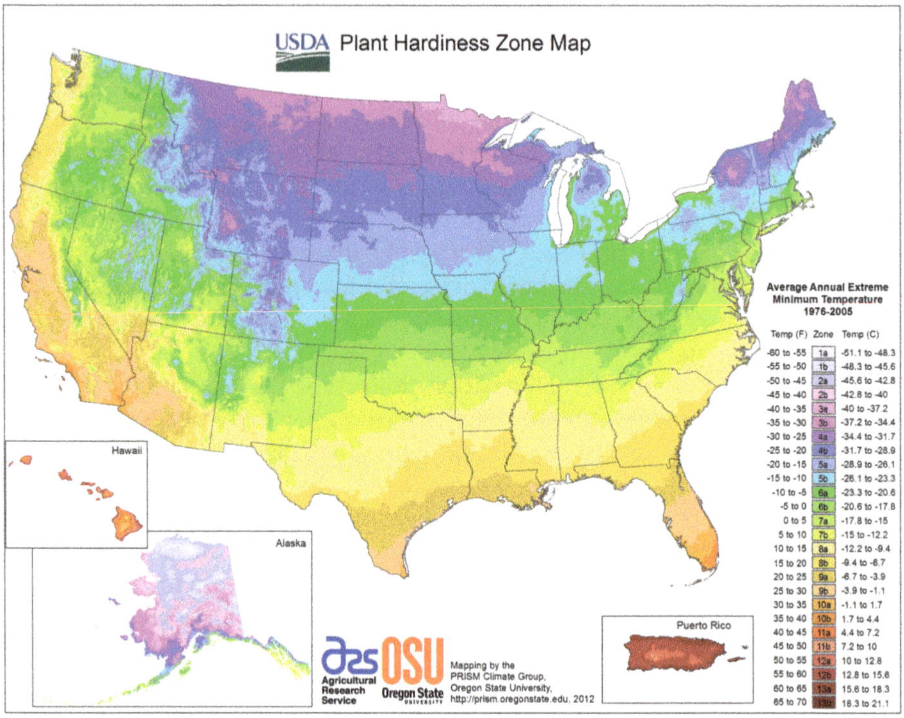


**Figure 4.9.** Plant hardiness zones in the United States. Reproduced from U.S. Department of Agriculture (2022). Image stated to be in the public domain.

which fuel ethanol is used almost exclusively in the form of E10 (i.e. 10% ethanol). It is clear, even from this simple calculation, that the use of any significant quantity of fuel ethanol produced from domestic corn in the United States would have a substantial impact on the availability of farmland for the production of food for humans and livestock. In fact, in order to satisfy this need, a significant fraction of agricultural land would need to be repurposed for corn production.

Although the electromagnetic energy associated with sunlight is transformed into the chemical energy associated with corn, it requires energy to grow the corn and process it into ethanol. This energy is needed for preparing the farmland, planting and harvesting the corn, producing fertilizer and related materials, processing the corn, producing ethanol, and transporting and distributing the final fuel. The energy balance or energy ratio is the ratio of the chemical energy associated with the ethanol to the energy required for agriculture, ethanol processing, and distribution. In the case of corn-based ethanol production in the United States, the energy balance has been estimated (Mekonnen *et al* 2018) to be about 2.1. This means that the chemical energy that is available from the bioethanol that is produced is about twice the energy that is input into the process of producing the ethanol.

The final important aspect of ethanol production from corn that needs to be considered is the overall environmental effect. While it is true that the carbon dioxide that is emitted during the combustion of the ethanol is equal to the carbon dioxide that is absorbed from the atmosphere during the growth of the feedstock, it is essential to consider the overall utilization of energy. This means that the energy input into the growth of the corn and the production of the ethanol needs to be considered in this overall assessment. At present (in the United States, at least) the energy that is used for operating farm equipment, for processing the corn, for the fermentation and distillation of the ethanol and its subsequent transport and distribution, comes mainly from the consumption of fossil fuels. This type of assessment is not straightforward and also varies with time; as technology develops, input energy sources change and even assumptions that go into the analysis can vary.

A number of such assessments for corn-based ethanol production in the United States have been reported over the years, and many of these have recently been reviewed by Scully *et al* (2021). Figure 4.10 shows the results of Scully *et al*'s overview of the carbon emissions during bioethanol production in the United States along with a breakdown of where those emissions occur. It is relevant to compare these results to the carbon dioxide emission for gasoline which are about 85 g $CO_2$ per MJ. Early studies (up to the mid-2000s) have divided the process into (a) corn production and transportation and (b) fuel production and transportation. The data show that between roughly 2000 and 2005, carbon dioxide emissions decreased, largely as a result of improved technology and a corresponding carbon dioxide reduction in ethanol production. Studies from the late 2000s onward have typically included an additional carbon dioxide source referred to as land-use change (Searchinger *et al* 2008). This refers to factors such as increased atmospheric carbon dioxide that results from the deforestation needed to provide additional land area for increased corn production. While there is some decrease in calculated carbon dioxide emission between 2009 and 2016, most analyses since that time are

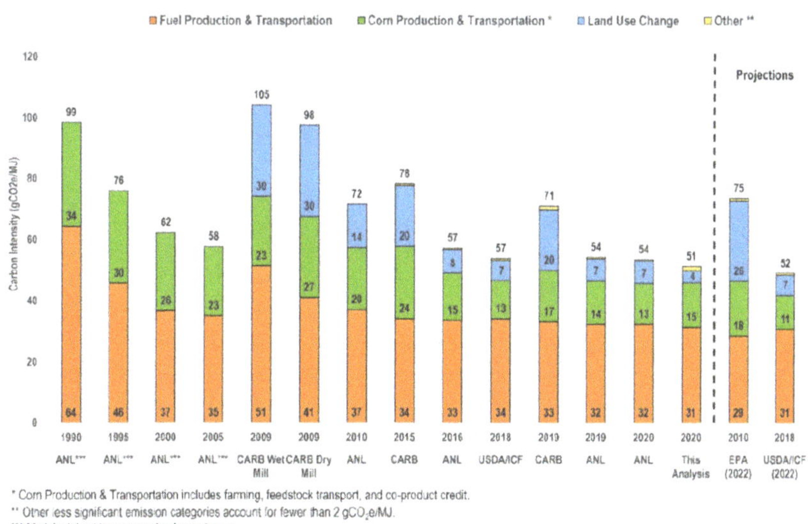


**Figure 4.10.** Carbon dioxide emissions for corn ethanol produced in the United States in g $CO_2$ per MJ. Reproduced from Scully *et al* (2021), copyright IOP Publishing Ltd. CC BY 4.0.

reasonably consistent and predict around 55 g $CO_2$ per MJ. This suggests a decrease in carbon emissions of about 35% compared to gasoline.

It is interesting to contrast the situation described above for ethanol produced from corn in the United States to the situation in Brazil, where ethanol is produced mainly from sugarcane. Table 4.2 shows that the ethanol yield from sugarcane is approximately twice that from corn. The table shows that the hardiness zones for sugarcane are zone 9 to zone 13. All of Brazil has a climate that is compatible with the growth of sugarcane. In the United States, on the other hand, sugarcane can only be grown in a few very southern regions, mostly along the Gulf Coast and in Southern California, as illustrated in figure 4.9.

By weight, sugarcane is the world's largest crop; close to $2 \times 10^9$ tonnes is produced annually and 40% of this is produced in Brazil (see figure 4.11). Brazilian sugarcane is used for several purposes. Principal among these are the production of fuel ethanol (about 55%), the production of sugar for the food industry (about 44%), and the production of alcoholic beverages (about 1%).

Because of its climate and its extensive sugarcane industry, Brazil is in an ideal position to utilize this alternative transportation fuel. In fact, bioethanol use in Brazil dates back to the mid-1970s when the government extensively promoted the use of domestic ethanol for use as a transportation fuel. This approach was a direct result of the 1970s oil crisis and was a move to secure some degree of energy independence. During this time, pure (or neat) ethanol (i.e. E100) was widely promoted, as illustrated in figure 4.5 by the significant number of E100 light-duty vehicles manufactured in Brazil from the late 1970s to the mid-1990s. The figure also

**Figure 4.11.** Sugarcane growing in Brazil for fuel ethanol production. This image (da Fonseca 2007) has been obtained by the author from the Wikimedia website where it was made available under a CC BY-SA 3.0 licence, https://creativecommons.org/licenses/by-sa/3.0/deed.en. It is included within this chapter on that basis. It is attributed to José Reynaldo da Fonseca.

shows that in the mid-1980s, E100 vehicles represented nearly 80% of the Brazilian automobile industry.

In the early 2000s, flex-fuel vehicle manufacture became important in Brazil as a result of the development of cost-effective technology that allowed the vehicle's engine to determine the ratio of ethanol to gasoline in the fuel and adjust the combustion parameters accordingly to optimize engine efficiency. Since about 2006, flex-fuel vehicles have dominated the Brazilian automobile industry. Figure 4.5 shows that at present, nearly 85% of vehicles produced in Brazil can operate on flex fuel or, in many cases, on any mixture of ethanol and gasoline up to E100.

The efficient and effective implementation of ethanol road vehicles in Brazil has led to an energy balance of around 6.0 for bioethanol (Mekonnen *et al* 2018). This about three times the energy recovery associated with corn ethanol production in the United States. In addition, Brazil uses only about 1% of its arable land to support its extensive ethanol fuel program. This significant difference between the sugarcane ethanol industry in Brazil and the corn ethanol industry in the United States can be attributed to several factors. Most importantly, these factors include the following:

- The greater yield of ethanol from sugarcane compared to corn.
- The much warmer climate in Brazil.
- A much less mechanized agricultural system.
- A dedicated effort to implement ethanol production technology.

A direct comparison of the carbon emissions produced by Brazilian sugarcane ethanol and United States corn ethanol is not straightforward. Studies over the years have reported quite variable results. A summary of data for various bioethanol production programs up to around 2013 is shown in figure 4.12. This comparison suggests a rather high value for the emissions from corn-based ethanol and a very low value for the emissions from sugarcane-based ethanol. While such comparisons are often quoted as a measure of Brazil's ethanol success, it is often difficult to compare results from different studies, because researchers often use different approaches and methodologies to determine the carbon footprints of different transportation technologies. This is exemplified by an inspection of the results of studies of United States ethanol production shown in figure 4.10. Mekonnen *et al* (2018) have compared ethanol production in both the United States and Brazil and have reported a value of 57.0 g $CO_2$ per MJ for United States corn ethanol and a value of 38.5 g $CO_2$ per MJ for Brazilian sugarcane ethanol. The value given for the United States is consistent with most contemporaneous results shown in figure 4.10. The authors, however, suggest a reduction of 12.1 g $CO_2$ per MJ (giving a net 44.9 g $CO_2$ per MJ) for the United States number to account for *distillers' grains and solubles* (DGSs). DGSs are residual components of the corn that are left over from the ethanol production process that can be used as livestock feed. The availability of these residual materials eliminates the carbon dioxide emissions associated with conventional feed production.

Thus, while most analyses indicate a reduction in carbon dioxide emission for bioethanol compared to gasoline, the exact amount of this reduction depends on the details of the ethanol production process as well as the method of analysis and is subject to considerable uncertainty. Studies, however, have consistently shown that

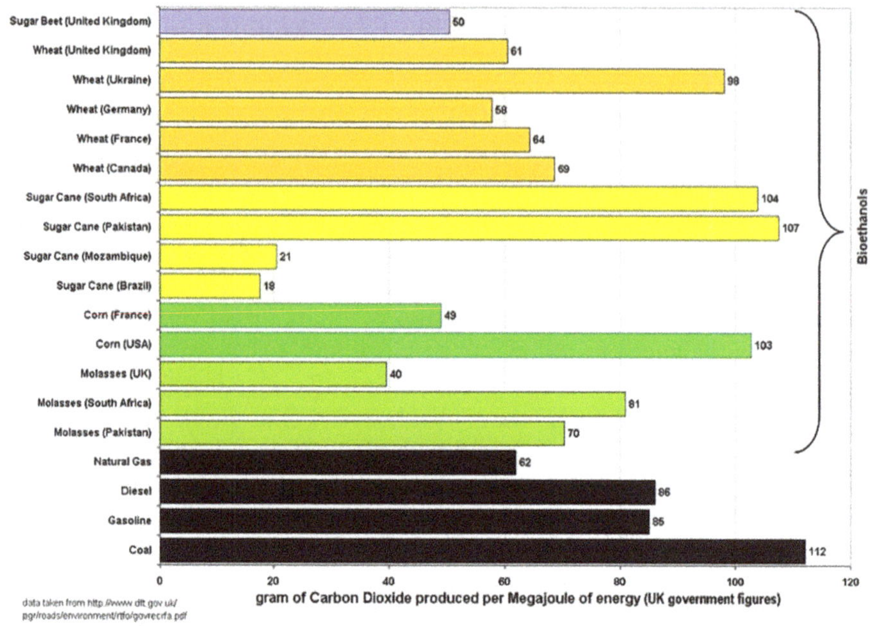


**Figure 4.12.** Grams of carbon dioxide produced per megajoule of energy generated from various fuels using current technologies. This image (Young 2008) has been obtained by the author from the Wikimedia website, where it is stated to have been released into the public domain. It is included within this chapter on that basis.

the implementation of ethanol production from cellulosic plant material has a greater impact on carbon dioxide reduction than ethanol production from sugars or starches. Ethanol production from cellulosic material as described above makes use of portions of plant material that do not contain sugar or starch, such as corn stalks, plants such as grasses, or wood. This approach to ethanol production opens up the possibility of utilizing a greater component of plant material or, more importantly, utilizing plants that can be grown on land that would not be suitable for the growth of food crops.

Some plant species that have been utilized for cellulosic ethanol production are summarized in table 4.3. The table shows that all three of these species are compatible with hardiness zones that cover most of the United States. Switchgrass, in particular, can be grown in fairly cold climates that are found in much of Canada, Russia, and China. Its annual yields are similar to the productivity of corn and even that of sugarcane.

As in the case of corn-based and sugarcane-based ethanol production, an exact analysis of greenhouse gas emission compared to that of gasoline is not straightforward. However, one study by Wang *et al* (2007) has estimated an 86% reduction in carbon dioxide emission for cellulosic ethanol compared to gasoline. This is a substantially greater reduction than has been estimated for United States corn ethanol production and is comparable to even the most advantageous estimates for Brazilian sugarcane ethanol production. Further work by Wang *et al* (2011) has

**Table 4.3.** Properties of some potential feedstocks for cellulosic ethanol production. Data adapted from Sanderson (2006) and references therein.

| Feedstock | Hardiness zones | Annual yield ($10^5$ L km$^{-2}$) |
|---|---|---|
| *Miscanthus* | 5–9 | 7.3 |
| Poplar | 5–9 | 3.1–7.6 |
| Switchgrass | 3–9 | 3.7–6.0 |

shown that both agricultural activities and fuel production related to cellulosic ethanol produce carbon dioxide emissions that are small or negligible compared to those of corn ethanol. Most importantly, however, is the negligible land-use change contribution to greenhouse gas emissions for cellulosic ethanol. This results from the ability to grow cellulosic ethanol crops on land that would otherwise not be suitable for agriculture. Including a credit for distillers' grains and solubles, Wang *et al* (2011) shows that cellulosic ethanol production is very close to carbon neutral.

While the production of ethanol from cellulose resolves the potential land-use and greenhouse gas emission problems associated with ethanol production from sugars or starches, an economically viable approach to the large-scale implementation of this technology still needs to be developed.

## 4.3 Biodiesel

While ethanol or methanol can replace gasoline in spark ignition internal combustion engines, biodiesel serves as a direct replacement for fossil-fuel-based diesel fuel in compression ignition engines. Biodiesel is produced from vegetable oil or animal fats as described below and consists of short-chain alkyl esters. Unprocessed vegetable oil (commonly referred to as straight vegetable oil (SVO)) can also be used as a replacement for diesel fuel, although significant engine modifications are required. The use of SVO as a fuel is discussed in further detail later in this section. It is also important to note that while the use of biodiesel as a transportation fuel is considered in the present book, home heating oil is essentially the same as diesel fuel, and the development of biodiesel production technologies to satisfy transportation needs may also provide a source of sustainable fuel for space heating applications.

### 4.3.1 Methods for the production of biodiesel

Biodiesel is a methyl, ethyl, or propyl ester produced by the transesterification of vegetable oil or animal fat. Transesterification is a process by which an organic group R' of an ester is exchanged with an organic group R' of an alcohol. This process is schematically illustrated in figure 4.13. Methanol is frequently used as the alcohol feedstock. This is because, at present, methanol is the least expensive alcohol available. Today, methanol is mostly produced from fossil fuel sources, which negates many of the environmental benefits of biodiesel. Future production methods could utilize methanol produced in a sustainable manner, as discussed in the previous chapter. Ethanol, as discussed above, can also be used as the alcohol feedstock.

**Figure 4.13.** Schematic representation of transesterification in which an organic group R' of an ester is exchanged with an organic group R' of an alcohol. This image (Benjah-bmm27 2007) has been obtained by the author from the Wikimedia website, where it is stated to have been released into the public domain. It is included within this chapter on that basis.

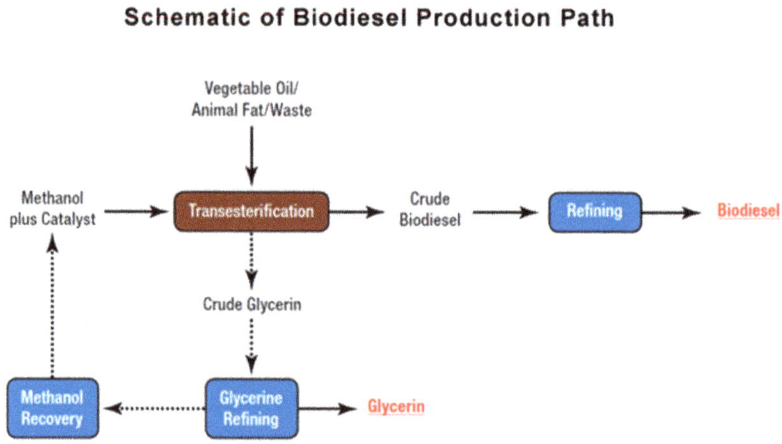


**Figure 4.14.** Schematic diagram of the steps involved in biodiesel production. Reproduced from U.S. Department of Energy (2022). Image stated to be in the public domain.

The reaction of oil or fat with alcohol is performed in the presence of a catalyst, usually sodium hydroxide (NaOH) or potassium hydroxide (KOH). The basic steps in the process of producing biodiesel are shown in figure 4.14. In a typical transesterification process, 1 tonne (1000 kg) of oil or fat is reacted with 100 kg of alcohol to yield 1 tonne (1000 kg) of biodiesel and 100 kg of glycerol ($C_3H_8O_3$, also known as glycerin or glycerine). The biodiesel that is produced is mixed with the glycerol and also contains a number of contaminants including soap, salts, residual methanol, and water (see e.g. Raman *et al* (2019)). As the figure shows, the biodiesel and glycerol are separated, and each is further refined. The purified biodiesel can be used as a substitute for petroleum-based diesel.

The glycerol ($C_3H_8O_3$) that is a by-product of biodiesel production has a variety of commercial uses. It is a component in numerous food products, pharmaceuticals, soaps, and personal care products. It is also used to produce nitroglycerin, an essential ingredient of many explosives. Unfortunately, the glycerol that results from biodiesel production is of relatively low purity and must be purified for most applications. This purification process is energy intensive. While glycerol utilization negatively impacts biodiesel production from an energetic standpoint, it may have a positive impact from an economic standpoint. Although this economic impact was once considered relevant, increases in biodiesel production in recent years has resulted in greater quantities of glycerol than are commercially needed. Ongoing research activities are considering additional viable options for the commercial utilization of glycerol.

A variety of oils and fats have been used as feedstocks for biodiesel production. Some of the most common materials that fall into this category are briefly described below.

*4.3.1.1 Virgin vegetable oils from land-based crops*

Dedicated crops such as sugarcane and corn are used for the production of ethanol from their sugar and starch contents; in a similar way, crops can be grown specifically for oil to be used as feedstock for the production of biodiesel. Table 4.4 lists some common plants that are used to produce vegetable oil for biodiesel production. It is clear that the plants with the highest yields are tropical (i.e. coconut palms) or at least require a fairly warm climate (i.e. palm oil trees). Under favorable climatic conditions, plants may have higher yields due to the possibility of producing more than one crop per year. It is interesting to note that some plants, such as jatropha, are potentially valuable sources of oil for biodiesel production but are toxic for human consumption. Chufa (*Cyperus esculentus*) is a plant in the sedge family which has good yield and is hardy (zones 3–9) throughout most of the United States and Canada.

Since the late 1970s, work on the development of commercially viable approaches to biodiesel production has been undertaken in several countries, including Brazil, South Africa, the United States, and several European countries. Rapeseed, soy, and sunflower have been popular feedstocks for this work. Following the approach described above for ethanol as a replacement for gasoline, we can look at the situation for domestic biodiesel production and use in the United States. Current diesel fuel consumption (primarily of petroleum-based diesel) is around $1.8 \times 10^{11}$ L per year (Dunlap 2019). Using soy (the most common biodiesel crop in the United States) with an (optimistic) yield of $92 \times 10^3$ L km$^{-2}$, we see that the land area required is $(1.8 \times 10^{11}\ \mathrm{L})/(92 \times 10^3\ \mathrm{L\ km^{-2}}) = 2 \times 10^6\ \mathrm{km}^2$. Comparing this area to the total arable land area in the continental United States given previously, i.e. $2.7 \times 10^6$ km$^2$, shows that this approach cannot supply all diesel fuel requirements without substantial interference with food agriculture.

**Table 4.4.** Properties of some potential oil feedstocks for biodiesel production.

| Feedstock | Minimum hardiness zone | Annual yield ($10^3$ L km$^{-2}$) |
|---|---|---|
| Palm | 7 | 475 |
| Coconut | 11 | 215 |
| Jatropha | 10 | 187 |
| Chufa | 3 | 163 |
| Rapeseed | 7 | 95 |
| Soy | 2 | 55–92 |
| Peanut | 6 | 84 |
| Sunflower | 2 | 77 |

#### *4.3.1.2 Waste vegetable oil*

The use of waste vegetable oil, primarily from the food industry, would seem to be an environmentally desirable approach to biodiesel production. However, primary vegetable oil production in the United States is about $1.3 \times 10^{10}$ L per year. In addition, the recovery of much of the waste vegetable oil from the food industry may be difficult and/or expensive. Thus, at most, biodiesel prepared from waste vegetable oil could replace a few percent of the diesel consumed. However, if the recovery and processing of waste oil could be accomplished in an economically viable manner, then it could provide a small contribution toward meeting future transportation energy needs.

#### *4.3.1.3 Animal fats*

As in the case of producing biodiesel from the vegetable oil of specifically grown plant species, it might be possible to consider raising dedicated animal species for the production of biodiesel from the resulting animal fat. However, this approach is not economically (or energetically) viable, and it is most appropriate to utilize waste fats and oils from the animal-based food industry as feedstock for biodiesel production. The primary production of animal-derived fats, including oils obtained from aquatic species, amounts to about half that of primary vegetable oil production. Thus, as for waste vegetable oil, residual animal fat could make only a small contribution to total diesel needs.

#### *4.3.1.4 Sewage*

Biodiesel can also be produced from sewage sludge. The utilization of sewage to produce fuel results in a reduction in greenhouse gas emissions compared to the use of petroleum-based diesel and also benefits the environment by avoiding the decomposition of sewage sludge in landfills where it emits methane. Research in this area is in its fairly early stages, but encouraging results have already been reported in the scientific literature (e.g. Kumar *et al* (2020)).

#### *4.3.1.5 Algae*

Algae can produce oil for use as a feedstock for biodiesel production. This approach has several attractive characteristics that could mitigate the problems encountered using terrestrial plants to produce biodiesel. First, algae can be grown in dedicated ponds or other facilities or in coastal regions that could not be used for food agriculture. Second, algae have a very high yield compared to other oil sources for biodiesel. For example, estimates of annual algae biodiesel yield are between $5.87 \times 10^6$ and $1.37 \times 10^7$ L km$^{-2}$ (Atabani *et al* 2012), which is up to about 30 times the yield of the best terrestrial crops. The reasons for this high yield are that algae grow very quickly, and they can contain up to 60% oil, compared to about 3% for soybeans. It is also beneficial that algae can be grown using waste products such as sewage and that the by-products of algae biodiesel production can be used as livestock feed. Using even the minimum expected biodiesel yield from algae indicates that a land (or water) area of $(1.8 \times 10^{11}\ \text{L})/(5.87 \times 10^6\ \text{L km}^{-2}) = 3 \times 10^4\ \text{km}^2$ would be sufficient to replace traditional diesel usage in the United States. This land area would certainly be quite viable (especially since it does not need to be arable land).

The most common approaches to the growth of algae for biodiesel production are open ponds and photobioreactors. Open ponds for algae growth are often of the 'raceway' design illustrated in figure 4.15. Water is circulated by means of one or more paddle wheels. These systems utilize less power and are more cost effective than the photobioreactor design described below. Since the growth of the algae relies on incident sunlight, the effective depth of the pond is limited by the opacity of the growth in the water. One major concern, however, is the possibility of contamination from the environment, particularly from birds, insects, and dust.

Algae can be grown in a system of plastic or glass pipes known as a photobioreactor. An example of a photobioreactor is illustrated in figure 4.16. The photobioreactor is exposed to sunlight, and water containing nutrients and carbon dioxide is circulated through the closed pipe system. Although the photobioreactor allows for more control of growing conditions and eliminates concerns about contamination, it is less cost effective than the simple open pond arrangement. In addition, scaling up the photoreactor for commercial production creates challenges due to light attenuation and increased carbon dioxide requirements.

Research into algae biodiesel began slowly in the 1990s and accelerated (at least for a time) after 2000. While algae biodiesel appears attractive compared to biodiesel

**Figure 4.15.** A raceway open pond for the cultivation of microalgae operated by Ingrepro Renewables BV in Amsterdam, The Netherlands. This image (JanB46 2011) has been obtained by the author from the Wikimedia website where it was made available under a CC BY-SA 3.0 licence, https://creativecommons.org/licenses/by-sa/3.0/deed.en. It is included within this chapter on that basis. It is attributed to JanB46.

**Figure 4.16.** A tubular glass photobioreactor for the cultivation of microalgae operated by IGV GmbH in Nuthetal, Germany. This image (IGV Biotech 2013) has been obtained by the author from the Wikimedia website where it was made available under a CC BY-SA 3.0 licence, https://creativecommons.org/licenses/by-sa/3.0/deed.en. It is included within this chapter on that basis. It is attributed to IGV Biotech.

derived from terrestrial plants, it has not achieved great commercial success. The main obstacles to the development of algae fuel seem to be primarily economic and also the anticipated time scale for commercialization. Since 2009, ExxonMobil has been involved in a research program dedicated to the development of algae biofuels with the biotechnology company Synthetic Genomics. In 2013, however (Carroll 2013), ExxonMobil speculated that commercialization would take at least 25 years. Many smaller research organizations have moved away from algae biofuels because of concerns over near-term profitability.

### 4.3.2 The use of biodiesel as a transportation fuel

Mixtures of biodiesel and petroleum-derived diesel are labeled in the same way as ethanol and methanol blends with gasoline. For example, B20 is a mixture of 20% biodiesel and 80% petroleum-based diesel, while B100 is pure biodiesel. In order to decrease dependence on petroleum resources, biodiesel is a common additive to traditional diesel fuel in many locations. Biodiesel is available throughout most of the United States, Canada, and the European Union, as well as other locations across the globe. B20 and B100 are the most commonly available mixtures. It is generally believed that mixtures up to B20 can be used in conventional diesel engines with little or no modifications. However, some vehicle manufacturers will not honor warranties for vehicles that have that have been damaged as a result of the use of biodiesel-containing fuel. The manufacture of biodiesel blends is not as standardized as the manufacture of ethanol–gasoline blends, largely because the composition of biodiesel is not unique (as is the case for ethanol). Thus, the exact properties of biodiesel may vary from manufacturer to manufacturer. Contaminants can be present in biodiesel that is not properly produced or stored. Perhaps the greatest concern is the presence of water, which can cause corrosion of components in the fuel delivery system as well as the pistons. In addition, the presence of water in the fuel can cause gelling of the fuel at low temperatures, resulting in the clogging of fuel delivery system components. Finally, microbes that may be associated with water contamination can cause the deterioration of filter elements. A small amount of petroleum-based diesel (e.g. ~1%) is sometimes blended with the biodiesel to retard the growth of microorganisms.

Some properties of typical biodiesel (B100) and typical regular-grade petroleum-based diesel fuel are shown in table 4.5. The table shows that biodiesel has a lower heat of combustion per unit volume, but a slightly higher density compared to petroleum diesel. The flash point is the minimum temperature at which the fuel will ignite in the presence of a spark. Although this value is not directly related to engine performance, it is an indication of fuel storage safety and shows that biodiesel is superior in this respect.

Three properties of diesel fuels that require particular consideration are lubricity, viscosity, and efficiency.

#### *4.3.2.1 Lubricity*

In the past, traditional diesel fuels contained sulfur, and this was responsible for some of the environmental concerns related to fossil fuel use, as described in chapter 1. Many contemporary diesel fuels are designed to have low sulfur contents to mitigate this problem. The main difficulty with this approach is that the sulfur compounds in petroleum-based diesel are largely responsible for its lubricating properties. Biodiesel is virtually sulfur-free, but has desirable lubricating properties compared to many petroleum-based diesel fuels. For this reason, small amounts of biodiesel are added to low-sulfur petroleum-based diesel in order to enhance its lubricating properties without introducing sulfur pollution into the environment. The higher lubricity of biodiesel may increase the usable lives of fuel injectors and the associated injection pumps, as these rely on the fuel itself to provide lubrication. However, some studies (e.g. Clark *et al* (1984)) have indicated that after two to four years fuel injectors may suffer deterioration and that this may be due to corrosion caused by moisture absorbed by the biodiesel fuel (Monyem and Van Gerpen 2001).

**Table 4.5.** Typical values of biodiesel (B100) compared with petroleum-based diesel.

| Property | Biodiesel | Petroleum diesel |
|---|---|---|
| Density (kg $m^{-3}$) | 880 | 850 |
| Heat of combustion (MJ $L^{-1}$) | 29.8 | 35.7 |
| Autoignition temperature (°C) | 374 | 210 |
| Flash point (°C) | 130 | 52 |

#### *4.3.2.2 Viscosity*

The viscosity of a fuel is an important parameter in determining its utilization in an engine. The fuel used in a diesel engine must be viscous enough to provide lubrication to the fuel injection components, as described above. However, its viscosity must be low enough to allow it to flow efficiently at operational temperatures. At 40 °C, the viscosity of biodiesel is between 4.2 and 4.6 cSt. A centistoke (cSt) is a unit of kinematic viscosity, i.e. the fluid's internal resistance to flow under gravitational forces. By comparison, the viscosity of petroleum-based diesel fuel is between 2.5 and 3.2 cSt. The high viscosity of biodiesel can cause it to clog fuel filters and fuel injection systems. In addition, the viscosity of a fluid typically increases with decreasing temperature, meaning that engines operating on high percentages of biodiesel can have fuel supply difficulties at low ambient temperatures.

#### *4.3.2.3 Efficiency*

Although the energy content of biodiesel is somewhat lower than that of petroleum-based diesel, as determined from its heat of combustion, this is partially offset by a typically higher engine efficiency. This is analogous to the situation described above for bioethanol compared to gasoline. The greater efficiency of biodiesel combustion is, in part, due to its better lubricity, but is primarily from more complete combustion. The combustion properties of fuel used in a compression ignition engine are generally measured by the so-called cetane rating (or cetane number). This rating is analogous to the octane rating used to measure the properties of a fuel in a spark ignition internal combustion engine. The cetane rating is based on an assessment of the ignition delay of the fuel in a compression ignition engine. The cetane rating is inversely related to the ignition delay, so fuels with lower cetane ratings have longer ignition delays in compression ignition engines. This means that low-cetane-rating fuels are not suitable for high-speed engines. The cetane rating of a given fuel is defined as the ratio of cetane and heptamethylnonane that would have the same ignition delay. The scale used for the cetane rating defines a cetane rating of 100 for pure cetane and 15 for pure heptamethylnonane. Analogous to the octane scale for spark ignition fuels, a cetane rating of 100 would indicate a fuel with the same ignition delay as pure cetane, but would not indicate that the fuel actually contained cetane. Many jurisdictions impose minimum cetane ratings on transportation fuels. Minimum cetane ratings are often around 40 or so. Typical petroleum-based diesel cetane ratings are in the range of 45–50, and those of premium diesel products are often around 60. By comparison, biodiesel derived from vegetable oil has a cetane rating of between 46 and 52, and biodiesel derived from animal fat has a cetane rating of between 56 and 60. This, on average, shows improved combustion for biodiesel compared to standard grade petroleum-based diesel.

### 4.3.3 Biodiesel production

Biodiesel utilization as a transportation fuel is not as advanced as, for example, the use of ethanol. It is also the case that its industry standards and mixture specifications are not as well established. Nonetheless, biodiesel use has increased considerably

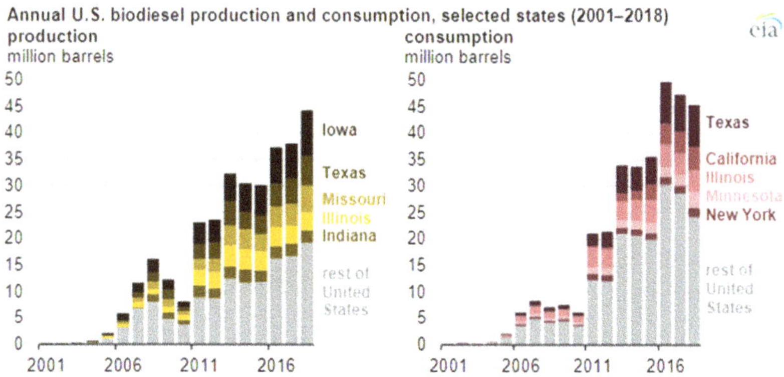


**Figure 4.17.** Annual biodiesel production and consumption in various regions of the United States from 2001 to 2018. Reproduced from EIA (2020). Image stated to be in the public domain.

since the early 2000s. As an example, biodiesel production and consumption in the United States from 2001 to 2018 are shown in figure 4.17. In recent years, this has amounted to about 4% of total United States diesel fuel use. Global biodiesel production in 2020 was about $46 \times 10^9$ L (compared to about $115 \times 10^9$ L of bioethanol). About 50% of the biodiesel comes from the European Union, 22% from the United States, and the remainder mostly from Brazil and Argentina.

### 4.3.4 Environmental aspects of biodiesel use

As discussed previously, the greenhouse gas reduction that results from the replacement of gasoline with bioethanol is highly dependent on a number of factors and is generally not as significant as might be anticipated. This situation is the same for biodiesel. The development of biodiesel as a transportation fuel is not as advanced as the development of bioethanol. Similarly, environmental assessments of biodiesel use are not as extensive or detailed as they are for bioethanol. A summary of some research on this subject is shown in figure 4.18. Typically, biodiesel produced from virgin oils show a carbon dioxide emission reduction (compared to petroleum-based diesel) in the range of 30%–55%. The figure also shows that the carbon dioxide reduction is substantially greater for waste oil and fat than for virgin oils. It should be noted that these results assume that there is no contribution from land-use change. As seen for ethanol production, however, this may be an important factor in an overall environmental assessment.

A 2010 report by the United States Environmental Protection Agency (EPA 2010) on renewable fuels provided an assessment of biodiesel produced from soy oil and waste fat. This report included land-use change and found a 57% carbon dioxide reduction from soy oil biodiesel and an 86% carbon dioxide reduction from waste fat biodiesel, which are reasonably consistent with the results shown in figure 4.18. The EPA has set a standard of a 20% reduction in carbon emission compared to a conventional fossil fuel equivalent in order for a biofuel to be considered to be a renewable fuel. In 2012, the EPA ruled that palm-oil-based biodiesel did not meet the

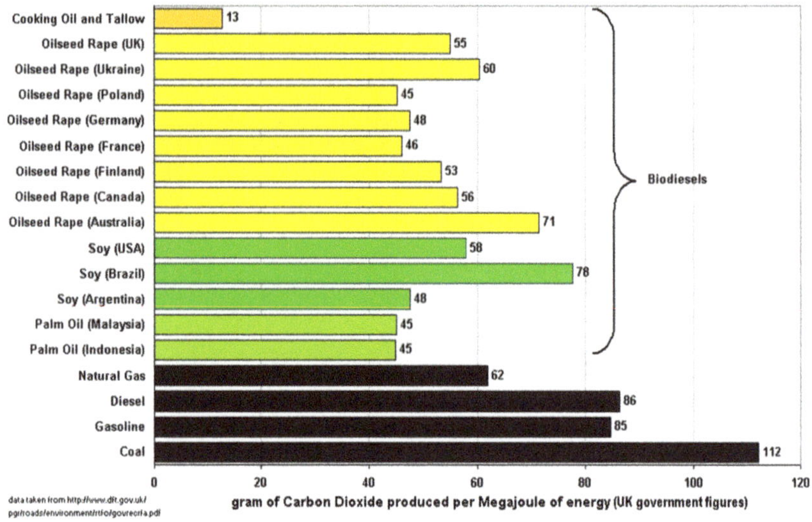


**Figure 4.18.** Carbon dioxide emissions in grams of $CO_2$ per MJ of energy produced of biodiesels produced from different feedstocks and a comparison with traditional fossil fuels. This image (Young 2008) has been obtained by the author from the Wikimedia website, where it is stated to have been released into the public domain. It is included within this article on that basis. Public domain by Mike Young.

requirements for a renewable fuel (Mongabay 2012). This failure to satisfy EPA standards for a renewable fuel stemmed largely from the carbon emissions associated with the deforestation practices that are typically used for oil palm agriculture.

### 4.3.5 Straight vegetable oil as a fuel

Fuel viscosity is a concern for the use of biodiesel as a road vehicle fuel. It is an even greater concern for the use of SVO. SVO may be virgin vegetable oil as produced from one of the sources described above, or it may be waste vegetable oil from the food industry (e.g. leftover oil from restaurants). The problem with SVO viscosity is illustrated in figure 4.19. As the figure shows, this problem is greatly exacerbated by low temperatures. One of the major effects of the transesterification process that is used to produce biodiesel from vegetable oil is the substantial reduction in viscosity which results from the breaking down of long hydrocarbon chains into smaller ester molecules.

A second potential problem with the use of SVO as a fuel is its reactivity with oxygen. A consequence of this reactivity is carbon buildup (commonly called coking) in the engine. The results of investigations into coking in diesel engines as a function of the compositions of mixtures of SVO and petroleum-based diesel are illustrated in figure 4.20.

Despite the potential difficulties of using SVO as a fuel in diesel engines, the conversion of vintage diesel passenger vehicles to run on vegetable oil is popular among environmental enthusiasts. Mercedes-Benz diesel vehicles, as shown in figure 4.21, have frequently been the starting point for such conversions, generally

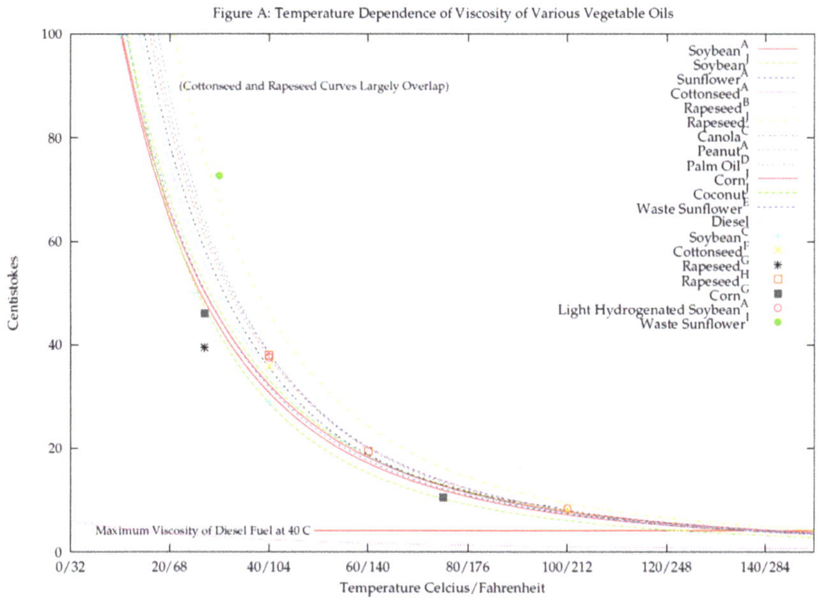


**Figure 4.19.** Kinematic viscosity of various vegetable oils as a function of temperature. The temperature dependence of the kinematic viscosity of petroleum-based diesel fuel is shown for comparison. Reproduced from [Bunkum, 2022]. CC BY-SA 3.0 US, https://creativecommons.org/licenses/by/3.0/us/.

with the goal of operating the vehicle on inexpensive waste vegetable oil. Typical modifications to make this feasible include:

- ensuring all plastic and rubber components in the fuel system are compatible with vegetable oil;
- replacing fuel filters so that they can accommodate the higher viscosity of vegetable oil;
- adding pre-heaters to the fuel supply system to lower the viscosity of the fuel before it gets to the engine.

While vehicles that have undergone such modification operate well, the modifications do not mitigate the problem of coking. Studies have shown that the long-term operation of diesel engines on SVO leads to engine deterioration and maintenance issues (Sidibé *et al* 2010). For this reason, as well as the limited availability of waste oil and the lack of any large-scale collection infrastructure, it is unlikely that vehicles taking this approach to renewable fuel would be commercially viable.

## 4.4 Renewable diesel

Biodiesel is a form of diesel fuel that is prepared by the transesterification of oils or fats. The term 'renewable diesel' is used in a manner that is distinct from biodiesel and refers to diesel fuel that is prepared from oils or fats by methods other than transesterification, as described below. Renewable diesel is chemically very similar to petroleum-based diesel and is, therefore, compatible with conventional diesel

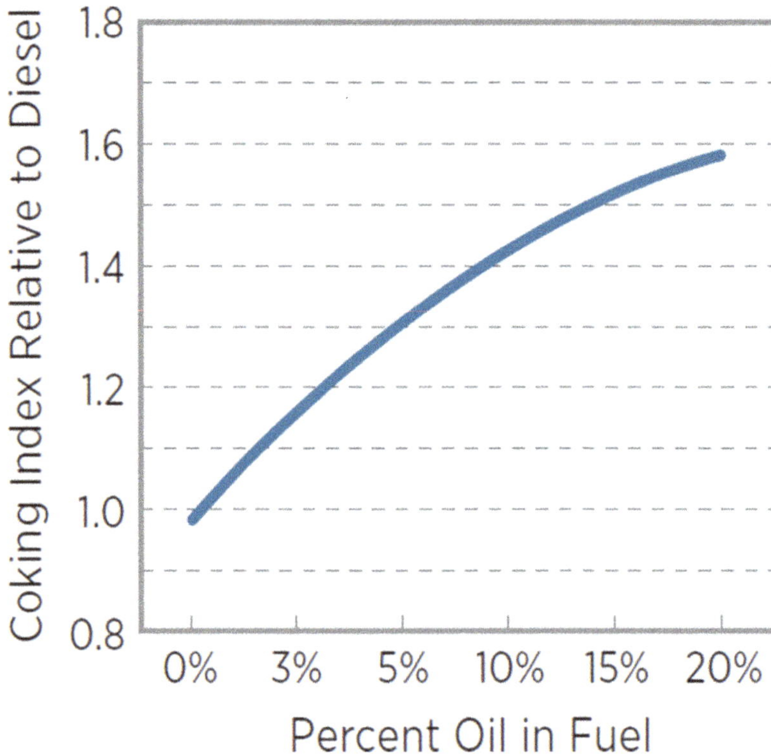


**Figure 4.20.** Coking (carbon buildup) index relative to petroleum-based diesel for mixtures of diesel with SVO. Reproduced from [U.S. Department of Energy (2022). Image stated to be in the public domain.

engines without any modification. Renewable diesel can be blended with petroleum-based diesel, and such blends are referred to as, for example, R20 for 20% renewable diesel with 80% petroleum-based diesel or R100 for 100% renewable diesel.

Renewable diesel is prepared from the same feedstock as biodiesel, i.e. virgin plant oils or waste cooking oils and animal fats. Renewable diesel is produced from feedstock by a reaction with hydrogen in the presence of a catalyst at elevated temperatures and pressures in a process referred to as hydrotreatment (Snåre *et al* 2008). This process is sometimes referred to as deoxygenation, as it serves to remove the oxygen from oils, thereby leaving a pure hydrocarbon. This basic process can be described as follows:

$$CH_xO_{(x-1)} + (x/2)H_2 \rightarrow CH_2 + (x-1)H_2O,$$

where a unit of oil containing one carbon atom reacts with enough hydrogen to produce water from its oxygen content. This process is in contrast to the trans-esterification process used to produce biodiesel, in which the resulting biofuel contains oxygen. The oxygen is a major factor in the adverse effects of biodiesel fuel on engine components. Other impurities that can be detrimental to the fuel properties, e.g. sulfur, nitrogen, and chlorine, can be removed from the oil during hydrotreatment.

**Figure 4.21.** Vintage Mercedes-Benz Diesel automobile modified to use biodiesel as a fuel. This image (Mejidori 2007) has been obtained by the author from the Wikimedia website, where it is stated to have been released into the public domain. It is included within this chapter on that basis.

Renewable diesel has the same (or very nearly the same) chemical composition as petroleum-based diesel and can, therefore, be used as a direct substitute for its fossil fuel equivalent in unmodified diesel engines either as R100 or as blends. In addition, renewable diesel has shown good low-temperature performance. Neste, a Finnish company with major renewable diesel refineries in the Netherlands and Singapore, is a major supplier of renewable diesel worldwide and reports cetane ratings between 75 and 85 for its product, which are higher than typical petroleum-based diesel ratings. This high cetane rating leads to efficient combustion and reduces vehicle maintenance.

Renewable diesel production has grown substantially in recent years. In 2019, about $3.4 \times 10^9$ L of renewable diesel was consumed in the United States. Most of this was used in California due to the economic benefits of renewable fuels resulting from state energy policies. More than half of this renewable diesel was imported, mostly from Asian suppliers, although domestic production is expected to increase in the coming years.

Studies have typically shown a 50%–80% reduction in carbon emissions for renewable diesel compared to petroleum-based diesel. This is comparable to or better than the reduction seen for typical terrestrial feedstocks used to produce biodiesel. However, as with biodiesel, land-use changes, in particular, deforestation, are a major concern for extensive renewable diesel utilization.

Initiatives for the production of other renewable hydrocarbon fuels (to replace gasoline and aviation fuel, for example) are underway and are discussed further in the next chapter.

## 4.5 Biogas and biomethane

Biogas is produced from organic feedstock by anaerobic decomposition (that is decomposition without oxygen). A wide variety of raw materials can be used as feedstocks for biogas production, including manure, agricultural waste, sewage, food waste, and municipal waste. The composition of the resulting biogas depends on the details of the feedstock and the method of preparation. Typical biogas consists of methane and a smaller amount of carbon dioxide. It may also contain small amounts of nitrogen, hydrogen, and other impurity gases. Table 4.6 gives the typical composition ranges of different components of a biogas.

### 4.5.1 The production of biogas and biomethane

The decomposition of feedstock typically takes place by anaerobic digestion inside a closed system, commonly referred to as a bioreactor or anaerobic digester. A diagram of a basic bioreactor design is shown in figure 4.22. This design processes materials with low contents of solids (typically less than 15%). Other designs utilize materials with high solids contents. Digesters may be designed for either batch processing, as shown in figure 4.22, or continuous flow processing. The decomposition process that produces biogas from biomass feedstock proceeds in four steps, as described below.

#### *4.5.1.1 Hydrolysis*

Typically, biomass material that is used for the production of biogas is comprised of large organic polymer molecules such as carbohydrates. The first step in anaerobic digestion is to break down these large organic molecules. This process is known as bacterial hydrolysis, in which bacteria break apart complex organic polymers into their component parts such as simple sugars and amino acids. Some hydrogen and ammonia may also be produced at this stage of the digestion. Other by-products of the hydrolysis must be catabolized (that is broken down further) before the methanogenesis stage. This catabolism takes place through acidogenesis and acetogenesis as described below.

**Table 4.6.** Typical composition of biogas produced by anaerobic digestion. Data adapted from https://web.archive.org/web/20100106022729/http://www.kolumbus.fi/suomen.biokaasukeskus/en/enperus.html.

| Gas | Composition range (%) |
|---|---|
| Methane | 55–75 |
| Carbon dioxide | 25–45 |
| Nitrogen | 1–5 |
| Hydrogen | 0–3 |
| Hydrogen sulfide | 0.1–0.5 |
| Carbon monoxide | 0–0.3 |

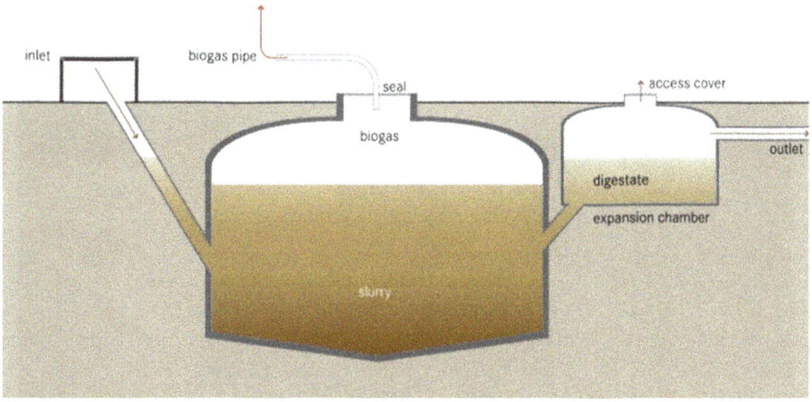


**Figure 4.22.** Diagram of a bioreactor for producing biogas. Reproduced with permission from Tilley *et al* (2014), copyright Eawag-Sandec.

*4.5.1.2 Acidogenesis*

The simple sugars and amino acids that have been produced by the hydrolysis of the feedstock are further broken down by acidogenic bacteria. This bacterial process converts the sugars and amino acids into simpler molecules such as carbon dioxide, hydrogen, hydrogen sulfide, and organic acids.

*4.5.1.3 Acetogenesis*

The organic acids that result from acidogenesis are converted by bacterial processes into acetic acid. Additional hydrogen, ammonia, and carbon dioxide can also be produced at this stage.

*4.5.1.4 Methanogenesis*

The final step of biogas production is methanogenesis. At this stage, the biomass consists of acetic acid with the additional hydrogen, hydrogen sulfide, ammonia, and carbon dioxide that were formed in previous digestion processes. Bacteria known as methanogens convert these materials into methane and carbon dioxide. Small concentrations of some other compounds, as described above in table 4.6, remain. A commercial biogas production facility in Germany is illustrated in figure 4.23.

Although biogas is primarily methane (as is natural gas) and has many of the same applications as natural gas, it cannot be used for many of these applications in its raw state. The main problem with using biogas in its raw state is caused by the impurities that it inevitably contains. Principal among these is carbon dioxide. However, other impurities, such as hydrogen sulfide and water vapor, are particularly troublesome. Hydrogen sulfide is particularly corrosive to metals and is not an acceptable component of fuel for internal combustion engines. For vehicle use, biogas must be purified (or as it is generally called, upgraded). There are five basic approaches to upgrading. These are water washing, membrane separation, pressure swing absorption, amine gas treatment, and selexol absorption. Water washing is the most common method used for commercial biogas processing. In this approach,

**Figure 4.23.** Biogas plant production facility in Sundern, Germany. This image (Thzorro77 2020) has been obtained by the author from the Wikimedia website where it was made available under a CC BY-SA 4.0 licence, https://creativecommons.org/licenses/by-sa/4.0/deed.en. It is included within this chapter on that basis. It is attributed to Thzorro77.

biogas at high pressure flows through a column where it scrubbed of impurities by a counter-flowing stream of water. Membrane separation has also become common in recent years. Upgraded biogas has methane concentrations of 98% or higher and is consistent with the purity and other characteristics of fossil-fuel-produced methane. It is, therefore, sometimes referred to as renewable natural gas.

### 4.5.2 The applications of biomethane

After upgrading, biogas (now mostly biomethane) can be used interchangeably with natural gas for most applications (EPA 2020). These applications include injection into the natural gas grid for domestic and industrial use, electricity generation in natural-gas-fired generating stations (this includes heat as well in co-generation schemes), and vehicle use. For vehicle use, biomethane is compressed and can replace compressed natural gas (CNG) for vehicle fuel. Biomethane is compressed to pressures of around 200–245 atm. At these pressures, the energy density per unit volume of compressed methane is about 25% that of gasoline. The combustion of methane in an internal combustion engine is described by the reaction

$$CH_4 + 2O_2 \rightarrow CO_2 + 2H_2O.$$

Thus, carbon dioxide is produced as a by-product. In an ideal situation, the combustion of biomethane is carbon neutral, as the carbon dioxide that is released during combustion is equivalent to the carbon dioxide that is sequestered from the atmosphere during the production of the biogas feedstock. The use of natural gas as a 'clean' alternative to other fossil fuels has long been promoted, even before the possibility of producing biomethane. In fact, the combustion of methane produces 50 g of $CO_2$ per MJ of energy compared to approximately 66 g $CO_2$ per MJ and approximately 112 g $CO_2$ per MJ for gasoline and coal, respectively.

The use of biomethane produced from biogas in vehicle engines follows from the fairly long history of natural-gas-powered vehicles. These vehicles fall into two basic categories: those which are conversions of gasoline-powered vehicles and those which have been manufactured specifically to use compressed natural gas as a fuel. In principle, any recent model of gasoline-powered vehicle can be converted to use compressed natural gas as a fuel. Conversions typically cost in the range of about 6000 USD, the major portion of which is usually due to the cylinders used to hold the high-pressure compressed gas. In vehicles that have been converted from gasoline vehicles, the gas storage tanks are typically located in the trunk and eliminate cargo space. In addition to having a lower volumetric energy capacity than gasoline and therefore requiring a larger tank volume for the same range, high-pressure gas cylinders must be of a suitable structural shape, while gasoline tanks can take on irregular shapes in order to best be integrated into the vehicle design. Vehicles may be converted from gasoline to compressed natural gas in the interest of reducing greenhouse gas emissions as well as other pollutants. They may also be converted for economic reasons. Although the conversion is costly, natural gas is less expensive than gasoline (per MJ), and vehicles that are driven a large number of kilometers can provide economic benefits.

Vehicles that are manufactured to operate on compressed natural gas are either dedicated natural gas vehicles or bi-fuel vehicles that typically operate on either natural gas or gasoline. In the latter case, the two fuels are stored in separate tanks (one compressed gas and one liquid) and this is analogous to the Mazda RX-8 Hydrogen RE (see section 3.5) which can operate on two fuels, compressed hydrogen and gasoline.

One of the earliest mass-produced compressed natural gas vehicles was the Honda Civic GX, which became available in the late 1990s and was manufactured until 2015. This vehicle is shown in figure 4.24 with the high-pressure compressed natural

**Figure 4.24.** 2009 Honda Civic GX showing the 'NGV' (Natural Gas Vehicle) logo. This image (IFCAR 2009) has been obtained by the author from the Wikimedia website, where it is stated to have been released into the public domain. It is included within this chapter on that basis.

gas filling hose connected. An interesting example of a compressed natural gas vehicle is the Fiat Siena Tetrafuel. This vehicle, which is produced in Betim (Brazil) and Córdoba (Argentina), has a 1.4 L engine that can run on pure gasoline, E25 or E100 from a (fluid) fuel tank, or compressed natural gas from high-pressure cylinders. Figure 4.25 shows the Tetrafuel and the location of the compressed gas cylinders.

**Figure 4.25.** 2008 Fiat Siena Tetrafuel 1.4. Top: Rear view of vehicle. Bottom: Compressed natural gas cylinders in trunk. This image (Mariordo 2008) has been obtained by the author from the Wikimedia website where it was made available under a CC BY-SA 3.0 licence, https://creativecommons.org/licenses/by-sa/3.0/deed.en. It is included within this chapter on that basis. It is attributed to Mariordo.

At present, there are approximately 30 million vehicles worldwide that operate on compressed natural gas. The majority of these are in China, Iran, India, Pakistan, Argentina, and Brazil. Worldwide, there are over 30 000 natural gas filling stations, mostly in Asia. In North America, there are currently less than 1000. Natural gas filling stations can be conveniently located anywhere where there is a natural gas distribution network. In other locations, compressed natural gas can be delivered by truck. Fueling is typically somewhat slower than fueling with gasoline or diesel, but is much faster than battery recharge times for battery electric vehicles. So-called 'fast fill' stations can typically provide compressed natural gas at a rate of 1–2 GGE per minute. GGE is sometimes used as a measure of alternative fuel volume and means 'gasoline gallon equivalent,' i.e. the volume of alternative fuel that has the same energy capacity as one gallon of gasoline. At the rate stated above, filling a typical passenger fuel tank with compressed natural gas would require around 10 min. In recent years, home fueling has become possible. Residences or business with natural gas service can have a unit mounted in a garage or on an exterior wall that can provide compressed natural gas for vehicle use. Such an arrangement, like home battery electric vehicle charging, is convenient albeit slow. Typical home filling units provide about 0.4 GGE per hour.

The use of biomethane in place of fossil-fuel-derived natural gas as a vehicle fuel is, in principle, relatively straightforward. Vehicles designed to operate on natural gas can also operate on biomethane, as the composition and properties are the same. Although the number of compressed natural gas filling stations worldwide is not large, the natural gas distribution infrastructure is in place in many parts of the world and the development of a filling station network would be fairly straightforward. Biomethane can be injected into the natural gas pipeline to supplement fossil fuel natural gas. In fact, this is commonly done using biogas that has been upgraded to natural gas standards. In the future, as biomethane production is increased, the percentage of biomethane relative to fossil fuel natural gas can be increased. In a sustainable situation, biomethane would replace natural gas.

In order to quantitatively evaluate the possible contribution biomethane could make to transportation, it is necessary to begin by looking at the current production levels and uses of fossil-fuel-derived natural gas. The current world production of natural gas is about $4 \times 10^{12}$ m$^3$ (Dunlap 2019). With an energy content of $3.8 \times 10^7$ J m$^{-3}$, the global production of natural gas provides about $1.5 \times 10^{20}$ J annually, or about a quarter of the world's primary use of about $6.1 \times 10^{20}$ J per year. The percentage of natural gas use in the United States is slightly higher than the world average at about $3.3 \times 10^{19}$ J per year, which about 34% of a total of $9.8 \times 10^{19}$ J (see figure 4.26). We can use the energy requirements of light-duty road vehicles as an example of transportation needs. As summarized in chapter 1, transportation in the United States accounts for 37% of primary energy use, and light-duty vehicles account for about 62% of this. Therefore, light-duty vehicles account for about 23% of primary United States energy use or about $(0.23) \times (9.8 \times 10^{19}\ \text{J}) = 2.3 \times 10^{19}$ J per year. This indicates that if all light-duty vehicles in the United States ran on compressed natural gas, this would utilize about two thirds of all natural gas production in the country.

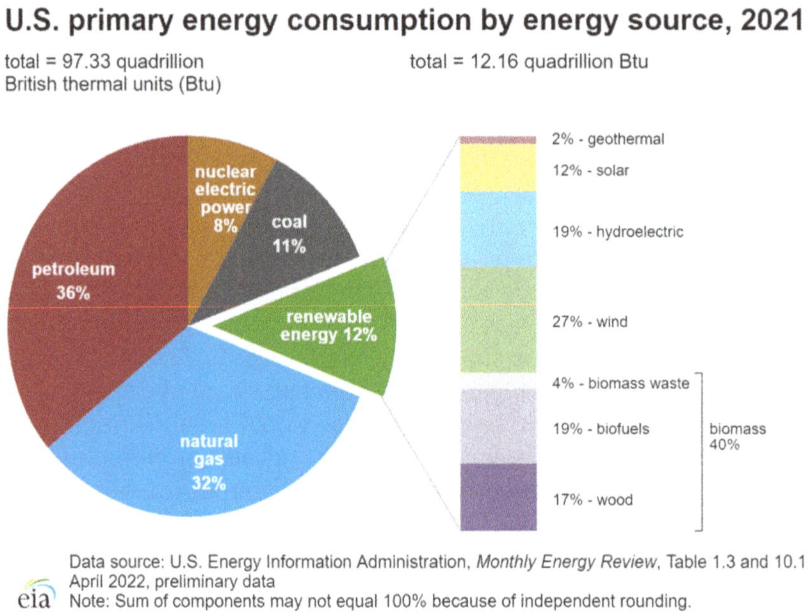


**Figure 4.26.** Breakdown of primary energy sources in the United States in 2021. Reproduced from EIA (2022). Image stated to be in the public domain.

At present, natural gas is used for electricity generation, industrial processes, residential heating, commercial heating, and transportation. Figure 4.27 shows a breakdown of these natural gas uses in the United States. Transportation accounts for only about 3% of the energy produced by natural gas. If biomethane were to replace natural gas as an energy source, it would need to accommodate all applications. Roughly speaking, biomethane production would have to accommodate all of natural gas's current applications and also accommodate the energy needs of transportation. In other words, biomethane production would have to be about 65% greater than current natural gas production. The actual situation is not likely to be as simple. It is not clear that biomethane would replace all natural gas applications in the future. However, it is clear that future biomethane production would have to be at least comparable to current natural gas use in order to satisfy transportation requirements.

We can look at biomethane use from a more global perspective. The International Energy Agency (IEA 2020) has predicted that biogas and upgraded biogas (i.e. biomethane) production will reach about $1.36 \times 10^{19}$ J per year by around 2040. This is only about 10% of the current natural gas consumption worldwide, which means that the expected future production of biomethane could, at best, satisfy only a small fraction of transportation needs.

### 4.5.3 The environmental aspects of biomethane

As previously noted, the combustion of methane produces less carbon dioxide per unit energy than the combustion of gasoline, 50 g of $CO_2$ per MJ compared to

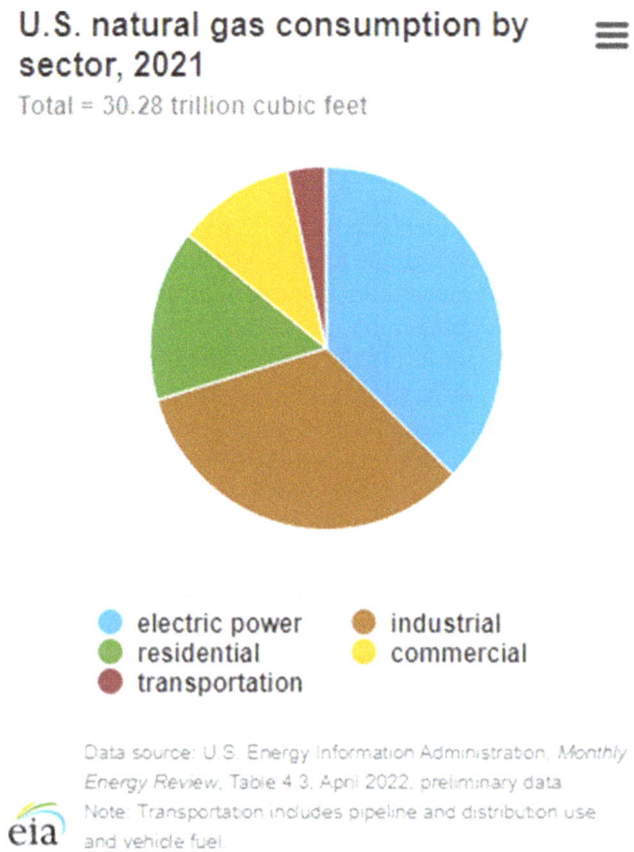


**Figure 4.27.** Breakdown of natural gas use in the United States in 2021. Reproduced from EIA (2022). Image stated to be in the public domain.

66 g $CO_2$ per MJ. The basic reason for this is that methane ($CH_4$) has a higher ratio of hydrogen to carbon (i.e. four to one) than that of gasoline (which is about 2.25 to one). The oxidation of carbon produces carbon dioxide, while the oxidation of hydrogen produces water. So, in simple terms, we can think of the combustion of a hydrocarbon as producing not only the energy due to the oxidation of the carbon component but also the energy due to the oxidation of the hydrogen component. The more energy is obtained from hydrogen, the less carbon dioxide is produced. Methane contains fewer impurities and burns more thoroughly than gasoline, so it produces fewer pollutants such as sulfur compounds. Like gasoline, it still produces nitrogen compounds, which result from the oxidation of nitrogen in the air.

The carbon dioxide that is emitted during the combustion of biomethane is compensated for by the carbon dioxide that is absorbed from the atmosphere during the growth of the organic material from which the biogas is produced. However, this simple assessment does not fully assess the carbon balance of biomethane production and use. As in the case of bioethanol, as discussed above, a quantitative assessment of the carbon dioxide greenhouse gas reduction achieved by replacing gasoline with biomethane is not straightforward and is subject to considerable uncertainty. In the case of biomethane (rather than bioethanol), such an assessment is considerably less relevant, as other factors need to be considered. The problem with large-scale methane utilization results from methane's properties as a greenhouse gas. As shown in table 1.9, methane is 25 times as effective a greenhouse gas as carbon dioxide. During the processing, distribution, and utilization of methane, some amount of gas is inevitably lost due to leaks. A recent analysis of the natural gas network (Alvarez *et al* 2018) concluded that about 2.3% of the natural gas that was extracted from wells was lost into the atmosphere. These authors assessed that this loss of methane had a net greenhouse effect that was comparable to that of the carbon dioxide produced by methane combustion. Thus, if, during the production, distribution, and utilization of biomethane, a comparable amount of gas is lost to the atmosphere to that lost in natural gas use, then any positive effects of greenhouse gas reduction that could be achieved would be negated. Therefore, perhaps the most important environmental consideration for the use of methane, either fossil fuel derived or biogas derived, is to ensure sufficient system integrity to minimize losses to the atmosphere.

A final point of relevance for biomethane is the origin of the material that is used to generate the biogas. While biogas may be produced using dedicated crops, it is preferable to utilize waste material for biogas production, as this eliminates potentially adverse environmental effects of land-use change. The potential benefits of biogas production depend significantly on the nature of the waste material used and what might otherwise be done with the material. Municipal solid waste may be directed to landfill sites. Manure or agricultural waste may be spread on farmland. In all of these cases, decomposing matter releases methane, and the use of waste material to produce biomethane may be environmentally preferable to the alternatives.

## 4.6 Biomethanol and biodimethyl ether

Methanol was first isolated in 1661 by Robert Boyle (1627–91). Its earliest application was probably as a fuel for space heating. Later, it was used as a feedstock for the production of various chemicals. Up until the early twentieth century, it was mainly produced by the destructive distillation of wood, hence one of its common names, wood alcohol. The destructive distillation of wood involves the heating of wood to high temperature in a relatively inert atmosphere (that is, with limited oxygen). The process is basically the same as the production of hydrogen by the pyrolysis of methane, as described in the last chapter. This approach to making biomethanol is effectively carbon neutral but, unfortunately, it is very inefficient. Methanol yields of 10–20 L per tonne of wood are typical. In order to produce biomethanol in a manner that is environmentally sustainable and economically competitive with the production of carbon-neutral e-methanol based on the reaction of hydrogen and carbon monoxide and/or carbon dioxide, as described in section 3.10, a more efficient approach is needed. Several practical technologies for the production of biomethanol have been utilized in recent years and some of the more promising are discussed below (see Gautam *et al* (2020), the International IRENA and Methanol Institute (2021), and Olah *et al* (2018)).

### 4.6.1 The production of biomethanol

#### *4.6.1.1 Biomethanol from the gasification of biomass*

The production of biomethanol from biomass closely follows the lines of its production from fossil fuels, as described in section 3.10. In the case of biomethanol, however, the feedstock is biomass rather than a fossil fuel (Kasmuri *et al* 2016). Typical examples of biomass are those that have been described above for the production of biomethane by anaerobic decomposition. These materials include, for example, agricultural waste and organic components of municipal solid waste, such as paper and food waste. The production of biomethanol from biomass follows several steps:

*Pretreatment of the feedstock*—In order to be effectively gasified, the feedstock must comprise reasonably uniformly sized small particles with an appropriate moisture content. The pretreatment process grinds the biomass feedstock, thereby increasing the surface-to-volume ratio of the particles to improve reactivity and allow for efficient material handling. This material is then dried to a moisture content of 15%–20%.

*Gasification*—Gasification is typically a two-step process consisting of pyrolysis and char conversion. Pyrolysis (see section 3.10) is the process of heating a material in an oxygen-deficient environment, resulting in partial oxidation. In this case, heating to a temperature in the range of 400 °C–600 °C is typical. Analogous to the process used for the production of methanol from coal, this process yields a gas that consists primarily of hydrogen, carbon monoxide, and carbon dioxide along with residual carbon (referred to as char). Char conversion consists of heating the by-products of pyrolysis to temperatures in the range of 1300 °C–1500 °C to convert the carbon to additional carbon monoxide. Thus, the gasification process produces a gas from biomass that is very similar to the syngas produced from fossil fuel carbon sources, i.e. a mixture of mostly hydrogen and carbon monoxide with some carbon dioxide which is often referred to as biosyngas.

*Biomethanol production*—The production of methanol from biosyngas follows the same lines as methanol production from fossil-fuel-derived syngas. The biosyngas is heated in the presence of an appropriate catalyst, and the hydrogen reacts with the carbon monoxide and carbon dioxide to produce methanol ($CH_3OH$):

$$2H_2 + CO \rightarrow CH_3OH$$

and

$$3H_2 + CO_2 \rightarrow CH_3OH + H_2O.$$

There are operational facilities that produce biomethanol from biomass in France, Germany, and the United States and a facility that produces biomethanol from municipal solid waste in Canada.

#### *4.6.1.2 Biomethanol from biogas*

As described in section 4.5, biogas is produced by the anaerobic digestion of organic matter. The biogas thus produced is primarily a mixture of methane and carbon dioxide, as shown in table 4.6, along with minor concentrations of impurity gases. This biomethane can be subjected to steam reforming as described in section 3.4.

Methane is reacted with water at high temperature to produce hydrogen and carbon monoxide:

$$CH_4 + H_2O \rightarrow 3H_2 + CO.$$

The carbon monoxide can be further reacted with water to yield additional hydrogen and carbon dioxide:

$$CO + H_2O \rightarrow CO_2 + H_2.$$

Thus, biogas can be used to produce a mixture of hydrogen, carbon dioxide, and carbon monoxide that is similar to biosyngas. Methanol can be produced from this biosyngas as described above. The efficiency of this conversion process depends on the ratios of the three components of the biosyngas as defined by the stoichiometric number $S$ (see section 3.10). There are currently facilities that produce methanol using this approach in the Netherlands, Germany, and the United States. In all cases, however, a mixture of biomethane from biogas and fossil fuel natural gas is used as the feedstock.

*4.6.1.3 Biomethanol from wood pulp*

During the process of producing paper from wood, methanol is produced as a by-product. Methanol is given off when wood is heated in the presence of chemicals such as sodium hydroxide and sodium sulfide that break down the wood fibers to produce wood pulp. The amount of methanol that is produced depends on the details of the cooking process, and this relationship has recently been reviewed by Zhu *et al* (2020). The methanol that is produced as a by-product of paper production contains a number of impurities. However, it is sufficiently flammable that it is generally used directly to produce combined heat and power for the paper mill. This methanol can also be upgraded and used as a transportation fuel or as a feedstock for biodiesel production. The first commercialization of this approach has been undertaken by Södra, Sweden's largest forest owner's association, at their pulp mill in Mönsterås, Sweden. This facility has the capacity to produce over 5000 t of biomethanol per year to be used as feedstock for biodiesel.

Another by-product of the paper industry is black liquor. This is a solution of lignin residues, hemicellulose, and inorganic chemicals used in pulp processing (see figure 4.28). Most black liquor that is produced by the pulp and paper industry is used by the industry for energy production, at least in the United States. Inorganic components are first separated and recycled for use in further pulp processing. The organic components can be burned to produce heat or to create steam for electricity generation. Alternately, black liquor can be further processed to form a possible transportation fuel. Black liquor can be gasified, as described above, to produce biosyngas. This biosyngas can be used for the production of biomethanol, as described above, or biodimethyl ether, as described below, for direct use as a fuel.

The production of biomethanol typically utilizes waste biomass, either directly or in the form of biomethane from biogas. Thus, the production of biomethanol competes for the same resources as the production of biomethane, and the discussion above related to the possible contribution of biomethane to transportation fuel needs is relevant.

Figure 4.28. A sample of black liquor produced by the pulp and paper industry. Reproduced from U.S. Department of Agriculture (2020). Image stated to be in the public domain.

### 4.6.2 The use of biomethanol as a transportation fuel

Section 3.10 described the use of e-methanol as a fuel. Biomethanol can also be used in the same ways, i.e. as a fuel for spark ignition internal combustion engines or reformed methanol fuel cells. Biomethanol can also contribute to the production of biodiesel. As noted in section 4.3, biodiesel is prepared by the transesterification of plant oil or animal fat. This transesterification process exchanges an organic group of an ester with an organic group of an alcohol. Methanol is the most commonly used alcohol for this process. Ironically, most biodiesel is produced using methanol that is derived from fossil fuel sources. Typically, 100 kg of methanol yields 1000 kg of biodiesel. However, the use of fossil-fuel-derived methanol does not completely free the production of biodiesel from the availability of fossil fuels or from their contribution to greenhouse gas emissions.

### 4.6.3 Biodimethyl ether

Dimethyl ether has been discussed in the previous chapter as a possible substitute for diesel fuel in compression ignition engines. That chapter described the production of dimethyl ether from e-methanol. Figure 4.29 shows the various routes by which dimethyl ether can be produced from fossil fuels, from biomass, or from e-methanol. Biodimethyl ether can be produced directly from gasified organic material or by the dehydration of biomethanol.

Research into the utilization of dimethyl ether as a transportation fuel is in its relatively early stages. An overview of work up to the mid-2010s has been provided by Patten and McWha (2015). Some relevant results include the details of dimethyl ether combustion characteristics. Dimethyl ether has a high cetane rating, typically in the range of 55–60. This means that the combustion process is very efficient despite its lower energy density compared to traditional diesel fuel. This efficient combustion produces fewer pollutants, such as particulate matter, and since it is inherently sulfur-free, it produces no sulfur compounds. An investigation of dimethyl ether as a replacement for diesel fuel in trucks by Oak Ridge National Laboratory, Volvo North America, and The Pennsylvania State University (Szybist *et al* 2014) showed promising results, as noted in the previous chapter.

## 4.7 An overview of biofuels

As illustrated in figure 4.30, there has been substantial growth in the use of biofuels since the early 2000s. The significant growth in North and South America has largely been the result of increased bioethanol and biodiesel production. Much of this growth has been driven by the perception that biofuels are indefinitely renewable and are an environmentally viable carbon-free approach to energy, particularly for transportation purposes. However, the actual advantages of biofuels as a future energy source for transportation need to be considered very carefully.

There are a number of factors that need to be considered in assessing the viability of different biofuels. For example, relevant factors might include cost, efficiency, convenience, safety, availability, land-use change, and greenhouse gas reduction.

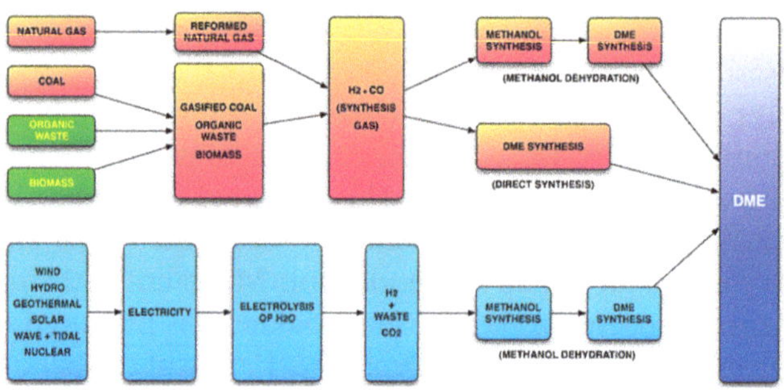


**Figure 4.29.** Routes to the production of dimethyl ether. This image (Moonduckzuma 2009) has been obtained by the author from the Wikimedia website, where it is stated to have been released into the public domain. It is included within this chapter on that basis.

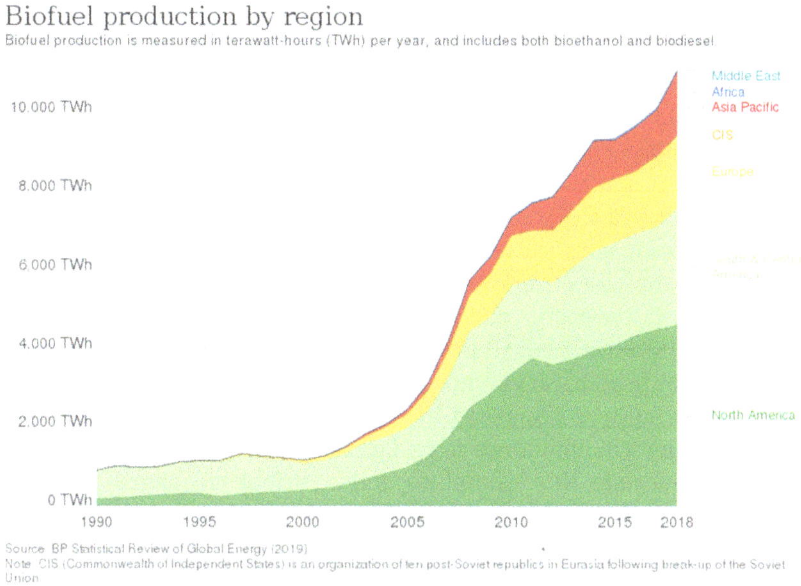


**Figure 4.30.** Biofuel production in different regions of the world from 1990 to 2018. Reproduced from Our World in Data (2022). CC BY 4.0.

Many of these are difficult to assess quantitatively; some approaches to biofuel resource development are in their very early stages and for these, such assessments may be premature. However, some general statements are more or less applicable to most forms of biofuel, and these are summarized below.

In many cases, biofuels are a convenient replacement for traditional fossil fuels, such as gasoline and diesel, that are used in internal combustion engines. As such, biofuels are compatible with current vehicle technology with minor modifications. It is also generally straightforward to integrate biofuels into existing infrastructure used for the distribution of transportation fuels. Specifically, for ethanol and methanol, existing gasoline infrastructure is appropriate with minor modifications. For biodiesel and renewable diesel, the existing diesel distribution system is appropriate. For biomethane, natural gas pipelines are suitable, and for dimethyl ether, infrastructure designed for the distribution of propane or liquefied petroleum gas is most appropriate. In fact, in many cases, partial (or even complete) integration of biofuels into the fossil fuel network has already been undertaken. For example, flex-fuel filling stations and stations in countries such as Brazil have already integrated ethanol into the gasoline supply. Similarly, biodiesel or renewable diesel is a common additive to fossil fuel diesel in many locations and biomethane (of suitable purity) can be injected into existing natural gas pipelines.

Biofuels are produced from biomass, and this biomass can roughly be divided into two categories: that which is grown for the specific purpose of producing fuel and that which is a waste product of other activities. In the first case, the utilization of agricultural land to produce biofuel feedstock can conflict with agricultural activities that produce food crops for human or livestock consumption. Alternately, the development of new agricultural land to accommodate the production of feedstock for biofuels can create adverse environmental effects, for example, due to deforestation. In this respect, the development of technologies for cellulosic ethanol production or biodiesel production from algae can provide an environmentally viable approach to biofuels.

The use of waste materials such as agricultural waste or municipal solid waste provides further possible feedstocks for biofuel production. As noted previously in this chapter, an appropriate infrastructure for the collection of materials needs to be in place. In addition, in many cases, the available supplies may not be sufficient to meet a significant fraction of the energy needs. However, the utilization of waste material for energy production, whether this is related to transportation energy, electricity generation, or space heating, has potential environmental benefits. Waste material that is not used for energy production must be eliminated by some other means. These methods include landfills or spreading agricultural waste on farmland. In either case, decaying material can release methane into the atmosphere and contribute to global greenhouse gas concentrations. Therefore, the utilization of waste-derived fuel for energy production can be an approach that can minimize an otherwise adverse environmental effect.

The overall environmental effect of biofuel production and utilization is a complex problem, and its careful consideration is not straightforward. Net greenhouse gas emissions and land-use change factors need to be evaluated. In this respect, the development of appropriate technologies is needed to ensure the environmental viability and sustainability of biofuel utilization.

## References and additional reading


Alvarez R A *et al* 2018 Assessment of methane emissions from the U.S. oil and gas supply chain *Science* **361** 186–8

Atabani A E, Silitonga A S, Badruddin I A, Mahlia T M I, Masjuki H H and Mekhilef S 2012 A comprehensive review on biodiesel as an alternative energy resource and its characteristics *Renew. Sustain. Energy Rev.* **16** 2070–93

Benjah-bmm27 2007 Transesterification: An alcohol-terminated oligomer and an ester-terminated oligomer condense to form an ester linkage, with loss of an alcohol. https://commons.wikimedia.org/wiki/File:Transesterification.png

Bunkum.us 2022 Kinematic Viscosity of Vegetable Oils at Different Temperatures https://bunkum.us/svo/viscosity.html

Carroll J 2013 Exxon at least 25 years away from making fuel from algae *Bloomberg* (8 March 2013) available at www.bloomberg.com/news/articles/2013-03-08/exxon-at-least-25-years-away-from-making-fuel-from-algae

Clark S J, Wagner L, Schrock M D and Piennaar P G 1984 Methyl and ethyl soybean esters as renewable fuels for diesel engines *J. Am. Oil Chem. Soc.* **61** 1632–8

da Fonseca J R 2007 Faz S Sofia canavial 090607 REFON https://commons.wikimedia.org/wiki/File:Faz_S_Sofia_canavial_090607_REFON.JPG

Dave7 2005 Ford Model T https://www.flickr.com/photos/21612624@N00/537880855

Department of Transportation (United Kingdom) 2013 *Renewable Transport Fuel Obligation (RTFO) order* available at www.gov.uk/government/collections/renewable-transport-fuels-obligation-rtfo-orders

Dogdibegovic E, Fukuyama Y and Tucker M C 2020 Ethanol internal reforming in solid oxide fuel cells: A path toward high performance metal-supported cells for vehicular applications *J. Power Sources* **449** 227598

Dunlap R A 2019 *Sustainable Energy* 2nd edn (Boston, MA: Cengage)

Duran-Ortiz M 2018 Carro Prius flex no Palácio do Planalto https://flickr.com/photos/30998987@N03/45433799935

EIA 2018 'Six states account for more than 70% of U.S. fuel ethanol production' (15 August 2018) available at www.eia.gov/todayinenergy/detail.php?id=36892

EIA 2020 'EIA now estimates biodiesel production and consumption by state' (24 July 2020) available at www.eia.gov/todayinenergy/detail.php?id=44496

EIA 2021a 'How much gasoline does the United States consume?' (9 March 2021) available at www.eia.gov/tools/faqs/faq.php?id=23&t=10#:~:text=How%20much%20gasoline%20does%20the,8.03%20million%20barrels%20per%20day

EIA 2021b 'U.S. energy facts' (14 May 2021) available at www.eia.gov/energyexplained/us-energy-facts/

EIA 2021c 'Natural gas explained' (26 May 2021) available at www.eia.gov/energyexplained/natural-gas/use-of-natural-gas.php#:~:text=Natural%20gas%20accounted%20for%2040,as%20lease%20and%20plant%20fuel

EPA 2010 Renewable Fuel Standard Program (RFS2) Regulatory Impact Analysis EPA-420-R-10-006 (February 2010) available at https://web.archive.org/web/20110202202402/http://epa.gov/otaq/renewablefuels/420r10006.pdf

EPA 2020 *An overview of Renewable Natural Gas from Biogas* EPA 456-R-20-001 (July 2020) available at www.epa.gov/sites/production/files/2020-07/documents/lmop_rng_document.pdf

Fagundes D 2018 Toyota confirma produção do primeiro híbrido flex no Brasil em 2019 (Toyota confirms production of the first flex hybrid in Brazil in 2019) available at https://motor1.uol.com.br/news/297507/toyota-primeiro-hibrido-flex-brasil/

Farrell A E, Plevin R J, Turner B T, Jones A D, O'Hare M and Kammen D M 2006 Ethanol can contribute to energy and environmental goals *Science* **311** 506–8

Gautam P, Neha, Upadhyay S N and Dubey S K 2020 Bio-methanol as a renewable fuel from waste biomass: current trends and future perspective *Fuel* **273** 117783

Hou T, Zhang S, Chen Y, Wang D and Cai W 2015 Hydrogen production from ethanol reforming: catalysts and reaction mechanism *Renew. Sustain. Energy Rev.* **44** 132–48

Huang H, Ramaswamy S, Al-Dajani W, Tschirner U and Cairncross R A 2009 Effect of biomass species and plant size on cellulosic ethanol: a comparative process and economic analysis *Biomass Bioenergy* **33** 234–46

IEA 2020 *Outlook for Biogas and Biomethane—Prospects for Organic Growth* (Paris: International Energy Agency) available at https://iea.blob.core.windows.net/assets/03aeb10c-c38c-4d10-bcec-de92e9ab815f/Outlook_for_biogas_and_biomethane.pdf

IFCAR 2009 2009 Honda Civic NGV with "Phil" refueler photographed at the 2009 Washington DC Auto Show https://commons.wikimedia.org/wiki/File:2009_Honda_Civic_NGV--DC.jpg

IGV Biotech 2013 Photobioreactor PBR 4000 G IGV Biotech https://commons.wikimedia.org/wiki/File:Photobioreactor_PBR_4000_G_IGV_Biotech.jpg

IRENA and Methanol Institute 2021 *Innovation Outlook: Renewable Methanol* (Abu Dhabi: International Renewable Energy Agency) available at www.methanol.org/wp-content/uploads/2020/04/IRENA_Innovation_Renewable_Methanol_2021.pdf

JanB46 2011 Raceway pond used for the cultivation of microalgae https://commons.wikimedia.org/wiki/File:Microalgenkwekerij_te_Heure_bij_Borculo.jpg

Jones S T, Peterson C L and Thompson J C 2001 (St. Joseph, MI: American Society ofAgricultural and Biological Engineers) Used vegetable oil fuel blend comparisons using injector coking in a DI diesel engine. Presented at *2001 ASAE Annual Int. Meeting, Sacramento, CA*, SAE paper No. 01-6051

Kasmuri N, Kamarudin S K, Abdullah S, Hassimi A H and Ayub M S 2016 Potential of biomass for biomethanol production *Int. J. Appl. Eng. Res.* **11** 10016–9

Kumar L R, Yellapu S K, Tyagi R D and Drogui P 2020 Cost, energy and GHG emission assessment for microbial biodiesel production through valorization of municipal sludge and crude glycerol *Bioresour. Technol.* **297** 122404

Mariordo 2008 New Fiat Siena Tetrafuel 1.4 model 2008 https://commons.wikimedia.org/wiki/File:SAO_09_2008_Fiat_Siena_TetraFuel_2_views_v1.jpg

Mejidori 2007 Biodiesel 3 https://commons.wikimedia.org/wiki/File:Biodiesel_3.jpg

Mekonnen M M, Romanelli T L, Ray C, Hoekstra A Y, Liska A and Neale C M U 2018 Water, energy, and carbon footprints of bioethanol from the U.S. and Brazil *Environ. Sci. Technol.* **52** 14508–18

Mongabay 2012 Palm oil does not meet U.S. renewable fuels standard, rules EPA mongabay.com (27 January 2012) available at https://news.mongabay.com/2012/01/palm-oil-does-not-meet-u-s-renewable-fuels-standard-rules-epa/

Monyem A and Van Gerpen J 2001 The effect of biodiesel oxidation on engine performance and emissions *Biomass Bioenergy* **20** 317–25

Moonduckzuma 2009 A chart showing production methods of dimethyl ether https://commons.wikimedia.org/wiki/File:Dme_production_dialg_090423.jpg

Olah G A, Goeppert A and Surya Prakash S K 2018 *Beyond Oil and Gas: The Methanol Economy* 3rd edn (Weinheim: Wiley-VHC)

Ortiz M R D 2010 Ford Escape E85 Flex Plug-in Hybrid views and badging WAS 2010 https://commons.wikimedia.org/wiki/File:Ford_Escape_E85_Flex_Plug-in_Hybrid_views_and_badging_WAS_2010.jpg

Ortiz M R D 2018 Historical trend of Brazilian production of light-duty vehicles by type of fuel from 1979 to 2017 https://commons.wikimedia.org/wiki/File:Historical_trend_ethanol_and_flex_veh_in_Brazil_1979_2008.png

Our World In Data 2022 Biofuel production by region https://ourworldindata.org/grapher/biofuels-production-by-region

Patten J and McWha T 2015 *Dimethyl Ether Fuel Literature Review* (Ottawa, ON: National Research Council of Canada Automotive and Surface Transportation) Technical ST-GV-TR-0032

Raman A A A, Tan H W and Buthiyappan A 2019 Two-step purification of glycerol as a value added by product from the biodiesel production process *Front. Chem.* **7** 774

Sanderson K 2006 US biofuels: a field in ferment *Nature* **444** 673–6

Scully M J, Norris G A, Alarcon Falconi T M and MacIntosh D L 2021 Carbon intensity of corn ethanol in the United States: state of the science *Environ. Res. Lett.* **16** 043001

Searchinger T *et al* 2008 Use of U.S. croplands for biofuels increases greenhouse gases through emissions from land-use change *Science* **319** 1238–40

Sidibé S S, Blin J, Vaitilingom G and Azoumah Y 2010 Use of crude filtered vegetable oil as a fuel in diesel engines state of the art: literature review *Renew. Sust. Energy Rev.* **14** 2748–59

Snåre M, Kubičková I, Mäki-Arvela P, Chichova D, Eränen K and Murzin D 2008 Catalytic deoxygenation of unsaturated renewable feedstocks for production of diesel fuel hydrocarbons *Fuel* **87** 933–45

Solomon B D, Barnes J R and Halvorson K E 2007 History, economics, and energy policy *Biomass Bioenergy* **31** 416–25

Szybist J P, McLaughlin S and Iyer S 2014 Emissions and performance benchmarking of a prototype dimethyl ether-fueled heavy-duty truck. *Oak Ridge National Laboratory Report* ORNL/TM-2014/59 February 2014 available at https://afdc.energy.gov/files/u/publication/ornl_dme_tm-2014-59.pdf

Thzorro77 2020 Anaerobic Digestion Biogas Plant with Composting for energetic and material recycling of Municipal Biowaste in Sundern, Germany https://commons.wikimedia.org/wiki/File:Biogasanlage_mit_Kompostierung.jpg

Tilley E, Ulrich L, Lüthi C, Reymond P and Zurbrügg C 2014 *Compendium of Sanitation Systems and Technologies* 2nd edn (Duebendorf: Swiss Federal Institute of Aquatic Science and Technology (Eawag)) https://www.iwa-network.org/wp-content/uploads/2016/06/Compendium-Sanitation-Systems-and-Technologies.pdf

U.S. Department of Agriculture 2012 USDA Plant Hardiness Zone Map https://pdi.scinet.usda.gov/phzm/vm/All_states_halfzones_title_legend_logos_300dpi.jpg

U.S. Department of Agriculture 2020 In the New Crops Research Unit at Peoria, Illinois, technician JoDean Sarins of Vision Paper, Inc., examines black liquor waste from kenaf pulping. https://www.ars.usda.gov/oc/images/photos/aug00/k8981-10

U.S. Department of Energy 2021 Alternative Fuels Data Center. Global Ethanol Production by Country or Region https://afdc.energy.gov/data/10331

U.S. Department of Energy 2022 Alternative Fuels Data Center. Biodiesel Production and Distribution https://afdc.energy.gov/fuels/biodiesel_production.html

U.S. Department of Energy 2022 Straight Vegetable Oil as a Diesel Fuel? https://afdc.energy.gov/files/u/publication/54762.pdf

Wang M, Wu M and Huo H 2007 Life-cycle energy and greenhouse gas emission impacts of different corn ethanol plant types *Environ. Res. Lett.* **2** 024001

Wang M Q, Han J, Haq Z, Tyner W E, Wua M and Elgowainy A 2011 Energy and greenhouse gas emission effects of corn and cellulosic ethanol with technology improvements and land use changes *Biomass Bioenergy* **35** 1885–96

Young M 2008 Excel Graph showing the Carbon Intensity of Bioethanols https://commons.wikimedia.org/wiki/File:BioethanolsCountryOfOrigin.jpg

Zhu J Y, Yoon S H, Liu P H and Chai X S 2020 Methanol formation during alkaline wood pulping *TAPPI J.* **83** 1–13 https://imisrise.tappi.org/TAPPI/Products/00/JUL/00JUL65.aspx

**IOP** Publishing

Transportation Technologies for a Sustainable Future

Renewable energy options for road, rail, marine and air transportation

**Richard A Dunlap**

# Chapter 5

# Other modes of transportation

## 5.1 Introduction

The previous three chapters have outlined the three basic approaches, i.e. batteries, hydrogen, and biofuels, for energy for road vehicles. Table 1.3 showed that road transportation accounts for about 85% of all transportation energy worldwide. However, this situation varies somewhat in different regions of the world. Figure 5.1 shows a breakdown of transportation energy for different locations. It is clear that non-road transportation is a more important contribution in many parts of the world than it is in the United States.

It is generally acknowledged that mass transport is a more efficient means of transportation per person than passenger automobiles. Table 5.1 summarizes the number of kilometers that one person can be transported by one kWh of energy using different transportation methods. As the figure indicates, passenger rail transportation is certainly favorable, from an energy standpoint, compared to bus, ship, and aircraft transportation. It can also be seen that mass transport is favorable in terms of energy per person, compared to light-duty passenger automobiles. In this chapter, we consider sustainable options for transportation modes other than road transportation, including the major forms of mass transportation shown in the table.

## 5.2 Rail transportation

### 5.2.1 Current railway locomotive designs

Railway trains are powered by a locomotive which can fall into one of several broad categories: freight electric, freight diesel, conventional electric, conventional diesel, urban electric, and high-speed electric. The primary propulsion system of virtually all current locomotives utilizes either an internal combustion engine operating on diesel fuel or an electric motor powered by an external source of electricity. In a few cases, as described below, electro-diesel locomotives may incorporate both sources

doi:10.1088/978-0-7503-5306-9ch5

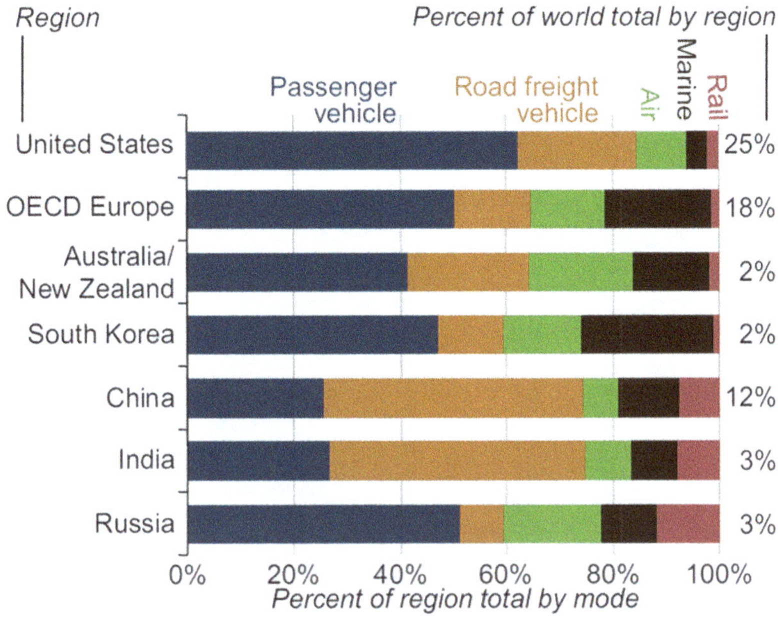


**Figure 5.1.** Breakdown of transportation energy for selected regions (for 2012). Reproduced from EIA (2016). Image stated to be in the public domain.

**Table 5.1.** The efficiencies of different modes of transportation showing typical power per person (kW), typical speed (km $h^{-1}$), and km traveled per kWh of energy per person. Data adapted from https://commons.wikimedia.org/wiki/File:EffizienzLeistungFahrzeuge.png.

| Mode | Power per person (kW) | Typical top speed (km $h^{-1}$) | km per kWh per person |
|---|---|---|---|
| Passenger automobile | 40 | 120 | 3 |
| Airplane | 200 | 700 | 3.5 |
| Ship | 5 | 25 | 5 |
| Bus | 15 | 80 | 5.3 |
| Passenger train | 8 | 100 | 12.5 |
| Electric scooter | 3 | 60 | 20 |
| Walking | 0.04 | 4 | 100 |
| Bicycle | 0.08 | 18 | 225 |

of power. The basic design characteristics of these types of locomotive are described below.

### *5.2.1.1 Diesel locomotives*

Diesel locomotives use a conventional internal combustion diesel engine. As in a road vehicle that utilizes an internal combustion engine, the engine of the locomotive

**Figure 5.2.** ČKD ČME3 diesel-electric locomotive manufactured in the Czech Republic. This is a common design for a diesel-electric freight locomotive. This image (Novoklimov 2006) has been obtained by the author from the Wikimedia website, where it is stated to have been released into the public domain. It is included within this chapter on that basis.

must be coupled to the drive wheels through a transmission. This allows the engine to continue running while the train is stopped. In comparison to a road vehicle, the amount of power that must be transferred from the engine to the wheels of a railway locomotive is very large, and this is a problem for the design of a reliable transmission. While a few diesel locomotive designs utilize hydraulic transmission, the vast majority of diesel locomotives are diesel-electric locomotives. This latter design uses the diesel engine to drive a generator which in turn powers an electric motor. The diesel engine can run efficiently at a relatively constant speed, and the power delivered to the drive wheels can be regulated by controlling the electric motor. A typical example of a diesel-electric locomotive used for freight trains is shown in figure 5.2.

*5.2.1.2 Electric locomotives and multiple-unit trains*

Electric trains utilize an electric motor supplied by power from either a third-rail (i.e. ground-level) system or an overhead catenary wire. Current designs can be generally categorized as conventional electric, freight electric, urban electric, and high-speed electric, as described below. This section also describes the electro-diesel locomotive.

*Conventional electric and freight electric rail transportation*—Conventional (passenger) electric and freight electric locomotives can replace traditional diesel-electric locomotives on intercity or rural rail routes. For safety reasons, power is typically provided by an overhead catenary (at least in rural areas, where tracks are at ground

**Figure 5.3.** Siemens ÖBB Class 1016 (ES 64 U) locomotive in Austria. This image (Murdockcrc 2011) has been obtained by the author from the Wikimedia website where it was made available under a CC BY 3.0 licence, https://creativecommons.org/licenses/by/3.0/deed.en. It is included within this chapter on that basis. It is attributed to Murdockcrc.

level). A typical example of a modern electric locomotive is shown in figure 5.3. This locomotive, the Siemens ES64U, is shown coupled to passenger cars in the image. The 'ES' in the designation stands for 'EuroSprinter,' the '64' represents the power output to the wheels (6400 kW), and the 'U' stands for the 'universal' version of the locomotive (usually used for passenger trains). An 'F' version (for freight) is also available. The typical top speed of the ES64U is 230 km $h^{-1}$ and that of the ES64F is 140 km $h^{-1}$.

*Urban electric rail transportation*—Urban rail transportation utilizes both third-rail systems (typically used on underground (subway) routes and elevated routes) as well as catenary systems (typically used on ground-level routes). While conventional locomotive-type systems are used for urban rail transportation, many trains are so-called multiple-unit trains. In this design, as shown in figure 5.4, individual units (called railcars) each have their own electric propulsion system. These cars can be used individually or can be coupled together in multiple-unit trains that can be controlled by a single driver.

*High-speed electric rail transportation*—High-speed electric trains are typically powered by an overhead catenary wire and are used for intercity passenger transport. Although there is no clearly defined speed at which a train is considered high-speed, new constructions are typically capable of at least 250 km $h^{-1}$, and older infrastructures with a speed capability in excess of 200 km $h^{-1}$ are generally considered high-speed. Most high-speed trains operate on standard-gauge tracks. However, the tracks are continuously welded and have large-radius turns and grade-separated intersections (i.e. at different heights) with roads and other rail tracks.

The Tōkaidō Shinkansen in Japan (commonly known as 'The Bullet Train') was world's first high-speed rail line and was opened commercially in 1964. A current high-speed train on this line is shown in figure 5.5.

**Figure 5.4.** A multiple-unit train consisting of four railcars on the métro de Marseille at the Bougainville terminal. This image (Fèvre F 2014) has been obtained by the author from the Wikimedia website where it was made available under a CC BY-SA 4.0 licence, https://creativecommons.org/licenses/by-sa/4.0/deed.en. It is included within this chapter on that basis. It is attributed to Florian Fèvre.

**Figure 5.5.** The Tōkaidō Shinkansen high-speed line in Japan. Mount Fuji appears in the background. This image (tansaisuketti 2010) has been obtained by the author from the Wikimedia website where it was made available under a CC BY-SA 3.0 licence, https://creativecommons.org/licenses/by-sa/3.0/deed.en. It is included within this chapter on that basis. It is attributed to tansaisuketti.

*Electro-diesel locomotives*—Electro-diesel locomotives can be powered either by an external third rail or overhead (catenary) power supply or by electricity produced on board by a generator that is run by a diesel engine. These locomotives are generally intended for operation on specific routes where the capability to operate on either external electricity or as a diesel-electric is appropriate. There are two notable situations in which this can occur. If a primarily electric locomotive has to traverse a section of non-electrified rail, then it can switch to diesel-electric operation. On the other hand, a primarily diesel-electric locomotive may enter a region where diesel operation is unacceptable, such as a tunnel or indoor train station where exhaust fumes need to be minimized. In such a case, the diesel-electric locomotive can switch to electric operation. While electro-diesel locomotives are, strictly speaking, hybrids, the term hybrid locomotive is typically used for battery–diesel locomotives, as described below.

### 5.2.2 Sustainable railway technologies

A number of alternative technologies that can reduce the dependence of rail transportation on fossil fuels are either in development or are in limited commercial use. Some of the more potentially viable (or interesting) are considered in this section.

#### *5.2.2.1 Battery locomotives and battery multiple-unit trains*

Battery locomotives and battery multiple-unit trains utilize batteries as a power source. Two typical designs are in use: those which utilize only batteries as a power source and those which utilize batteries but are also connected to an external power source, such as an overhead catenary line. In the first case, range is an important consideration, which generally precludes the use of battery locomotives for non-urban routes. In specialized situations, such as maintenance work or transport in areas that could potentially contain flammable gases (such as mines), electric locomotives can be a suitable replacement for diesel or conventional catenary electric trains. Battery-operated multiple units have found applications in urban transportation in cases where overhead catenary lines or third rails are not practical. In such cases, railcars often incorporate supercapacitors in addition to batteries. The supercapacitors can be quickly charged at rail stops and can maintain the battery charge between stops.

In the case of batteries incorporated into railway vehicles that have an external power connection, batteries can serve to provide propulsion in regions where there are gaps in the external power supply (which is analogous to the situation for electro-diesel locomotives). Batteries are also beneficial for the recovery of energy from regenerative braking. For urban routes with frequent stops, regenerative braking can recover as much as 45% of the energy expended.

#### *5.2.2.2 Battery–diesel hybrid locomotives*

Hybrid locomotives are analogous to battery–gasoline hybrid automobiles, since they combine power from batteries with power from an internal combustion engine. In the case of a hybrid locomotive, batteries can drive an electric motor directly or an internal combustion diesel engine can be used to run a generator to drive the motor. A typical example of a hybrid locomotive is illustrated in figure 5.6. This type of locomotive reduces carbon emissions compared to a conventional diesel-electric locomotive, as the (sometimes significant) regenerative braking energy can be recovered and stored.

**Figure 5.6.** RailPower GG20B ‘Green Goat’ hybrid battery/diesel locomotive manufactured in Saint-Jean-sur-Richelieu, Québec, Canada. This image (Flint Bryan 2007) has been obtained by the author from the Wikimedia website where it was made available under the GNU Free Documentation License, https://en.wikipedia.org/wiki/en:GNU_Free_Documentation_License. It is included within this chapter on that basis. It is attributed to Bryan Flint.

**Figure 5.7.** KuMoYa E995-1 ‘NE Train Smart Denchi-kun’ fuel cell–battery hybrid railcar. This image (Rsa 2011) has been obtained by the author from the Wikimedia website where it was made available under the GNU Free Documentation License, https://en.wikipedia.org/wiki/en:GNU_Free_Documentation_License. It is included within this chapter on that basis. It is attributed to Rsa.

### *5.2.2.3 Hydrogen fuel cell–battery hybrid rail transportation*

Hybrid hydrogen fuel cell–battery electric trains follow from the design of fuel cell automobiles discussed in chapter 3. At present, the development of such vehicles is in the very early stages of experimentation. Figure 5.7 shows the KuMoYa E995-1 experimental fuel cell–battery hybrid railcar. This vehicle was developed by JR East (the East Japan Railway Company) as a test bed for alternative railway transportation technologies. This version of the vehicle was tested in the early 2000s and utilized two 65 kW hydrogen fuel cells in conjunction with lithium-ion batteries to provide power for the railcar. The railcar had a top speed of about 100 km $h^{-1}$ and a range of 50–100 km.

In 2018, two Coradia iLint hydrogen fuel cell trains manufactured by the French company Alstom became operational on a 100 km line between Cuxhaven and Buxtehude, Germany (Agence France-Presse 2018). The manufacturer claims that the operating range between refueling procedures is 1000 km. Because the train's range is much greater than the length of its route, there is a single fueling station in Bremervörde, about midway between the end points of the route.

For hydrogen-powered railcars, the development of a hydrogen fueling network would likely be more straightforward than the more distributed system needed for hydrogen-powered road vehicles. However, compared to alternative approaches to sustainable railway transportation, fuel cell costs and the associated infrastructure development costs probably do not favor the extensive utilization of fuel cell technology.

#### *5.2.2.4 Biofuel locomotives*

Since conventional diesel-electric locomotives operate on fossil fuel diesel, biodiesel in the form of B100 (that is, pure biodiesel) or as a mixture with fossil fuel diesel should be a fairly direct replacement, requiring only minor engine modifications (see chapter 4 on biodiesel use in road vehicles). Since the mid-2000s, a number of railways have utilized biodiesel in some (or all) of their trains. Some notable examples include railways in Britain, France, the United States, and India, as described below.

Since 2007, the British Royal Train, which is the train officially used to convey senior members of the Royal Family, has operated on B100. In France, the Société Nationale des Chemins de Fer Français (SNCF) has operated an experimental B100 train between Granville and Paris (see figure 5.8). In the United States, Disneyland has operated its park trains on biodiesel produced from its own waste cooking oils (UPI 2009). Since 2007, the historic Mount Washington Cog Railway, which climbs the western slope of Mount Washington in New Hampshire, has included a biodiesel locomotive in its fleet (Kotrba 2013). Finally, the farmland around the rail line between Mumbai and Delhi in India is used to grow jatropha. These jatropha plants (as indicated in table 4.4) are used to produce biodiesel and this fuel is used in the train that runs on this rail line (Dunlap 2019).

#### *5.2.2.5 Maglev trains*

Maglev (for magnetic levitation) trains are a developing technology. These trains have been used commercially in Japan, Korea, and China since the early 2000s and future maglev infrastructure is also planned in these countries. Maglev trains utilize the interaction of magnetic fields to levitate the train above the track and to propel it along the track. These magnetic fields are produced by magnets on the train and tracks, and various designs have utilized conventional electromagnets, superconducting magnets, and/or permanent magnets. The levitation allows the train to move

**Figure 5.8.** SNCF Intercités train no. 3430 operating on B100 biodiesel from Granville (Normandy) to Paris at the Argentan Station in Orne. This image (Pépellin 2019) has been obtained by the author from the Wikimedia website where it was made available under a CC BY-SA 4.0 licence, https://creativecommons.org/licenses/by-sa/4.0/deed.en. It is included within this chapter on that basis. It is attributed to Florian Pépellin.

along the track without any significant friction, thereby allowing it to move efficiently and at high speed. Although maglev trains do not represent a new energy technology, they represent a means of improving railway efficiency and thereby reducing energy requirements. Levitation also has the advantage of minimizing noise and vibration.

A number of different designs have been used for experimental maglev trains (Boslaugh 2020). The design that has achieved the greatest commercial success utilizes electromagnetic suspension (EMS). A diagram of an EMS maglev train is shown in figure 5.9. In this design, a portion of the train containing magnetic coils wraps around the track. The train is lifted off the track by an attractive force between the magnets in the train and the magnetic rail. A complex feedback system is used to control the height of the train above the track at about 1 cm. Additional magnets on the sides of the train help to align the train in the track while it is in motion. A major advantage of this type of levitation system is that the train can be levitated at all speeds, as well as when it is at rest.

Propulsion for maglev trains can be provided by several methods. For EMS maglev trains, this is often done by a linear motor consisting of an arrangement of magnetic coils inside the train, as shown in figure 5.9. Figure 5.10 shows the design of a simple linear induction motor. This device is like a conventional rotary AC induction motor that is flattened out into a linear device. An alternating magnetic field is produced by applying three-phase alternating current to the windings.

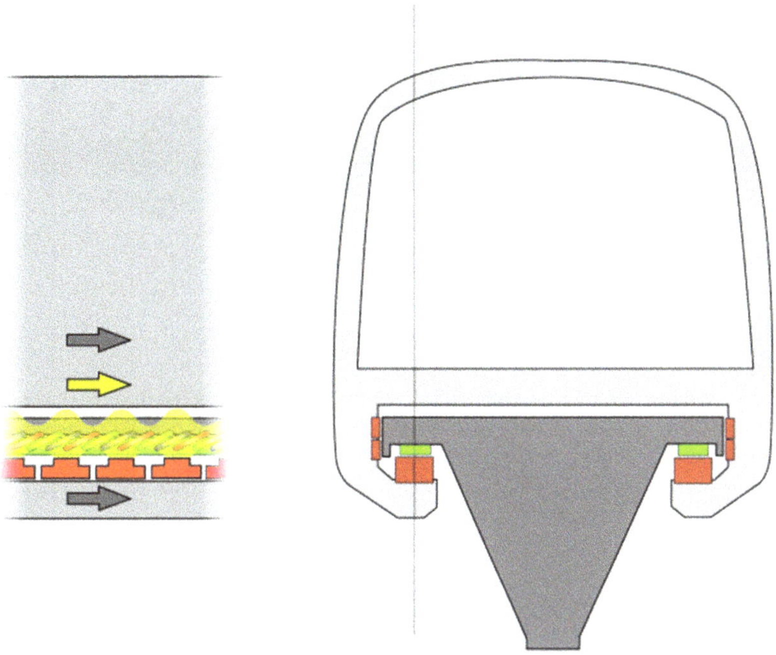

**Figure 5.9.** Diagram of an electromagnetic suspension (EMS) maglev train. The left portion of the figure shows a longitudinal section of the magnets in the train and the rail. The right portion of the figure shows a transverse cross section of the train on the tracks illustrating the wraparound train body with the levitation and positioning magnets. The gray vertical line in the right-hand image shows the location of the section in the left image. This image (Moralapostel 2006) has been obtained by the author from the Wikimedia website where it was made available under the GNU Free Documentation License, https://en.wikipedia.org/wiki/en:GNU_Free_Documentation_License. It is included within this chapter on that basis. It is attributed to Stefan_024.

Current is induced in a conductive (usually aluminum) plate that covers the core (not shown in the diagram). This induced current produces a magnetic field that interacts with the field from the winding and causes the plate to move linearly. The principle of the linear induction motor can be applied to a maglev train by placing the core and windings inside the train, facing downward towards the track. A conductive rail is attached to the racks adjacent to the inductive motor, and the magnetic interaction pushes the train forward. Linear induction motors are also used in some specialized situations to propel wheeled (non-levitating) trains.

Onboard power for maglev trains can be provided by several different means or a combination of methods. Power can be provided by onboard batteries or through electrical contact or induction from the track. Batteries can be an important source of backup power for some designs (e.g. EMS trains) to maintain levitation in the event of grid failure.

At present, there are six operational maglev systems: three in China, one in Japan, and two in South Korea. As examples of different system designs, we consider two well-known systems, the Shanghai Transrapid and the Tobu Kyuryo Line (commonly known as the Linimo line).

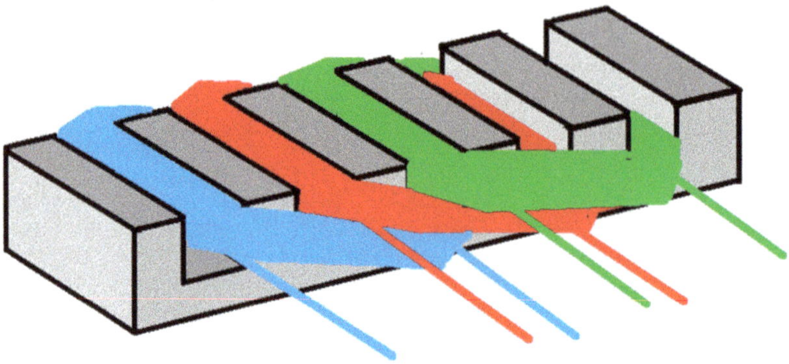

**Figure 5.10.** Basic operation of a three-phase linear induction motor showing the core and windings. The conductive plate which sits above the core is not shown. This image (GliderMaven 2015) has been obtained by the author from the Wikimedia website where it was made available under a CC BY-SA 4.0 licence, https://creativecommons.org/licenses/by-sa/4.0/deed.en. It is included within this chapter on that basis. It is attributed to GliderMaven.

**Figure 5.11.** The Shanghai Transrapid maglev train (EMS) leaving Pudong International Airport. This image (Needham 2006) has been obtained by the author from the Wikimedia website, where it is stated to have been released into the public domain. It is included within this chapter on that basis.

*Shanghai Transrapid*—The Shanghai Transrapid, as shown in figure 5.11, is the oldest maglev system still in operation; it has been in commercial operation since 2004. Transrapid connects the Shanghai Pudong International Airport and the outskirts of central Pudong, Shanghai. The train has a top speed of 430 km $h^{-1}$ and typically covers the 30.5 km route in seven to eight minutes at an average speed of about 245 km $h^{-1}$.

*Linimo*—The Linimo Line (more formally known as the Aichi High-Speed Transit Tobu Kyuryo Line) has operated commercially since 2005 in Aichi, Japan. This train, as shown in figure 5.12, is more analogous to an urban multiple-unit train than a high-speed intercity passenger train. The Linimo Line has nine

**Figure 5.12.** Linimo train approaching Banpaku Kaijo Station in Aichi, Japan. This image (Chris 73 2005) has been obtained by the author from the Wikimedia website where it was made available under a CC BY-SA 3.0 licence, https://creativecommons.org/licenses/by-sa/3.0/deed.en. It is included within this chapter on that basis. It is attributed to Chris 73.

stations along its 8.9 km route and the train has a top speed of about 100 km $h^{-1}$. As such, it is a low-noise alternative to conventional electric urban railway systems.

Maglev trains require dedicated tracks which are not compatible with traditional wheeled rail trains. The infrastructure cost associated with such tracks may be a limiting factor in the future commercial development of maglev technology. For example, the construction costs for the two maglev systems described above were 1.33 billion USD for the 30.5 km Shanghai system and 955 million USD for the 8.9 km Linimo system. However, although operating costs for maglev trains are greater than those of conventional electric trains, they can be competitive with short-run aircraft for routes of a few hundred kilometers.

There are several maglev systems in the construction stage at present. Most of these, although not all, are similar to the Linimo system and are relatively low-speed alternatives to conventional electric urban rail systems. The Japanese Chuo Shinkansen system is, by far, the most ambitious of the current maglev projects. The initial phase of this maglev line will connect Tokyo and Nagoya (a distance of 286 km) with a high-speed maglev line. The train is expected to have a maximum operating speed of 505 km $h^{-1}$. Long-term plans include an extension of the line from Nagoya to Osaka (an additional 152 km). Initial construction on the Tokyo–Nagoya line began in 2014. The completion date is, at present, somewhat uncertain; the Tokyo–Nagoya phase is expected to begin operation around 2027 and the Nagoya–Osaka phase is expected to begin operation between about 2037 and 2045.

### 5.2.3 The current and future use of rail transportation

Table 5.2 shows a breakdown of the energy used by different railway technologies worldwide. About 47% of all rail energy is electricity. If only passenger rail is considered, about 74% of the energy is electricity. There are significant differences in rail use in different regions. Table 5.3 shows the percentage of rail use (in terms of energy utilization) that provides passenger service (as opposed to freight) for different regions. The table also shows the percentage of rail energy that is provided by electricity. The most obvious feature in the table is the rail energy used in North America. There, the vast majority of rail use is for freight, and the vast majority of rail energy is provided by diesel. In other regions, railways are more often electrified and are more likely to be geared to passenger travel.

At present, rail transportation is, in a sense, the most environmentally sustainable mode of transportation. This is, in part, as a result of its efficiency, as indicated in table 5.1. However, it is primarily a result of the fact that a substantial fraction of world rail transportation is electric, i.e. it derives its energy from the electric grid (IEA 2019). Thus, the environmental sustainability of rail transportation is not only a function of rail transportation technology, but also significantly a function of the environmental sustainability of the electric grid. An overall analysis of the

**Table 5.2.** Breakdown of world rail energy for 2017 by rail technology. Data are adapted from figure 1.22 of IEA (2019).

| Technology | Percentage of total rail energy |
|---|---|
| Freight diesel | 44.2 |
| Conventional diesel | 8.9 |
| Freight electric | 21.8 |
| Conventional electric | 13.3 |
| Urban electric | 8.2 |
| High-speed electric | 3.6 |

**Table 5.3.** Total rail energy, percentage of passenger rail energy, and percentage of rail energy provided by electricity in 2017 for selected regions of the world. Data are adapted from figure 2.13 of IEA (2019).

| Region | Total rail energy ($\times 10^{11}$ MJ) | Percent rail energy for passengers | Percent electric rail energy |
|---|---|---|---|
| North America | 6.20 | 7 | 4 |
| China | 4.52 | 35 | 69 |
| Europe | 2.85 | 77 | 75 |
| Russia | 2.35 | 17 | 67 |
| Japan | 1.26 | 93 | 95 |
| India | 1.68 | 43 | 48 |

information as presented above suggests two major approaches to ensuring the optimal sustainability of rail transportation:

1. Ensure that grid energy is environmentally sustainable.
2. Reduce fossil fuel diesel consumption.

In the first case, overall energy approaches for the future were summarized in chapter 1 and will be considered further in the next chapter. However, a detailed analysis of this situation is beyond the scope of this book.

The reduction of fossil fuel diesel use for rail transportation depends on increasing the amount of electrified rail track and/or using alternative propulsion technologies on non-electrified lines. In the first case, the applicable technology is straightforward and well established. In densely populated regions in particular, where electrified railways are already common (such as Europe), an increase in the electrification of rail facilities may be possible, and this would result in an additional reduction in greenhouse gas emissions. If grid electricity is provided by renewable methods or nuclear energy, then this will translate fairly directly to reduced carbon dioxide emissions. Even in the case of grid electricity that is produced by fossil fuel consumption, the greater efficiency of conventional thermal generation combined with electric train efficiency compared to diesel-electric locomotive efficiency will translate into some level of carbon emissions reduction.

In the case of North American railways, the situation is somewhat more difficult to deal with. Electrification is a much more significant task because of the extensive use of diesel fuel. The length of non-electrified rail line in North America means that the infrastructure cost associated with converting from diesel to electricity may not be economically viable. Therefore, other approaches to the reduction of fossil fuel diesel use may be suitable. Some possible options are as follows:

- the conversion of diesel-electric locomotives to operate on biodiesel or renewable diesel;
- a reduction in diesel use due to the introduction of battery–diesel hybrid locomotives (this may also be done in conjunction with the replacement of fossil fuel diesel by biodiesel);
- the use of electro-diesel locomotives in conjunction with limited sections of electrified rail line;
- the use of battery electric locomotives in conjunction with limited sections of electrified rail line to provide propulsion and recharge batteries.

While all of these approaches will contribute to a reduction in greenhouse gas emissions, the exact magnitude of this reduction will depend on a number of factors and will be limited, to some extent, by the reliance of grid electricity on fossil fuel sources in North America.

## 5.3 Maritime transportation

Maritime transport involves the movement of people or goods over oceans, lakes, rivers, or canals. It may be undertaken for commercial, military, or recreational purposes. As indicated in table 1.3, it currently accounts for 4.1% of all primary transportation energy. Correspondingly, it produces between 3.5% and 4% of all carbon dioxide emissions related to the transportation sector (Walker *et al* 2019). Although greenhouse gas emissions from maritime transport were not a focus of the Paris Climate Accord, they are an important and potentially significant factor in future actions related to climate change. The International Maritime Organization (IMO 2015) has estimated that without intervention, maritime greenhouse gas emissions could increase by 50%–250% by 2050. Thus, from an environmental standpoint, it is important to consider the effect of energy used for maritime transportation. In this section, we consider the major modes of marine propulsion that are used to power commercial and military vessels.

### 5.3.1 Current marine propulsion systems

The earliest powered ships were developed in the 19th century and utilized reciprocating steam engines, generally fueled by wood (in the early part of the century) and later by coal. Around 1900, steam turbines were developed and because of their greater power-to-weight ratio, they replaced reciprocating steam engines for many applications. In the early part of the 20th century, fuel oil began to replace coal as the source of heat for steam engines. The most common types of propulsion system for modern ships are reviewed below.

*Diesel engines*—Reciprocating marine diesel engines can be categorized as slow-speed, medium-speed, or high-speed. These three categories are defined as engines that operate with a maximum speed of 300 rpm, 300–1000 rpm, and over 1000 rpm, respectively. Large ships typically use slow-speed or medium-speed engines, while smaller ships often use high-speed engines. In all cases, it is important to drive the propeller (or propellers) at an appropriate speed. In order to accomplish this, the engine can be coupled to the propeller(s) in several different ways. For slow-speed engines, direct coupling is sometimes appropriate. For medium-speed and high-speed engines, the engine speed is typically greater than the required propeller speed and some means of speed reduction is needed. A strictly mechanical approach is sometimes used, in which the engine is coupled to the propeller through a gearbox. In such a situation in larger ships, more than one engine may be coupled to each propeller shaft through a clutch. This allows the engines to be run independently and is beneficial for maintenance. A diesel-electric approach, along the same lines as the design used for diesel-electric locomotives, is also common.

Marine diesel engines run on a variety of fossil-fuel-derived liquid fuels. These marine fuels, commonly referred to as bunker fuels, fall into two general categories, heavy fuel oil and distillates (Marquard & Bahls 2015). Heavy fuel oil is often referred to as residual fuel, as it is the heavy liquid petroleum product that remains after crude oil is refined to extract products such as gasoline. Distillates are lighter hydrocarbons that are extracted from crude oil during the distillation process; the marine industry refers to them as marine gas oil. Marine fuels are typically mixtures of heavy fuel oil and distillates which are matched to the operating characteristics of the specific engine. Cargo ships with large engines usually operate on relatively pure heavy fuel oil, and these fuels account for about 77% of the marine fuel that is used. They tend to be high in sulfur content; current regulations allow for a maximum of 3.5% sulfur content. The combustion of high-sulfur fuels emits sulfur dioxide (in addition to carbon dioxide), which is a major source of acid rain. Marine fuels with a higher percentage of distillates are commonly used in smaller ships.

*Liquefied natural gas (LNG) engines*—In order to comply with relevant pollution standards, which have been largely established as a result of the work of the IMO (an agency of the United Nations), some ships utilize marine LNG engines (see figure 5.13). LNG engines are typically dual-fuel engines that can operate on either traditional marine diesel fuel, such as heavy fuel oil, or on LNG (Dorrington *et al* 2013, Wärtsilä Corporation 2012). LNG is the most environmentally sustainable of the fossil fuels, as it virtually sulfur-free and produces the least carbon dioxide per unit energy generated. Dual-fuel vessels have the option of utilizing less expensive marine diesel when they are far from land and then switch to less polluting LNG when they are in port.

*Gas turbines*—Gas turbines can operate on a variety of liquid or gaseous fuels. They are less polluting than traditional reciprocating marine diesel engines. However, compared to traditional diesel engines, gas turbines have lower efficiency, and gas turbine fuel is more costly than heavy fuel oil. Thus, gas turbines are more expensive to operate than traditional reciprocating diesel engines and they have had limited applications. Some large passenger ships utilize gas turbines in conjunction with diesel engines to increase maximum power output for operation at high speeds

**Figure 5.13.** The *Bit Viking*, a 177 m long tanker. In 2011, the *Bit Viking* was the first vessel to be converted to dual-fuel use. The liquefied natural gas is stored in the cylindrical tank on the deck. This image (Lek S 2019) has been obtained by the author from the Wikimedia website where it was made available under a CC BY-SA 4.0 licence, https://creativecommons.org/licenses/by-sa/4.0/deed.en. It is included within this chapter on that basis. It is attributed to Steven Lek.

or to substitute for the primary diesel engines in environmentally sensitive areas. Gas turbines are also sometimes used in military ships, where the available power is an important factor and the operating cost is secondary. In a few cases, excess heat from gas turbines is used to produce steam to run a steam turbine, thereby increasing total available power, as well as efficiency.

### 5.3.2 Sustainable marine propulsion systems

*5.3.2.1 Battery electric marine vessels*

The major concern for battery electric ships is the availability of charging stations. In contrast to the situation with battery electric road vehicles, for which significant numbers of charging stations exist along many routes, or battery electric rail vehicles, for which periodic grid connection through overhead catenary lines can be incorporated into the rail line infrastructure, charging options for ships are much more limited. While this probably precludes the possibility of long-distance ocean routes for battery-operated cargo or passenger ships, the possibility of using this technology on short ferry routes may be realistic.

The *MV Ampere* is the first commercial battery electric ferry. It has operated since 2015 on the Sognefjord, the longest and deepest fjord in Norway, between the villages of Lavik and Oppedal (Barry 2013). The 5.9 km crossing takes 20 min each way. The MV *Ampere* is a 79 m long vessel that can accommodate up to 120 automobiles and 360 passengers (see figure 5.14). Power is provided by two 450 kW electric motors supplied by 10 tonnes of lithium-ion batteries with a total storage capacity of 1000 kWh. The batteries can be recharged fully in 10 min from onshore

**Figure 5.14.** The battery electric MV *Ampere*, a ferry traversing the Sognefjord in Norway. This image (Cavernia 2019) has been obtained by the author from the Wikimedia website where it was made available under a CC BY-SA 4.0 licence, https://creativecommons.org/licenses/by-sa/4.0/deed.en. It is included within this chapter on that basis. It is attributed to Cavernia.

high-capacity batteries at each port while the ferry is unloading/loading. Onshore batteries can provide more charging current than a grid connection, and these onshore batteries can be recharged more slowly from the grid while the ferry is away from the dock. The MV *Ampere* has been a successful approach to providing an environmentally sustainable ferry service on this route.

In October 2020, the Maid of the Mist Tours in Niagara Falls, New York launched two battery electric tour boats (Kane 2020), the *James V. Glynn* and the *Nicola Tesla*, see figure 5.15. Each boat is 27 m in length and incorporates lithium-ion batteries with a total capacity of 316 kWh to power a 400 kW electric motor. Each boat has a capacity of about 600 passengers and provides a 20-minute tour of Niagara Falls. Batteries are recharged during the approximately 10 min docking period required for passenger disembarkation and boarding. An added benefit of this particular implementation of electric vessel technology is that the electricity used to recharge the batteries comes from the local hydroelectric generating station associated with Niagara Falls (see Dunlap 2019 pp 374–75 for a description of Niagara Falls Generating Station).

Other battery electric marine transportation projects include a tanker, a container ship, and a cruise ship. These projects are described briefly below.

The Asahi Tanker Co., Ltd is constructing two battery electric tankers for the short-range transport of bunker fuels (Kane 2021a). The ships will be 62 m in length and will utilize four electric motors with a total maximum output of 736 kW powered by lithium-ion batteries with a total capacity of 3.5 MWh. The ships are expected to be operational sometime in 2022.

The chemical company Yara International is operating the battery electric ship MV *Yara Birkeland* (Kane 2021b). The *Yara Birkeland* is an 80 m container ship

**Figure 5.15.** The *James V. Glynn*, one of two battery electric tour vessels operated by the Maid of the Mist Tours. This image (Bona1858 2020) has been obtained by the author from the Wikimedia website where it was made available under a CC BY-SA 4.0 licence, https://creativecommons.org/licenses/by-sa/4.0/deed.en. It is included within this chapter on that basis. It is attributed to Bona1858.

designed to traverse the route between Porsgrunn and Brevik in Norway (about 16 km). The ship has four electric motors with a total maximum output of 3.2 MW powered by lithium-ion batteries with a total capacity of 7 MWh. The ship is currently undergoing testing and it is anticipated that fully autonomous operation will be achieved within about two years.

The construction of the *Yangtze River Three Gorges 1*, a battery electric cruise ship is being undertaken by the Wuxi Saisiyi Electric Technology Company. This cruise ship utilizes lithium-ion batteries with a total capacity of 7.5 MWh (the equivalent of 75 Tesla Model S automobiles) and is intended for use in river cruises on the Yangtze River. Initial announcements suggested a completion date around the end of 2021 for the construction of this vessel, but the current status of the project has not been made known (Kane 2021c).

All of the projects described above involve the utilization of battery electric marine vessels on relatively short routes and between well-defined locations where appropriate charging infrastructure is located. Studies of the viability of battery electric ferries with rapid recharging from onshore batteries have suggested that this option is most suitable for vessels on routes with one-way crossing times of up to 30 min. While some of the projects described above may push this envelope somewhat, it is important to consider the necessary operational time and the required battery charging time in the context of battery capacity and vessel power requirements.

#### *5.3.2.2 Hydrogen marine vessels*

While a number of hydrogen-fuel-cell-powered marine vehicles have been proposed and a few experimental boats of this design have been built, there has been little actual progress in applying fuel cell technology to ship propulsion. The world's first hydrogen-fuel-cell-powered boat was the *Hydra*, shown in figure 5.16. The *Hydra* is a 12 m long passenger boat with a capacity of 22 that is propelled by a 5 kW alkaline

**Figure 5.16.** The 12 m long *Hydra* on the Karl Heine Chanal in Leipzig, Germany in 2000. This image (Cchhrriissttiiaann 2000) has been obtained by the author from the Wikimedia website, where it is stated to have been released into the public domain. It is included within this chapter on that basis.

fuel cell. Other hydrogen fuel cell watercraft to date have been similar to the *Hydra* or have been sail-powered craft that use hydrogen fuel cells to provide auxiliary electrical power. The viability of using hydrogen fuel cells to power larger ships has not been demonstrated and is not likely to be the first approach to developing sustainable marine transportation.

*5.3.2.3 Biofuel and electrofuel marine vessels*

Biofuels have been discussed in detail in the previous chapter with regard to road vehicles and earlier in the present chapter with regard to rail transportation. As engines in most of today's large ships are diesel engines, the substitution of biofuels for fossil fuels is a potentially viable option. As marine diesel engines operate on a variety of fossil fuels, there are a number of options for biofuels. Biofuels that can serve as direct replacements for marine diesel fuels in compression ignition engines with minor modifications have recently been reviewed by McGill *et al* (2013) and Hsieh and Felby (2017). The most likely possibilities for biofuels to replace fossil fuels in marine diesel engines are biodiesel, renewable diesel, methanol, and dimethyl ether. Fuels that can be used as direct replacements for fossil-fuel-based marine diesel without the need for engine modifications are generally referred to as 'drop-in' fuels. Here, we look at some of the most probable sustainable replacements for marine diesel.

*Biodiesel and renewable diesel*—Biodiesel and renewable diesel are the possible marine fuels that are most compatible with existing fossil fuels. In many cases, biodiesel and renewable diesel can be integrated into the existing fossil fuel diesel supplies, as has been discussed previously for road and rail transportation. Biodiesel is not considered to be a drop-in fuel, as some engine modifications are required for its use in diesel engines designed to operate on traditional fossil fuels. However, blends of marine diesel with up to 20% biodiesel are in common use in the shipping industry. Renewable diesel is generally considered a drop-in fuel, although its use to date for marine transportation has been somewhat limited. Since 2010, the United States Navy has used a 50:50 fossil fuel–renewable diesel blend in some of its ships. These renewable diesel tests have indicated that the 50:50 blend functions suitably as a drop-in fuel.

*Methanol*—One of the most interesting recent utilizations of the use of biofuels in marine transportation is the conversion of the ferry *Stena Germanica* for methanol use (figure 5.17). This ship was launched in 2001, at which time the ship was named the *Stena Hollandica*. In 2007, the vessel's length was increased from 188 m to 240 m by the insertion of a 52 m section in the middle of the ship. In 2010, her registry was changed from the Netherlands to Germany and the ship's name was changed to *Stena Germanica*. Most recently, and of most significance to the present discussion, were the 2015 modifications that converted the engine to dual-fuel use. In addition to the marine gas oil that the engine used originally, the modifications allowed it to also operate on methanol. Dual-fuel engines using biofuel or e-fuel as well as a fossil fuel can be highly beneficial during a transition period. The use of methanol as a fuel was expected to reduce carbon dioxide emissions by 25%, sulfur emissions by 99%, and nitrogen-compound emissions by 60%.

**Figure 5.17.** The *Stena Germanica*, a 240 m long methanol-powered ferry that operates between Gothenburg, Sweden and Kiel, Germany (about 430 km). This image (Sharma 2017) has been obtained by the author from the Wikimedia website where it was made available under a CC BY-SA 4.0 licence, https://creativecommons.org/licenses/by-sa/4.0/deed.en. It is included within this chapter on that basis. It is attributed to Gaurav Sharma.

Since methanol is a liquid fuel, the infrastructure related to its transport and storage may be more convenient than that for (e.g.) dimethyl ether (IRENA and Methanol Institute 2021, Kowalewicz 1993). However, its lower energy density (17.9 MJ $L^{-1}$) compared to typical diesel fuels (38.6 MJ $L^{-1}$) means that this fuel occupies larger volumes. This translates into either larger fuel tanks, meaning reduced passenger or cargo volumes, or limited range. Shorter routes, such as that traversed by the *Stena Germanica*, might be a suitable application for marine methanol use.

A final point concerns the origin of the methanol fuel. As noted in chapter 3, most of the methanol produced today comes from natural gas, and this is the source of the methanol that is used by the *Stena Germanica*. The long-term sustainability of this approach will require the implementation of processes to produce methanol from sustainable hydrogen (i.e. e-methanol, as discussed in chapter 3) or from biological feedstock (as discussed in chapter 4). Other e-fuels (or electrofuels) as discussed in the section below on aviation fuels may contribute to marine transportation at some point in the future.

#### *5.3.2.4 Ammonia-powered marine vessels*

In recent years, interest has grown in the use of ammonia as a fuel in internal combustion engines for marine applications (see DNV GL (2020)). As mentioned in chapter 3, ammonia may be produced from hydrogen and atmospheric nitrogen by the Haber–Bosch process:

$$3H_2 + N_2 \rightarrow 2NH_3. \tag{5.1}$$

Combustion of the ammonia produces only nitrogen and water:

$$4NH_3 + 3O_2 \rightarrow 2N_2 + 6H_2O + 22.5\ \text{MJ/kg} \tag{5.2}$$

and has a heat of combustion of 22.5 MJ $kg^{-1}$. If the hydrogen utilized in equation (5.1) is green hydrogen, as defined in chapter 3 (i.e. it does not contribute carbon to the atmosphere), then the ammonia produced by the process in equation (5.1) and the by-products of the combustion in equation (5.2) are also carbon free.

The advantages and disadvantages of ammonia as a fuel for internal combustion engine road vehicles were discussed in section 3.9. Concerns related to the size and complexity of ammonia engines compared to those using more traditional alternative fuels such as methanol or ethanol may be less serious for large marine vessels than for light-duty road vehicles. For example, a large marine vessel can more easily accommodate a dual-fuel system and/or onboard cracking facilities than a typical road vehicle. However, the toxicity of ammonia is a major safety concern. As noted previously, fuel storage and delivery systems must be carefully monitored for leaks. It is also crucial to deal with any unburned fuel that may be expelled through the exhaust ports. For this reason, exhaust gas must be treated to remove any residual ammonia before being released to the environment.

Figure 5.18 shows the basic layout of an ammonia engine. The important features here that are relevant to the use of ammonia as a fuel include the ammonia cracker (lower left), which provides a small quantity of hydrogen as a combustion promoter, and the exhaust aftertreatment device, which removes any residual unburned ammonia from the exhaust gas before it is released to the atmosphere.

There are two basic approaches to an ammonia-fueled engine for marine applications—an internal combustion engine and a gas turbine. Following on from figure 5.18, more detailed diagrams for these two possibilities are shown in figure 5.19. Both designs incorporate an ammonia cracker to provide the hydrogen combustion promoter and an exhaust gas scrubber (SCR) to deal with residual unburned ammonia. Depending on engine design and performance, the scrubber

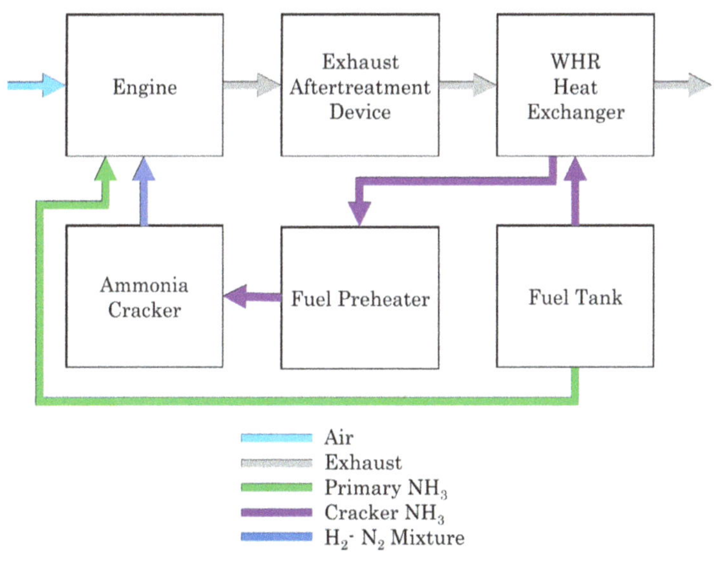


**Figure 5.18.** Layout of a marine ammonia power train. Reproduced from Imhoff *et al* (2021). CC BY 4.0.

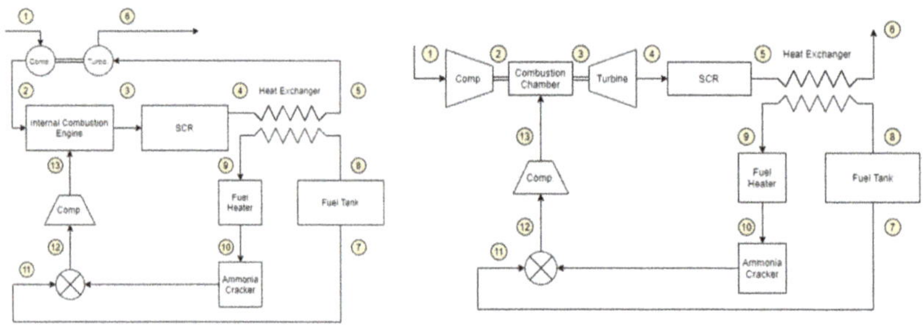


**Figure 5.19.** A diagram of an ammonia internal combustion engine for marine applications (left) and a diagram of an ammonia gas turbine engine (right). Reproduced from Imhoff *et al* (2021). CC BY 4.0.

may also be necessary to reduce $NO_x$ emissions, which are present in the exhaust from the oxidation of atmospheric nitrogen.

The development of viable ammonia-fueled engines is in its early stages, although several significant programs have been initiated. For example, MAN Energy Solutions (2022) is developing a two-stroke internal combustion engine operating on ammonia for marine applications. Along similar lines, Wärtsilä Corporation (2020) is working on a four-stroke internal combustion marine engine. Mitsubishi is developing a large ammonia-fueled gas turbine (Patel 2021). Although this turbine is aimed at applications in small to medium-size generating stations, its power output (40 MW) is comparable to those of many current diesel engines used in tankers and other large marine vessels. A parallel approach is the development of dual-fuel engines using ammonia in conjunction with another renewable fuel (e-methanol, for example). In the short term, dual-fuel ammonia–LNG engines can bridge the gap to all renewable technologies.

#### *5.3.2.5 Wind-powered marine vessels*

Wind power was the first non-human means of marine propulsion. In fact, sailing ships were the principal means of ocean transportation up until the late 19th century. A large sailing ship, such as that illustrated in figure 5.20, could generate up to about 8 MW of power in a strong wind. By comparison, the RMS *Titanic* (as an example) could produce about 34 MW.

The use of the wind as a source of energy for commercial ocean-going vessels has been considered again in recent years. One design that has attracted some recent interest utilizes Flettner rotors instead of traditional sails. An example (the *E-Ship 1*) is shown in figure 5.21. The Flettner rotor (named after its developer, the German engineer Anton Flettner (1885–1961)), is a rotor sail that makes use of the Magnus effect to create a propulsive force. The Magnus effect was first described in detail in 1852 by the German physicist Heinrich Gustav Magnus (1802–70). However, the basic principle of the Magnus effect had been described nearly two centuries earlier by Sir Isaac Newton (1643–1727). The Magnus effect refers to the force that acts on a rotating object that is moving relative to a fluid (in this case, air), as illustrated in figure 5.22. There is a net force on the object perpendicular to both the axis of rotation and the relative velocity of the fluid. In the case of the *E-Ship 1*, the vertical

**Figure 5.20.** The *Preußen*, a 147 m long full-rigged five-masted sailing ship that was operational and was used to carry cargo from 1902 to 1910. The *Preußen* was the largest traditional sailing ship ever constructed and had no additional source of energy for propulsion. Reproduced from [Nautical Photo Agency (n.d.)]. Image stated to be in the public domain.

**Figure 5.21.** The 130 m long *E-Ship 1* has been operational since 2010 and uses four Flettner rotors, each 27 m high and 4 m in diameter, to harness wind energy. It is interesting to note that the *E-ship 1* is owned by Enercon GmbH, a major wind turbine manufacturer in Germany and is used to transport wind turbine components from Emden, Germany to customers around the world. This image (Lpele 2016) has been obtained by the author from the Wikimedia website where it was made available under a CC BY-SA 4.0 licence, https://creativecommons.org/licenses/by-sa/4.0/deed.en. It is included within this chapter on that basis. It is attributed to Lpele.

cylinders are rotated by a motor along their cylindrical axis. Air moving perpendicular to the ship produces a net force that drives the ship forward. If the wind approaches the vessel from the opposite side, the rotors must be rotated in the opposite direction in order to move the ship forward (rather than backward). Current commercial ship designs that incorporate a means of harnessing energy from the wind have another primary means of propulsion, and the wind power can be used as a means of reducing fuel consumption. Analyses of vessels such as the

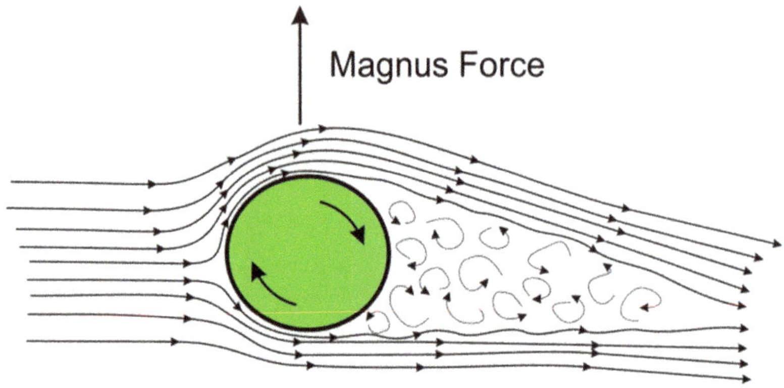


**Figure 5.22.** Illustration of the Magnus effect showing the net force on a rotating object in a flowing fluid. This image (Rdurkacz 2013) has been obtained by the author from the Wikimedia website where it was made available under a CC BY-SA 3.0 licence, https://creativecommons.org/licenses/by-sa/3.0/deed.en. It is included within this chapter on that basis. It is attributed to Rdurkacz.

*E-Ship 1* have indicated that the power that can be harnessed from the wind using Flettner rotors can reduce fuel consumption and thus carbon emissions by up to about 25% (MI News Network 2021).

*5.3.2.6 Nuclear-powered marine vessels*

Nuclear-powered marine propulsion systems were first developed by the United States Navy in the 1950s. The advantage (from a military standpoint, at least) of nuclear-powered ships (rather than fossil-fuel-powered vessels) stems from the fact that nuclear fuel has a very high energy density (compared to the energy density of chemical fuels, such as diesel fuel) and from the fact that nuclear reactors do not require a constant supply of oxygen (as is the case for a diesel engine). These advantages translate into direct benefits, such as

- long intervals (typically many years) between refueling
- no loss of space for fuel tanks, combustion air intake structures, or exhaust stacks
- the ability of submarines to remain submerged for extended periods of time, as oxygen is not needed for combustion

Most nuclear reactors used for ship propulsion are of the pressurized water design. Energy produced by the reactor heats water under pressure in the primary circuit to temperatures of around 250 °C to 300 °C. The pressure prevents the water from boiling. The hot water in the primary circuit transfers heat through a heat exchanger to low-pressure water in the secondary circuit. This water boils to produce steam which is then used to drive a steam turbine.

The ship can be propelled by one of two methods: (a) by coupling the steam turbine output through a gearbox to the propeller or (b) using the steam turbine to run an electrical generator which is then used to drive an electric motor connected to the propeller. The mechanical drive system is commonly used in British, Russian, and United States ships, while the electrical drive system is commonly used in Chinese and French ships. Submarines often have a single nuclear reactor. Surface ships often have two nuclear reactors and aircraft carriers can have up to eight.

The basic physics of the operation of a marine nuclear reactor is the same as a conventional pressurized water thermal neutron fission reactor used for power generation. There are, however, some practical differences that stem from the fact that the power output of a marine reactor is typically much less than that of a commercial power reactor. Typical marine reactors might produce a couple of hundred megawatts of (thermal) output, while a large commercial power reactor might produce around 2500 MW (thermal). This smaller size has some implications for reactor design, most importantly for the level of enrichment of the fuel. Naturally occurring uranium is about 99.3% uranium-238 and about 0.7% uranium-235. In a conventional nuclear reactor, it is the uranium-235 that produces energy by undergoing fission (that is, breaking up into lighter nuclei). A chain reaction is maintained when neutrons produced by the fission of one uranium-235 nucleus go on to react with other uranium-235 nuclei and induce further fission. The natural abundance of uranium-235 is too low to allow this to occur in most thermal neutron reactor designs, so fission reactors utilize enriched uranium, i.e. uranium that has artificially had its fraction of uranium-235 nuclei increased. In a typical commercial pressurized water power reactor, uranium fuel enriched to about 3.2% uranium-235 is used. Because the core of a maritime reactor is much smaller than that of a commercial power reactor, the fuel must have an even higher fraction of uranium-235 in order to maintain the chain reaction. While some nuclear marine reactors have utilized low-enrichment fuel, enrichments are more typically anywhere from about 20% to as much as 97% uranium-235. An added benefit of this high level of enrichment is that the fuel (per unit mass) contains more nuclei that can undergo fission and produce energy; in other words, its energy density is very high. This means that nuclear-powered ships can go for long periods between refueling operations. Modern aircraft carriers can have a core life of as much as 50 years. More modern reactor designs, such as molten salt reactors, are under consideration for future nuclear ship designs.

The first nuclear-powered ship was the United States Navy submarine USS *Nautilus*, which was commissioned in 1955 and operated until 1980. It is currently on exhibit at the US Submarine Force Museum and Library, Groton, CT (figure 5.23).

Nuclear power is very compatible with the needs of a military submarine, as it allows the vessel to remain submerged more or less indefinitely. In addition to those constructed for the United States Navy, nuclear submarines have also been built by Britain, China, France, India, and Russia. By 1989, the number of operational nuclear submarines was in excess of 400. At that time there were more nuclear reactors (albeit small ones) at sea than there were in commercial generating stations. In more recent years, the number of nuclear submarines has decreased to about 160 worldwide (World Nuclear Association 2021).

Nuclear power has also been utilized in a number of other military vessels including destroyers, cruisers, and communication ships. The most successful

**Figure 5.23.** The 98 m long USS *Nautilus* at the US Submarine Force Museum and Library, Groton, CT in 2008. This image (Victor-ny 2008) has been obtained by the author from the Wikimedia website where it was made available under a CC BY-SA 3.0 licence, https://creativecommons.org/licenses/by-sa/3.0/deed.en. It is included within this chapter on that basis. It is attributed to Victor-ny.

application of marine nuclear power, after submarines, has probably been in aircraft carriers.

The Soviet Union (and later Russia) has been active in utilizing nuclear power for icebreakers. This power source is beneficial for icebreakers in the Arctic, as it allows them to operate for extended periods of time in isolated regions. Eleven nuclear icebreakers have been built, of which six are still in service; an early example is illustrated in figure 5.24. Two additional vessels are planned for construction in the near future.

The use of nuclear power in commercial ships has been very limited. The only notable example of a commercial nuclear-powered ship is the 182 m long NS *Savannah* (NS for 'Nuclear Ship') shown in figure 5.25. The *Savannah* was operational from 1962 to 1972 and was intended to demonstrate the technical feasibility of constructing a nuclear-powered cargo ship. The ship, however, was designed with more luxury features than typical cargo ships, as well as less cargo volume. As such, the ship was not economically viable. This lack of viability was also related to the construction and operating costs associated with the nuclear reactor. The original cost of constructing the *Savannah* was 46.9 million USD, of which 28.3 million USD was the cost of the reactor and the nuclear fuel. The Savannah was designated as a National Historic Landmark in 1991 and will likely, at some point in the future, be opened as a museum.

Nuclear-powered ships have characteristics (such as refueling requirements) that make them attractive for military and some specialized applications. However, it is

**Figure 5.24.** The Russian nuclear icebreaker *Arktika*. The 148 m long *Arktika* was in service from 1975 to 2008 and in 1977 became the first surface vessel to reach the North Pole. This image (Abarinov 2006) has been obtained by the author from the Wikimedia website, where it is stated to have been released into the public domain. It is included within this chapter on that basis.

**Figure 5.25.** The NS *Savannah* in Baltimore, MD, in 2012. This image (Acroterion 2012) has been obtained by the author from the Wikimedia website where it was made available under a CC BY-SA 3.0 licence, https://creativecommons.org/licenses/by-sa/3.0/deed.en. It is included within this chapter on that basis. It is attributed to Acroterion.

unlikely that they will ever be a significant component of overall marine transportation. Specifically, the application of nuclear power to commercial shipping is unlikely to be economically competitive with other options that have been discussed above. In addition, the possibility of using nuclear reactors to power passenger vessels may not meet with public acceptance.

Many of the factors that adversely affect the viability of marine nuclear power are also potential drawbacks of the application of nuclear power to electricity generation. These include both sustainability and environmental issues. While nuclear power (i.e. fission power) is, in principle, carbon free, it is not a renewable source (Dunlap 2019). There is a limited amount of available uranium, and this poses problems for the longevity of commercial nuclear power reactors. New technologies, such as fast breeder reactors (that make use of the uranium-238) or thorium reactors (that utilize more plentiful thorium, rather than uranium) may substantially extend the useable life of nuclear fission as a power source. However, the practicality, or in some cases even the scientific feasibility of such technologies, has not been satisfactorily demonstrated for land-based reactors. Their applicability to small marine reactors is even more uncertain.

Other factors that adversely affect the possibility of nuclear technology in general include waste disposal, reactor safety, and radioactive material security. For marine nuclear reactors, the disposal of radioactive material at the end of the vessel's life is not a straightforward process. The fuel that remains in the reactor is removed and reprocessed or disposed of as for spent fuel from land-based reactors. Low-level activity associated with the reactor materials must, in principle, be removed from the vessel and buried in secure shallow landfills. Unfortunately, in many cases, particularly in Russia, decommissioned submarines are sealed with radioactive material inside and stored whole. Although the safety record of nuclear-powered ships has generally been good, there have been a number of reactor accidents that have resulted in loss of life for crew members as well as radioactive material leakage to the environment. However, the overall potential dangers are somewhat mitigated by the small sizes of marine reactors compared to those of commercial power reactors. Finally, the security of radioactive material is an important concern for ship reactors, and the high level of enrichment that is used in many reactors is of particular relevance in this respect.

## 5.4 Air transportation

### 5.4.1 Current aviation fuels

Table 5.1 shows that for mass passenger transportation, aircraft require the greatest power per person. At present, air transportation accounts for about 12% of transportation energy and virtually all air transportation uses jet fuel or avgas (described below). Jet fuel is a mixture of fossil-fuel-derived hydrocarbons that is similar to kerosene and has an average of 5–16 carbon atoms per molecule. By comparison, gasoline has 4–12 carbon atoms per molecule, and diesel fuel has 12–20. Jet fuel is suitable for use in turbine engines, as well as compression ignition reciprocating engines. The precise composition of jet fuel varies, and fuels are categorized on the basis of performance characteristics more than on composition. There are four fuel standards that are in common use for commercial aircraft (Shell 2021). *Jet A* is used in the United States and in some Canadian airports. *Jet A-1* is used in most of the rest of the world. Both of these fuels are kerosene based and have similar, but not identical properties. *Jet B* is a naphtha–kerosene based fuel that is more flammable than *Jet A* or *Jet A-1* and has enhanced performance at low temperatures. It is only used in cold climates, such as in northern Canada, Alaska, and sometimes Russia. *TS-1* is a Russian fuel standard that is also suitable for low temperatures It is used in Russia and former Soviet states.

A variety of different jet fuels are used for military applications (Shell 2021) and these are designated as *JP* (Jet Propellant) fuels. *JP-8* is similar to *Jet A-1* and *JP-4* is similar to *Jet B*. Other military fuels are designed for specific applications.

Avgas (which is short for aviation gasoline) is designed for use in spark ignition internal combustion engines. It is similar to gasoline that is used in road vehicles, with a couple of important differences. Avgas typically has a higher octane rating than gasoline sold for light-duty vehicle use. Octane ratings of around 100 are common for avgas. Avgas traditionally contained lead additives and, although lead additives have been banned in the United States for road vehicle gasoline since 1996, many grades of avgas still contain lead. Note that leaded gasoline is also still available for use in off-road vehicles, marine engines, farm equipment, and race cars. Newer avgas grades are available that do not contain lead, but many aircraft currently in operation require lead-containing fuel.

While we would expect that the carbon emissions per unit energy generated would be similar to those of the fossil fuels used in (e.g.) road transportation, the problem of aircraft emissions is not as simple as looking at the fuel consumption for different modes of transportation. Aircraft release carbon dioxide as well as other pollutants such as carbon monoxide, nitrogen oxides, water vapor, and particulates, and these are released high in the atmosphere. The reactions between these pollutants and the constituents of the upper atmosphere are not well known, but there is evidence that certain reactions can lead to increased global warming compared to ground-level greenhouse gas emissions (Penner *et al* 1999). Thus, it is particularly important to look at ways of minimizing carbon emissions from air transportation. Air transportation is perhaps the most demanding mode of transportation from the standpoint of implementing environmentally sustainable options. This is a result of the requirement for a source of propulsion energy that has a very high energy density. We can understand this need by looking at a Ragone plot for the relationship between energy density and power density for different energy storage mechanisms. In chapter 2, we saw the relationship between specific energy and specific power for the potential power sources for transportation, namely batteries, capacitors, and fuel cells. We can include fossil fuels that are appropriate for use in internal combustion engines or gas turbines on such a plot, as shown in figure 5.26. It is obvious that no other potential transportation energy source can compete with fuels such as gasoline for combined specific energy and specific power. Supercapacitors can provide similar specific power but fall short by more than two orders of magnitude in terms of specific energy. Fuel cells are comparable in terms of specific energy but provide less specific power by two orders of magnitude. Thus, the replacement of fossil fuels for aviation energy is a substantial challenge. The following sections summarize some of the approaches that have been taken in recent years to mitigate greenhouse gas emissions from air transportation.

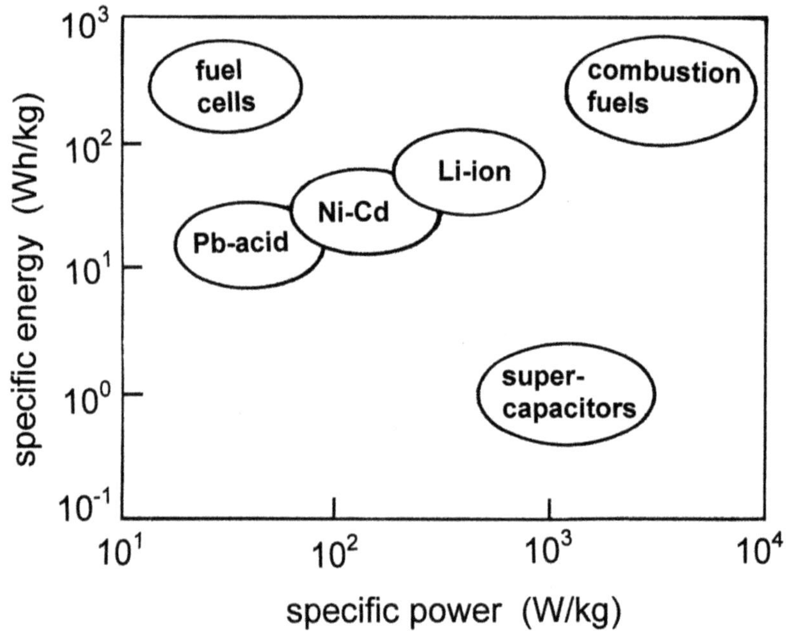


**Figure 5.26.** Ragone plot showing the relationship of different energy storage methods for transportation technology. Combustion fuels include both fossil fuels and biofuels, as well as hydrogen, as used in internal combustion piston engines and jet engines.

### 5.4.2 Battery electric aircraft

Several different designs of aircraft that use electric motors to drive propellers have been constructed. These designs include battery electric aircraft (including both unmanned drones and aircraft capable of carrying one or more persons), fuel cell aircraft, and solar aircraft. Battery electric aircraft that can carry passengers are discussed in this section and fuel cell aircraft are discussed in the next section. Solar-powered aircraft that use photovoltaic cells for power have been tested experimentally but are not likely to ever contribute in any significant way to our transportation needs.

The prospect of constructing an aircraft that utilizes batteries and an electric motor for propulsion is somewhat uncertain on the basis of the information contained in the Ragone plot in figure 5.26. Even for the best available batteries, both the specific energy and the specific power are substantially less than that provided by the combustion of fossil fuels. In fact, the possibility of a battery-powered aircraft with the ability to carry at least one passenger did not become a reality until the early 1970s with the development of nickel–cadmium batteries. Clearly, even on the basis of the characteristics of lithium-ion batteries, we would expect that small, short-range battery electric aircraft would be the most successful.

In 1973, the first experimental battery-powered aircraft constructed by Fred Militky and Heino Brditschka flew for 14 min with one person on board. A notable experimental battery electric aircraft, the E-Fan, as shown in figure 5.27,

**Figure 5.27.** Airbus E-fan 1.1 battery electric aircraft in 2014. Reproduced form Sieker (2014) Flickr (CC BY-SA 2.0).

was developed by Airbus in 2013. The E-Fan utilized two ducted, variable-pitch fans driven by two electric motors totaling 60 kW of power. On 9 July 2015, the E-Fan 1.1 crossed the English Channel from Lydd Airport to Calais–Dunkerque Airport. Airbus originally planned to produce two production variants of the E-Fan: the E-Fan 2.0, a two-seat version to be used for pilot training, and the E-Fan 4.0, a four-seat variant to be used for touring. In 2017, Airbus canceled its production plans for the E-Fan.

In 2020, the first commercially available battery electric aircraft certified by EASA (the European Union Aviation Safety Agency) became available. This craft is the Velis Electro manufactured by Pipistrel in Ajdovščina, Slovenia, as shown in figure 5.28 (Pipistrel 2021). The electric motor produces 57.6 kW for 90 s during takeoff and 49.2 kW in continuous operation. The lithium-ion batteries have a capacity of 24.8 kWh, giving the aircraft an estimated range of 200 km. The total permitted payload is 172 kg.

While the availability of the Velis Electro provides evidence of the applicability of battery electric power to small, short-range aircraft that can carry two passengers, it is a long way from large-scale commercial air travel. A somewhat more ambitious project being undertaken by the battery electric aircraft company Eviation is the development of a small battery electric commuter aircraft named *Alice* (see figure 5.29) that can carry two pilots and up to nine passengers on flights of up to 815 km. As of 2022, test flights of a prototype aircraft have been undertaken and commercial availability is planned for around 2027.

Another approach to electric aircraft in the short term is the implementation of hybrid technology for air travel. Typical approaches to this type of design combine a

**Figure 5.28.** The Velis Electro manufactured by Pipistrel in Ajdovščina, Slovenia first became commercially available in 2020. This image (Airjuice Photography 2020) has been obtained by the author from the Wikimedia website where it was made available under a CC BY-SA 4.0 licence, https://creativecommons.org/licenses/by-sa/4.0/deed.en. It is included within this chapter on that basis. It is attributed to Airjuice Photography.

**Figure 5.29.** Eviation Alice prototype. This image (Blume 2019) has been obtained by the author from the Wikimedia website where it was made available under a CC BY-SA 4.0 licence, https://creativecommons.org/licenses/by-sa/4.0/deed.en. It is included within this chapter on that basis. It is attributed to Matti Blume.

battery electric drive train with a conventional diesel or avgas piston engine to drive one or more propellers. Electric power would be used for takeoffs and landings in regions where a reduction in the noise and pollution produced by internal combustion engines would be most beneficial. The traditional engines would be used for cruising to increase the total range of the aircraft. Aircraft carrying 10–20 passengers on flights of less than about 1000 km are probably the most likely to make use of this technology.

### 5.4.3 Hydrogen-powered aircraft

As for the application of hydrogen as a fuel in road vehicles, hydrogen can be used in a fuel cell to provide electricity, or it can be used in an engine to provide energy by combustion. An inspection of the Ragone plot in figure 5.26 suggests that hydrogen fuel cells (probably in the form of a fuel cell–battery hybrid) may be better suited to larger aircraft than batteries alone.

#### *5.4.3.1 Fuel cell aircraft*

A small number of experimental aircraft that utilize hydrogen fuel cells connected to an electric motor have been tested. The first such craft that had the ability to carry a person was the Antares DLR-H2 (see figure 5.30) developed in 2009 by the Deutsches Zentrum für Luft- und Raumfahrt e.V. (DLR 2009). The fuel cell system had an output of 30 kW and the plane had a maximum payload of 265 kg.

A further step towards the commercialization of fuel cell aircraft technology came with the development of the ZeroAvia HyFlyer I in 2020 (Henderson 2021). This aircraft is a six-seat Piper Malibu that was modified to operate as a fuel cell–battery hybrid. It includes a PowerCell MS-100 100 kW fuel cell and lithium-ion batteries rated at 75 kWh. The electric motor drives a single propeller and has a maximum output of 250 kW. Comparing the motor power with the fuel cell output makes it clear that a significant portion of the power used during periods of high demand (such as takeoff) comes from the batteries (Harris 2021). ZeroAvia has announced that it

**Figure 5.30.** Antares DLR-H2. The Antares DLR-H2 research aircraft is the world's first piloted aircraft to be powered exclusively by fuel cells, ensuring that it takes off, flies, and lands without emitting any carbon dioxide. Reproduced from https://commons.wikimedia.org/wiki/File:Antares_DLR-H2_in_flight.jpg. CC BY 3.0 by German Aerospace Center.

expects to market a fuel-cell-powered aircraft that will carry up to 20 passengers with a range of about 650 km by 2023 and that an aircraft that will carry 80 passengers with a range of over 900 km will be available by 2026. Much more development will be certainly needed over the next few years to meet those expectations.

#### *5.4.3.2 Hydrogen internal combustion engine aircraft*

The Ragone plot in figure 5.26 suggests that extracting energy from the combustion of hydrogen would more easily meet the energy and power requirements of an aircraft than the use of hydrogen in a fuel cell and suggests the possibility that this technology could be applicable to larger, long-range aircraft. The use of hydrogen as a fuel in an internal combustion engine for road transportation has been discussed in detail in section 3.5. The concerns related to the use of hydrogen as a fuel for air transportation may be quite different than those for road transportation. For example, the hydrogen fueling infrastructure at selected airports would be much simpler than the distributed fueling infrastructure with many small stations that would be required for road transportation.

Perhaps the first notable experiment with hydrogen as a fuel for a jet engine was the TU-155 (see figure 5.31) constructed by Aviakor in the former Soviet Union. This aircraft was a Tupolev TU-154 that was modified for use as a testbed for alternative aviation fuels. It flew from 1988 to 1991. Its early flights utilized liquid hydrogen as a fuel. Later flights used liquefied natural gas.

As discussed in detail in chapter 3, hydrogen in the liquid state or as a compressed gas has a lower specific energy than liquid hydrocarbon fuels, such as diesel fuel or gasoline. It also requires more complex handling and an infrastructure which is not compatible with liquid biofuels, as discussed below. At present, virtually all commercial hydrogen is produced by the steam reforming of fossil fuel natural

**Figure 5.31.** The Tupolev Tu-155 alternative fuel testbed aircraft. Reproduced from Wilson (2012) Flickr (CC BY-SA 2.0)

gas. In order to develop a sustainable air transportation system utilizing hydrogen-fueled jet engines, the infrastructure for the production of carbon-free sustainable hydrogen must also be developed (IATA 2015a). In addition, hydrogen combustion produces more significant amounts of water vapor than traditional jet fuels, about 2.6 times as much per unit energy produced. The details of how this additional water vapor affects climate require additional study. These factors would seem to put hydrogen at a disadvantage compared to sustainable aviation fuels or electrofuels, as discussed below. However, Airbus has recently announced plans to develop three hydrogen-powered commercial aircraft (Tidey 2020), and future investigations of all reasonable aviation fuel options should be undertaken.

### 5.4.4 Alternative aviation fuels

Alternative aviation fuels are hydrocarbon-based fuels that are either liquid or gaseous and are designed to minimize the release of greenhouse gases. They are used to obtain energy through combustion in either a jet engine or a piston internal combustion engine. There are several ways of categorizing alternative fuels. Biofuels are those that are made from biological feedstock and include fuels such as biodiesel, as discussed in section 4.3. Electrofuels (or e-fuels) are those that are made from feedstock such as hydrogen that is produced electrolytically and include fuels such as e-methanol, as discussed in section 3.10 (Transport & Environment 2017). Sustainable aviation fuels (or SAFs) are those fuels that are based on feedstock that does not compete with the use of agricultural land used for human or animal food production. These fuels would include, for example, renewable diesel, as discussed in section 4.4.

Developing and utilizing suitable sustainable aviation fuels is a significant technical challenge. It is also a challenge in terms of developing a suitable distribution infrastructure, overcoming economic concerns, and achieving certification with the relevant regulatory bodies (IATA 2015b). In the short term, at least, it is important that alternative aviation fuels are drop-in fuels. This means that existing infrastructure, including aircraft engines, can continue to be utilized without modification. Importantly, drop-in fuels are also suitable for use in blends with existing fossil fuels, meaning that a transition to sustainable aviation fuels can occur over time as infrastructure for the production of alternative fuels is developed.

In recent years, much emphasis has been placed on the development and integration of drop-in sustainable aviation fuels. In fact, the use of biofuels that do not fit into the sustainable category (primarily because of competition with food production for land and the resulting land-use change component of the resulting environmental impact) has been viewed unfavorably in aviation fuel studies (Transport & Environment 2016). The future implementation of electrofuels is likely to follow, as the technology and infrastructure for these fuels has already been developed. In this section, we look at some features of sustainable aviation fuels and electrofuels for future aviation use.

#### *5.4.4.1 Sustainable aviation fuels*

At present, the organization that provides guidelines for aviation fuels is ASTM International (formerly, the American Society for Testing and Materials). In 2011,

**Table 5.4.** Pathways to sustainable aviation fuel production as certified by ASTM International.

| Process | Standard abbreviation | Viable sustainable feedstock | Approved blending ratio |
|---|---|---|---|
| Fischer–Tropsch hydroprocessed synthesized paraffinic kerosene | FT-SPK | Solid biomass | 50% |
| Synthesized paraffinic kerosene produced from hydroprocessed esters and fatty acids | HEFA-SPK | Waste oil, animal fat, algae | 50% |
| Synthesized kerosene isoparaffins produced from hydroprocessed fermented sugars | HFS-SIP | Biomass from sugar production | 10% |
| Fischer–Tropsch hydroprocessed synthesized paraffinic kerosene with aromatics | FT-SPK/A | Solid biomass | 50% |
| Alcohol-to-jet synthetic paraffinic kerosene | ATJ-SPK | Biomass from ethanol or isobutanol production | 50% |
| Catalytic hydrothermolysis jet fuel | CHJ | Fats, oils | 50% |

ASTM certified three pathways for the production of sustainable aviation fuel. Since then, they have certified three additional pathways. These pathways utilize biological feedstock in such a way as to satisfy the requirements for sustainability, principally that land-use change does not adversely affect food sources. The six pathways are as shown in table 5.4.

While the technical details of these processes are outside the scope of the present book, it is interesting to look briefly at the utilization of the Fischer–Tropsch process for the production of hydrocarbon-based sustainable fuels. The Fischer–Tropsch process was first discovered in 1925 by the German chemists Franz Joseph Emil Fischer (1877–1947) and Hans Tropsch (1889–1935). The Fischer–Tropsch process is a collection of chemical reactions that convert syngas, i.e. a mixture of hydrogen and carbon monoxide, into heavier hydrocarbons in the presence of certain metal catalysts, typically cobalt, or sometimes iron. The reactions typically take place at temperatures between 150 °C and 300 °C and pressures of several tens of atmospheres.

The Fischer–Tropsch process yields heavy hydrocarbons that are typically compounds in the alkane chain, with compositions of $C_nH_{2n+2}$ (see section 4.2). These hydrocarbons are created by the process

$$(2n + 1)H_2 + nCO \rightarrow C_nH_{2n+2} + nH_2O,$$

where the values of $n$ are usually in the range of 10–20. This is consistent with the requirements for jet fuel, which has a carbon content similar to that of kerosene, with 6–12 carbons per molecule. The exact composition of the resulting hydrocarbon can be controlled by the details of the Fischer–Tropsch process. In this way, appropriate aviation fuels for specific applications can be obtained from biosyngas that is produced from feedstock derived from sources that do not compete with food agriculture. Such feedstock might be agricultural waste, manure, or municipal solid waste, as discussed in section 4.5.

It is interesting to note that ASTM has certified sustainable aviation fuels for use in commercial airline flights only as blends with fossil fuels. The first commercial use of a blend of biofuel and fossil fuel occurred on a 2008 Virgin Atlantic flight. Since then, substantial research has been conducted on the development of aviation biofuels. Since the establishment of the ASTM guidelines, work has concentrated on certified pathways for the production of sustainable aviation fuel; much of this work has dealt with algal or municipal waste feedstock. Numerous airlines have used blended fuel on commercial flights, but not on a routine basis. A major factor limiting the use of aviation biofuels is lack of availability. As of 2019, only five airports, namely Bergen, Brisbane, Los Angeles, Oslo, and Stockholm have biofuel available on a regular basis, although it is occasionally available at other locations. In 2019, the last year of air travel that was not interrupted by COVID-19 restrictions, 9.0 million liters of sustainable aviation fuel was used in the United States, compared to a total aviation fuel use of 69 billion liters. In other words, about 0.1% of the aviation fuel use in the United States was sustainable. In 2009, the International Air Transport Association (IATA 2021) committed to reducing carbon emissions from the airline industry by 50% by 2050, a task that will require substantial technological development to make it economically feasible.

#### *5.4.4.2 Electrofuels*

The previous discussions in this book have suggested that the direct use of hydrogen as a transportation fuel has to overcome a number of serious challenges in order to become economically and environmentally viable, as well as practical. However, if hydrogen is viewed merely as an electricity storage mechanism, then it can be used as a feedstock for the production of other fuels that are more easily utilized. This approach was described in some detail in chapter 3 (section 3.10), which described e-methanol production in Iceland. Efficiency, which ultimately translates into economy, may be a factor for the use of these fuels, particularly for general road vehicle use. However, they may prove suitable for aviation, for which other approaches, particularly for large long-range aircraft, are either impossible (such as the use of catenary electric connections) or uncertain (such as the use of battery electric aircraft).

In order for electrofuels to be viewed as beneficial to climate change mitigation, it is essential that the environmental impacts of all aspects of their production should be minimized. There are two fundamental components of electrofuel production, namely hydrogen and carbon dioxide, which need to be considered. First, the hydrogen feedstock must be made using carbon-free sources of electricity. Second, since the combustion of the electrofuel emits carbon dioxide into the atmosphere, it is essential that the carbon dioxide that is used to make the electrofuel is taken from the atmosphere. These conditions are not currently fulfilled. However, if electrofuels are ultimately expected to reduce greenhouse gas emissions, their development must include the establishment of the infrastructure required to produce them in an environmentally sustainable manner. A recent roadmap for the sustainable production of electrofuels has been published by Transport & Environment (2017), a European non-profit organization dedicated to the development of a transportation system that minimizes adverse impacts on health, the climate, and the environment.

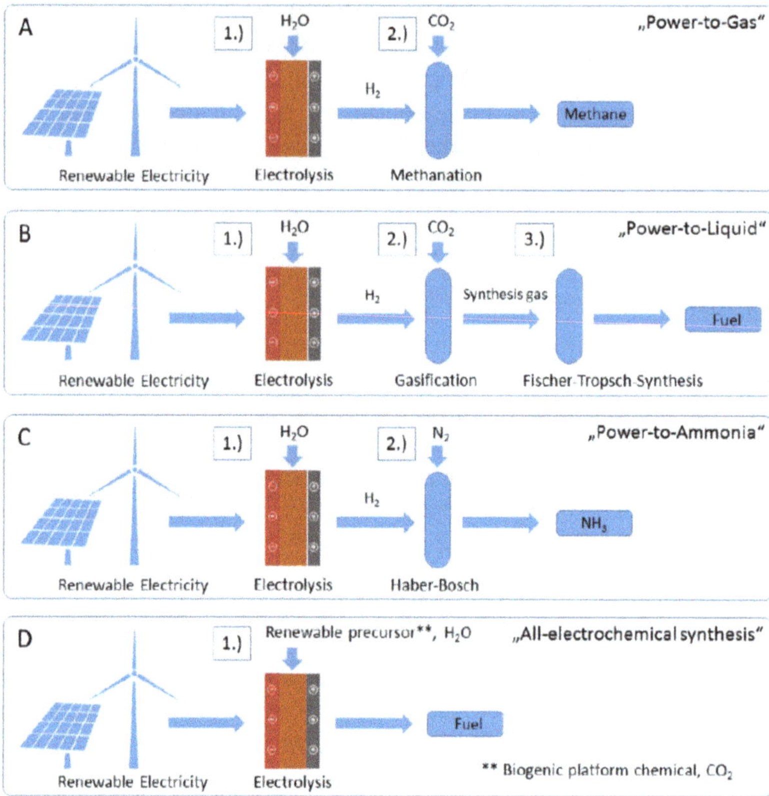


**Figure 5.32.** Pathways for electrofuel production. Reproduced from Goldmann *et al* (2018). CC BY 4.0.

Several pathways for the production of electrofuels from electrolytically produced hydrogen are illustrated in figure 5.32.

The simplest approach to electrofuel production (figure 5.31(a)) is the formation of methanol ($CH_4$) from electrolytically produced hydrogen ($H_2$) and carbon dioxide ($CO_2$) from the atmosphere by the process of methanation (Grond *et al* 2013). The overall process is:

$$4H_2 + CO_2 \rightarrow CH_4 + 2H_2O.$$

This process occurs in two steps, as follows:

$$H_2 + CO_2 \rightarrow CO + H_2O$$

and

$$3H_2 + CO \rightarrow CH_4 + H_2O.$$

This is sometimes referred to as 'power-to-gas.' The synthetic methane produced by this method can be used as a transportation fuel, typically either in the form of a compressed gas (as a replacement for compressed natural gas (CNG)) or as liquid (as a replacement for LNG).

More useful liquid fuels can be produced as shown in figure 5.31(b). Such processes are sometimes referred to as 'power-to-liquid.' The simplest of these reactions is the formation of methanol ($CH_3OH$), as previously discussed in section 3.10. This process begins with the reverse water–gas shift reaction:

$$3H_2 + CO_2 \rightarrow (2H_2 + CO) + H_2O.$$

The quantity in parentheses on the right-hand side of the reaction is syngas, and this is converted into methanol (in this case e-methanol) by the following reaction (Dieterich *et al* 2020):

$$2H_2 + CO \rightarrow CH_3OH.$$

As noted in figure 5.31 and as described above for sustainable aviation fuels, the syngas produced from hydrogen and carbon dioxide can be used to produce other fuels by the Fischer–Tropsch process. This process, as noted in the section above on sustainable aviation fuels, allows for the production of heavier hydrocarbons which can act as drop-in fuels for aircraft.

A recent analysis of electrofuels (Lehtveer *et al* 2019) has suggested that at present or in the near future, electrofuels are not likely to be sufficiently cost-effective to compete with alternative options. These authors particularly noted the need to improve the efficacy of electrofuel synthesis and electrolysis. In addition, the environmental sustainability of electrofuel utilization requires the implementation of infrastructure that can produce carbon-free electricity. This requires the large-scale development of renewable energy resources, particularly solar power and wind power.

## References

Abarinov 2006 Russian nuclear icebreaker "Arktika" https://commons.wikimedia.org/wiki/File:Russian_Nuclear_Icebreaker_Arktika.jpg

Acroterion 2012 NS Savannah https://commons.wikimedia.org/wiki/File:NS_Savannah_exterior_MD2.jpg

Agence France-Presse 2018 Germany launches world's first hydrogen-powered train *Guardian* (17 September 2018) available at: www.theguardian.com/environment/2018/sep/17/germany-launches-worlds-first-hydrogen-powered-train

Airjuice Photography 2020 Pipistrel Velis Electro https://commons.wikimedia.org/wiki/File:Pipistrel_Velis_Electro_9992.jpg

Barry K 2013 World's first electric car Ferry recharges in 10 min *Wired* (1 February 2013) available at: www.wired.com/2013/02/electric-ferry/

Blume M 2019 Eviation Alice at Paris Air Show 2019 https://commons.wikimedia.org/wiki/File:Eviation_Alice,_Paris_Air_Show_2019,_Le_Bourget_(SIAE8856).jpg

Bona1858 2020 The James V. Glynn, one of two new all-electric, emission-free vessels, was placed into service by Maid of the Mist in October 2020 https://commons.wikimedia.org/wiki/File:Maid_of_the_Mist_11_5_20.png

Boslaugh S E 2020 'Maglev' Encyclopedia Britannica (12 May 2020) available at: www.britannica.com/technology/maglev-train

Cavernia 2019 «Ampere» i Sognefjorden https://commons.wikimedia.org/wiki/File:9683611_Ampere_2015.jpg

Cchhrriissttiiaann 2000 This is a photo of the world's first certified Fuel Cell Boat (HYDRA) realised by Christian Machens https://commons.wikimedia.org/wiki/File:Die_Hydra_in_Leipzig_I.jpg

Chris 73 2005 Linimo approaching Banpaku Kaijo Station https://commons.wikimedia.org/wiki/File:Linimo_approaching_Banpaku_Kinen_Koen,_towards_Fujigaoka_Station.jpg

Dieterich V, Buttler A, Hanel A, Spliethoff H and Fendt S 2020 Power-to-liquid via synthesis of methanol, DME or Fischer–Tropsch-fuels: a review *Energy Environ. Sci.* **13** 3207–52

DLR 2009 *Antares DLR-H2 Take-Off with the Power of Fuel Cells* (Cologne: Deutsches Zentrum für Luft- und Raumfahrt) available at: www.dlr.de/tt/en/Portaldata/41/Resources/dokumente/ec/Antares-DinA5-V6_E_WEB.pdf

DNV GL 2020 *Ammonia as a Marine Fuel* (Group Technology & Research, White paper) www.dnv.com/Publications/ammonia-as-a-marine-fuel-191385

Dorrington G, Baxter G, Bil C, Subic A and Trivailo P 2013 Prospects for liquefied natural gas and other alternative fuels for future civil air transportation *Proc. 15th Australian Int. Aerospace Cong.* 116–25

Dunlap R A 2019 *Sustainable Energy* 2nd edn (Boston, MA: Cengage)

EIA 2016 *International Energy Outlook 2016* DOE/EIA-0484(2016) May 2016 available at: www.eia.gov/outlooks/ieo/pdf/0484(2016).pdf

Fèvre F 2014 MPM76 n°33 sur la ligne 2 du métro de Marseille, à son terminus Bougainville https://commons.wikimedia.org/wiki/File:MPM76_n%C2%B033_RTM_Bougainville.jpg

Flint Bryan 2007 Union Pacific locomotive Y2315 https://commons.wikimedia.org/wiki/File:UNION_PACIFIC_Y2513_20070228.jpg

GliderMaven 2015 Standard 3 phase linear induction motor core and windings https://commons.wikimedia.org/wiki/File:Three_phase_linear_induction_motor.gif

Goldmann A, Sauter W, Oettinger M, Kluge T, Schröder U, Seume J R, Friedrichs J and Dinkelacker F 2018 A study on electrofuels in aviation *Energies* **11** 392

Grond L, Schulze P and Holstein J 2013 *Systems Analyses Power to Gas* (Groningen: KEMA Nederland B.V. Energieweg: Groningen) available at: https://web.archive.org/web/20150124021520/http://dnv.com/binaries/DNV%20KEMA%20(2013)%20-%20Systems%20Analyses%20Power%20to%20Gas%20-%20Technology%20Review_tcm4-567461.pdf

Harris M 2021 ZeroAvia's hydrogen fuel cell plane ambitions clouded by technical challenges—aspirations remain sky high *TechCrunch* (14 April 2021) available at: https://techcrunch.com/2021/04/14/zeroavias-hydrogen-fuel-cell-plane-ambitions-clouded-by-technical-challenges/

Henderson C 2021 The hydrogen revolution in the skies *BBC Future Planet* (7 April 2021) available at: www.bbc.com/future/article/20210401-the-worlds-first-commercial-hydrogen-plane

Hsieh C C and Felby C 2017 *Biofuels for the marine shipping sector—an overview and analysis of sector infrastructure, fuel technologies and regulations* (IEA Bioenergy, Task 39) available at: www.ieabioenergy.com/wp-content/uploads/2018/02/Marine-biofuel-report-final-Oct-2017.pdf

IATA 2015a *Fact Sheet #7: Liquid Hydrogen as a Potential Low-carbon Fuel for Aviation* (Montreal, Geneva: International Air Transport Association) available at: www.iata.org/contentassets/d13875e9ed784f75bac90f000760e998/fact_sheet7-hydrogen-fact-sheet_072020.pdf

IATA 2015b *Sustainable Aviation Fuel Roadmap* (Montreal, Geneva: International Air Transport Association) available at: www.iata.org/contentassets/d13875e9ed784f75bac90f000760e998/safr-1-2015.pdf

IATA 2021 *Our actions for the environment* (Montreal, Geneva: International Air Transport Association) www.iata.org/en/programs/environment/

IEA 2019 *The Future of Rail—Opportunities for Energy and the Environment* (Paris: International Energy Agency) available at: https://iea.blob.core.windows.net/assets/fb7dc9e4-d5ff-4a22-ac07-ef3ca73ac680/The_Future_of_Rail.pdf

Imhoff T B, Gkantonas S and Mastorakos E 2021 Analysing the performance of ammonia powertrains in the marine environment *Energies* **14** 7447

IMO 2015 *Third IMO GHG Study 2014* (London: International Maritime Organization) available at: https://web.archive.org/web/20151019064411/http://imo.org/en/OurWork/Environment/PollutionPrevention/AirPollution/Documents/Third%20Greenhouse%20Gas%20Study/GHG3%20Executive%20Summary%20and%20Report.pdf

IRENA and Methanol Institute 2021 *Innovation Outlook: Renewable Methanol* (International Abu Dhabi, UAE: Renewable Energy Agency) available at: www.methanol.org/wp-content/uploads/2020/04/IRENA_Innovation_Renewable_Methanol_2021.pdf

Kane M 2020 Maid of the mist Ferries are electric and truly zero-emission *InsideEVs* (27 October 2020) available at: https://insideevs.com/news/450985/maid-of-the-mist-ferries-electric/

Kane M 2021a Corvus energy to supply battery for all-electric Asahi bunker tanker *InsideEVs* (01 February 2021) available at: https://insideevs.com/news/484730/corvus-energy-battery-electric-asahi-tanker/?utm_source=msn.com&utm_medium=referral&utm_campaign=msn-feed

Kane M 2021b See 7 MWh electric container ship *Yara Birkeland* on maiden voyage *InsideEVs* (22 November 2021) available at: www.msn.com/en-ca/lifestyle/travel/news/see-7-mwh-electric-container-ship-yara-birkeland-on-maiden-voyage/ar-AAQYDwT?ocid=msedgntp

Kane M 2021c China builds the world's largest electric cruise ship *InsideEVs* (10 January 2021) available at: https://insideevs.com/news/465115/china-world-largest-electric-cruise-ship/

Kotrba R 2013 Name that biodiesel train contest *Biodiesel Magazine* (29 May 2013) available at: www.biodieselmagazine.com/blog/article/2013/05/name-that-biodiesel-train-contest

Kowalewicz A 1993 Methanol as a fuel for spark ignition engines: a review and analysis *Proc. Inst. Mech. Eng.* D **207** 43–52

Lehtveer M, Brynolf S and Grahn M 2019 What future for electrofuels in transport? Analysis of cost competitiveness in global climate mitigation *Environ. Sci. Technol.* **53** 1690–7

Lek S 2019 Het schip de Bit Viking in de Nieuwe Waterweg tussen Maassluis en Rozenburghttps://commons.wikimedia.org/wiki/File:Bit_Viking_Nieuwe_Waterweg_2019.jpg

Lpele 2016 Navire Enercon de transport d'éoliennes https://commons.wikimedia.org/wiki/File:Bateau_enercon_P9240983_part.jpg

MAN Energy Solution 2022 www.man-es.com/discover/sustainability-is-a-mindset

Marquard & Bahls 2015 'Marine fuels' available at: www.marquard-bahls.com/en/news-info/glossary/detail/term/marine-fuels-bunker-fuels.html

McGill R, Remley W and Winther K 2013 *Alternative Fuels for Marine Applications A Report from the IEA Advanced Motor Fuels Implementing Agreement, Annex 41* (Paris: International Energy Agency) available at: www.iea-amf.org/app/webroot/files/file/Annex%20Reports/AMF_Annex_41.pdf

MI News Network 2021 'Sea cargo ship to be world's first vessel with tilting rotor sails arrives in Rotterdam' (14 January 2021) available at: www.marineinsight.com/shipping-news/sea-cargo-ship-to-be-worlds-first-vessel-with-tilting-rotor-sails-arrives-in-rotterdam/

Moralapostel 2006 Linearmotorgetriebene Magnetschwebebahn in Langstatorbauweise (Transrapid) https://commons.wikimedia.org/wiki/File:Magnetschwebebahn.svg

Murdockcrc 2011 An OEBB 1016 Taurus locomotivehttps://commons.wikimedia.org/wiki/File:OEBB_1016_LCD.jpg

Nautical Photo Agency (n.d.) Preussen (Ship) in full sail (State Library of Queensland) https://collections.slq.qld.gov.au/viewer/IE298608

Needham A 2006 A maglev train is coming out of the Pudong International Airport https://commons.wikimedia.org/wiki/File:A_maglev_train_coming_out,_Pudong_International_Airport,_Shanghai.jpg

Novoklimov 2006 ЧМЭ3 (ChME3,ČME3) diesel-electric locomotive https://commons.wikimedia.org/wiki/File:%D0%A7%D0%9C%D0%AD3_(ChME3,%C4%8CME3).jpg

Patel S 2021 'Mitsubishi power developing 100% ammonia-capable gas turbine' *Power—News & Technology for the Global Energy Industry* (3 March 2021) www.powermag.com/mitsubishi-power-developing-100-ammonia-capable-gas-turbine/

Penner J E, Lister D H, Griggs D J, Dokken D J and McFarland M (ed) 1999 *Aviation and the Global Atmosphere* (Geneva: IPCC)

Pépellin F 2019 Sight of the French SNCF Intercités train n° 3430 from Granville (Normandy) to Paris, stopped at Argentan station, in Orne https://commons.wikimedia.org/wiki/File:R%C3%A9giolis_Granville-Paris_en_gare_d%27Argentan_(ao%C3%BBt_2019).JPG

Pipistrel 2021 www.pipistrel.ad/pipistrel/entrainement/velis-electro

Rdurkacz 2013 Sketch of Magnus effect with streamlines and turbulent wake https://commons.wikimedia.org/wiki/File:Sketch_of_Magnus_effect_with_streamlines_and_turbulent_wake.svg

Rsa 2011 JR East KuMoYa E995-1 NE Train Smart Denchi-kun https://commons.wikimedia.org/wiki/File:JR_East_E995-1_omiya_20111015.jpg

Sharma G 2017 Stena Germanica https://commons.wikimedia.org/wiki/File:GS001962.jpg

Sieker B 2014 Airbus' all-new full-electric technology demonstrator in flight at ILA 2014 Berlin Air Show https://www.flickr.com/photos/98529529@N00/14088845198

Shell 2021 'Aviation fuel' available at: www.shell.com/business-customers/aviation/aviation-fuel.html

tansaisuketti 2010 N700 series Shinkansen in front of Mt. Fuji https://commons.wikimedia.org/wiki/File:Shinkansen_N700_with_Mount_Fuji.jpg

Tidey A 2020 Airbus unveils concepts for zero-emission planes powered by hydrogen *Euronews* (23 September 2020) available at: www.euronews.com/2020/09/21/airbus-unveils-concepts-for-zero-emission-planes-powered-by-hydrogen

Transport & Environment 2016 *Globiom: The Basis for Biofuel Policy Post-2020* (Brussels, Belgium: Transport & Environment) available at: https://www.transportenvironment.org/wp-content/uploads/2021/07/2016_04_TE_Globiom_paper_FINAL_0.pdf

Transport & Environment 2017 *What role is there for electrofuel technologies in European transport's low carbon future?* (Brussels, Belgium: Transport & Environment) available at: https://www.transportenvironment.org/wp-content/uploads/2021/07/2017_11_Cerulogy_study_What_role_electrofuels_final_0.pdf

UPI 2009 'Disneyland trains running on biodiesel' *UPI* (29 January 2009) available at: www.upi.com/Top_News/2009/01/29/Disneyland-trains-running-on-biodiesel/10151233252145/?ur3=1

Victor-ny 2008 The USS Nautilus https://commons.wikimedia.org/wiki/File:USS_Nautilus_SSN571.JPG

Walker T R *et al* 2019 Environmental effects of marine transportation *World Seas: An Environmental Evaluation (Second Edition) Volume III: Ecological Issues and Environmental Impacts* (Cambridge, MA: Academic), ch 27 pp 505–30

Wärtsilä Corporation 2012 Case: tanker ship Bit Viking LNG conversion for environmental sustainability. Available at: https://cdn.wartsila.com/docs/default-source/services-documents/learning-center/references/services-reference-tarbit-shipping.pdf?sfvrsn=600ee245_0

Wärtsilä Corporation 2020 World's first full scale ammonia engine test—an important step towards carbon free shipping www.wartsila.com/media/news/30-06-2020-world-s-first-full-scale-ammonia-engine-test-an-important-step-towards-carbon-free-shipping-2737809

Wilson A 2012 Tupolev Tu-155 (Tu-154) 'CCCP-85035' https://www.flickr.com/photos/ajw1970/8601902283

World Nuclear Association 2021 Nuclear powered ships. Available at: https://world-nuclear.org/information-library/non-power-nuclear-applications/transport/nuclear-powered-ships.aspx

# Chapter 6

# The future of transportation

## 6.1 Introduction

Mitigating climate change caused by the anthropogenic emission of greenhouse gases is certainly one of the greatest challenges facing humanity. These greenhouse gases, particularly carbon dioxide, come from a variety of anthropogenic sources. Figure 6.1 shows a breakdown of the greenhouse gases produced by human activities. Transportation is certainly a major component (at 16.2% of the total) of global emissions, and any plan that intends to mitigate the human contributions to climate change must particularly address transportation energy. Based on the information provided in chapter 1, figure 6.2 shows that: (a) among transport modes, road transportation is, by far, the most significant consumer of primary energy, and (b) light-duty vehicles are the single most significant component of transportation energy.

The sources of world transportation energy are described in figure 6.3, which includes a projection for the next two decades or so from the United States Energy Information Agency. It is clear from the figure as well as the previous discussions in chapters 2–5 that gasoline and diesel fuel are, by far, the most common sources of transportation energy, followed by jet fuel. In fact, in a sense, our transportation energy system in the past has been very simple. Virtually all transportation energy came from the same source, i.e. petroleum that was distilled to produce hydrocarbon fuels with different numbers of carbon atoms per molecule as appropriate for different applications. It is probable that in the future, a more diverse approach to transportation energy will be required to fulfill all our needs.

Renewable energy sources, as shown in figure 6.3, are only a small fraction of total transportation energy at present and are predicted to remain so in the near future. The electricity category includes both battery-operated vehicles as well as those that obtain electricity from an external source, such as a third rail or a catenary. Other liquids include biofuels such as ethanol and biodiesel. Biofuels are most commonly utilized as a component (up to 10%) in blends with fossil fuels (either gasoline or diesel) for road vehicles. Figure 6.4 shows the global greenhouse

doi:10.1088/978-0-7503-5306-9ch6  

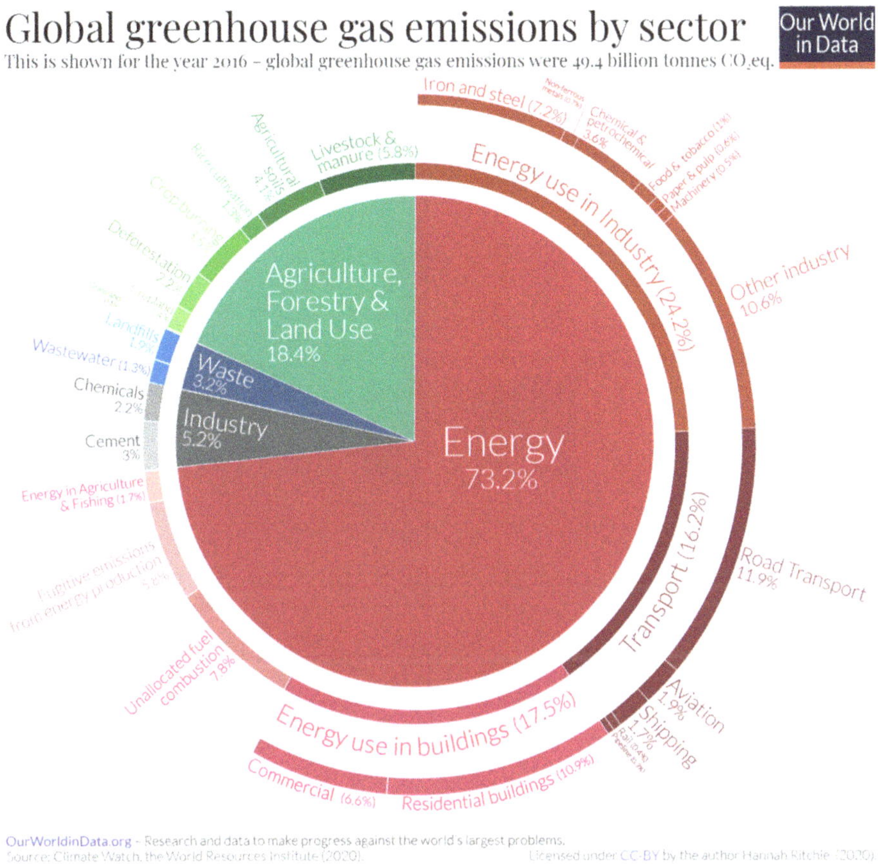


**Figure 6.1.** Breakdown of world greenhouse gas emissions by sector for 2016. Reproduced from Ritchie (2020). CC BY 4.0.

gas emissions for the period 1990–2019. Overall, there has been a consistent increase in transportation greenhouse gases, with the result that the current emissions are nearly twice those of thirty years ago. For road vehicles, the result of increased vehicle efficiency, the utilization of biofuel blends, and the increasing numbers of electric vehicles (hybrids, plug-in hybrids, and battery electric vehicles) has been more than balanced by an increase in the total number of vehicles on the roads, or more specifically, the total number of kilometers driven.

This unfortunate trend provides an incentive to look carefully at our transportation energy use and find strategies that will mitigate the effects of greenhouse gas emissions. The remainder of this chapter provides a summary of the technologies that have been discussed in previous chapters and an analysis of their suitability for providing satisfactory transportation services while simultaneously addressing environmental issues. Before we can assess these various energy technologies and establish their viability, it is beneficial to look in more detail at how transportation requirements fit into our general energy needs.

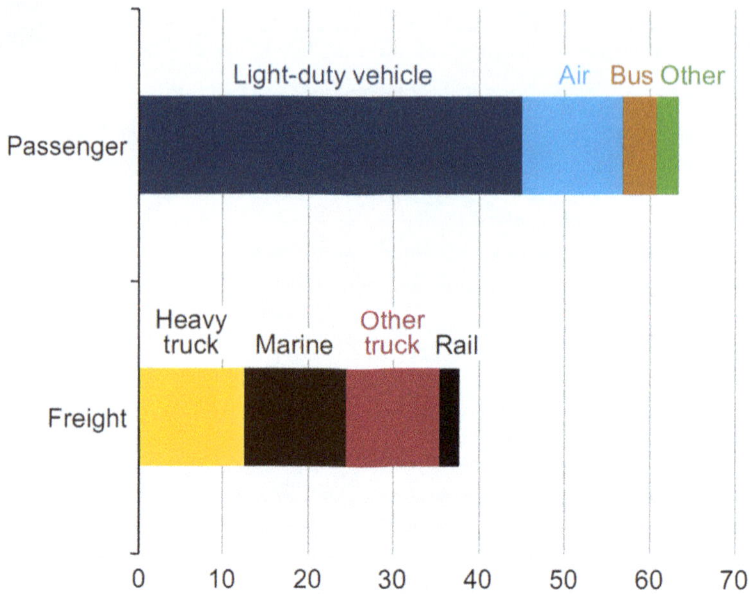


**Figure 6.2.** World transportation energy as a function of transportation mode in 2012. Reproduced from EIA (2016). Image stated to be in the public domain.

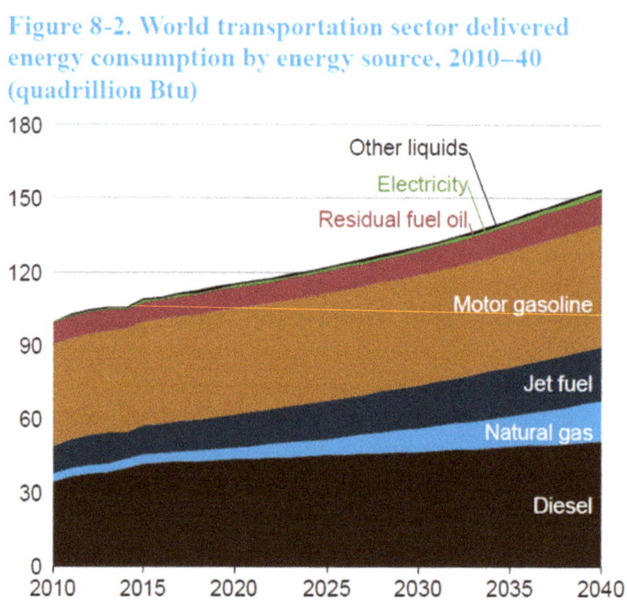


**Figure 6.3.** World transportation energy by source from 2010 and projected to 2040. Reproduced from EIA (2016). Image stated to be in the public domain.

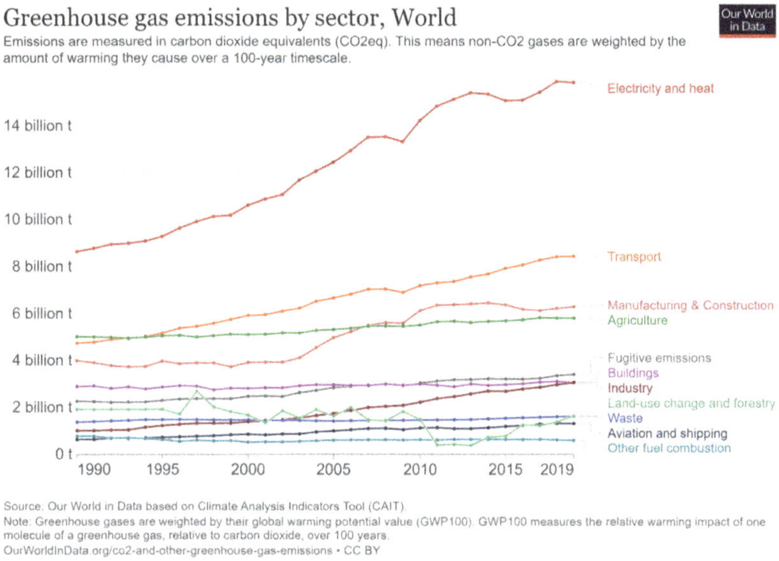


**Figure 6.4.** Global greenhouse gas emissions broken down by sector. Reproduced from Our World in Data (2022a). CC BY 4.0.

## 6.2 What we need to do

As discussed in chapter 1, it is essential that we decrease our dependence on fossil fuels and minimize greenhouse gas emissions by transitioning to low-carbon energy sources. While the depletion of fossil fuel resources will ultimately require this transition, it is important reduce carbon emissions as soon as possible in order to mitigate their adverse environmental effects. Determining the most appropriate plan of action requires us to ask two questions: What do we need to do? And how fast do we need to do it? This may be viewed in another way by considering the following points:

What is the ideal course of action?
What is an adequate course of action?
What is a feasible course of action?

The first question is, perhaps, the easiest to answer. It would have been ideal if society had had the insight fifty years ago to realize the severity of greenhouse gas emissions and had been proactive in reducing fossil fuel consumption at that time. Although a concern about the future availability of fossil fuels was raised in the early 1970s, as long as they were plentiful and inexpensive, there was little motivation to move to energy sources that were more expensive and less convenient.

Determining an adequate approach to deal with climate change is difficult. Scientifically, the results of our actions are not well defined. Figure 6.5 shows the results of models for different approaches to greenhouse gas emissions and indicates the ranges of annual global greenhouse gas emissions and mean global temperature changes. It is often assumed that it is necessary to limit the global temperature

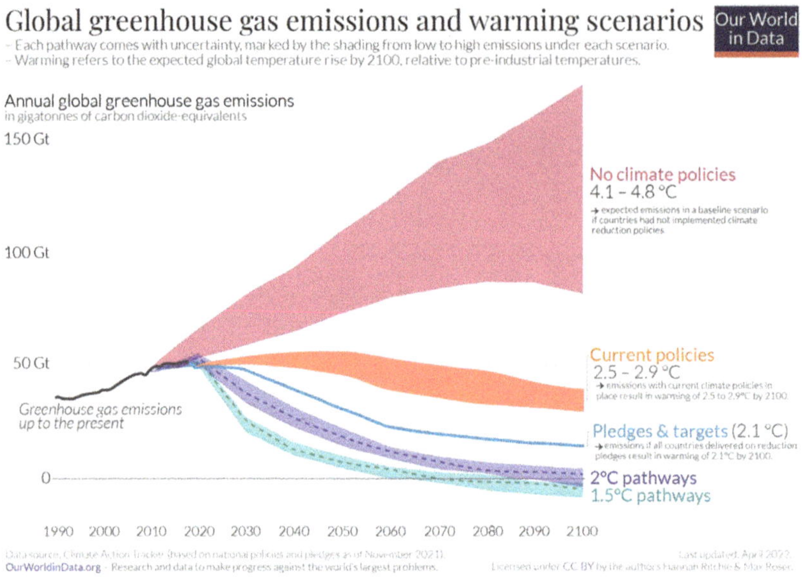


**Figure 6.5.** Projected global greenhouse gas emissions and resulting global temperature changes for different scenarios. The vertical axis shows the annual greenhouse gas emissions in units of $10^9$ tonnes of carbon dioxide equivalent. Thus, all greenhouse gases are included and are scaled by their effect relative to that of carbon dioxide as given in table 1.7. Reproduced from Ritchie *et al* (2020). CC BY 4.0.

increase to 2 °C above preindustrial levels in order to minimize severe adverse effects on humanity that result from conditions such as sea-level increase, drought, severe heat waves, and extreme weather conditions. Recent studies (IPCC 2018), however, have suggested that it may be necessary to limit the temperature increase to 1.5 °C in order to avoid unacceptable environmental changes. Figure 1.8 illustrates that global temperatures are already about 1.5 °C above the 1880 value; certainly, the adverse effects of climate change are already apparent.

Figure 6.5 shows that inaction will result in global temperature increases of 4.1 °C to 4.8 °C by the end of this century. Current polices will lead to temperature increases in the range of 2.7 °C to 3.1 °C. The Paris Agreement aims to address this issue and to implement a roadmap for future greenhouse gas emissions. Some analyses, as shown in figure 6.5, suggest that even the current 'pledges and targets' scenario as specified by the Paris Agreement misses the 2 °C mark and leads to a 2.4 °C increase over preindustrial global temperatures.

This brings us back to the question of what is feasible. Feasibility may be viewed from a technical standpoint, an economic standpoint, or a sociopolitical standpoint, and an analysis of what is actually feasible is a very complex task with a great deal of uncertainty. It is not known whether meeting 'pledges and targets' is a realistic expectation; neither is it known whether what we can actually accomplish is consistent with what is adequate.

Limiting the global temperature increase to 1.5 °C by the end of the century may not be ideal, but it may be the best we can do going forward under the present circumstances. As figure 6.5 shows, limiting global warming to 1.5 °C would require eliminating all net greenhouse gas emissions by 2070 followed by sequestering existing atmospheric greenhouse gas beyond that point. Even the 2 °C limit would require eliminating about 90% of greenhouse gas emissions by 2070.

Of course, eliminating all net greenhouse gas emissions might be achieved by a combination of reducing actual emissions combined with the sequestration of some existing atmospheric gases such as carbon dioxide. However, we can take a simple approach to dealing with the feasibility question by just looking at the carbon dioxide emissions that result from fossil fuel consumption and the actions necessary to replace all the energy currently produced using fossil fuels with energy from carbon-free sources (or at least low-carbon sources) by 2070. We can see the present (and past) distribution of global primary energy sources in figure 6.6. This figure shows the so-called 'substitution' primary energy. In this method of calculating the primary energy, the energy of fossil fuel sources is scaled in order to take into account the inherent lack of efficiency of their utilization. Since fossil fuels are inevitably burned to produce heat, such heat, unless it is used directly for (e.g.) space heating, is converted into mechanical energy through a heat engine. This conversion may take place in an power station, in which case the efficiency would be around 40%, or in an internal combustion engine, in which case the efficiency would be around 20%. On the other hand, renewable electricity from (say) a wind turbine can be used to power an electric vehicle at an efficiency of around 85%. (This characteristic of energy conversions is discussed in more detail in the context of transportation in section 6.3.) Thus, the substitution primary energy of low-efficiency fossil fuels is determined in

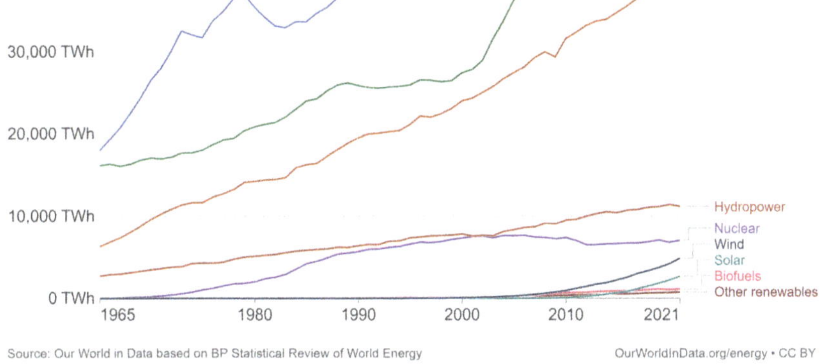


**Figure 6.6.** Primary world energy consumption from 1965 to 2021 for energy obtained from different sources. Reproduced from Our World in Data (2022b). CC BY 4.0.

such a way as to provide a direct measure of the amount of greenhouse gas emissions. The relationship between the distribution of primary energy sources as determined by the substitution method is compared to a direct measurement in figure 6.7.

Figure 6.6 illustrates several relevant features of our energy use. First, overall energy use has increased substantially over the past half century. In fact, adding the various energy components shows that the total energy has increased from around 40 000 TWh in 1965 to around 160 000 TWh in 2020. Despite the impressive increase (relatively speaking) in the use of wind energy (see figure 1.16) and solar power (see figure 1.17) over this period, the largest absolute increases (by far) have been for the fossil fuels—coal, oil, and natural gas. It is, however, interesting to note the decrease in fossil fuel use in 2020 (compared to 2019), particularly for oil. This decrease resulted from the decrease in the use of (primarily) transportation energy that was caused by the COVID-19 pandemic.

In order to understand the viability of transitioning all of our energy use from fossil fuels to low-carbon sources by (say) 2070, it is necessary to look at the infrastructure that must be developed. Figure 1.4 shows a model of projected primary energy use for the remainder of the 21st century. The most likely scenario is that the annual energy consumption in 2070 will be about twice the current consumption, i.e. about 1200 EJ (or about 330 000 TWh). A further discussion of the relationship between transportation technologies and overall energy needs is presented in the next section. In any case, low-carbon energy sources will need to replace the current fossil fuel primary energy (i.e. about 84% of the substitution primary energy, as shown in figure 6.6) and also provide the additional annual energy needed by 2070. The total new low-carbon infrastructure will, therefore, need to provide about 300 000 TWh (equivalent to 1080 EJ) per year.

It is more convenient to view this low-carbon energy in terms of the necessary generating capacity. Dividing 300 000 TWh by the number of hours in a year (8760 h $y^{-1}$) gives an average power of 34 TW. We can now look at some

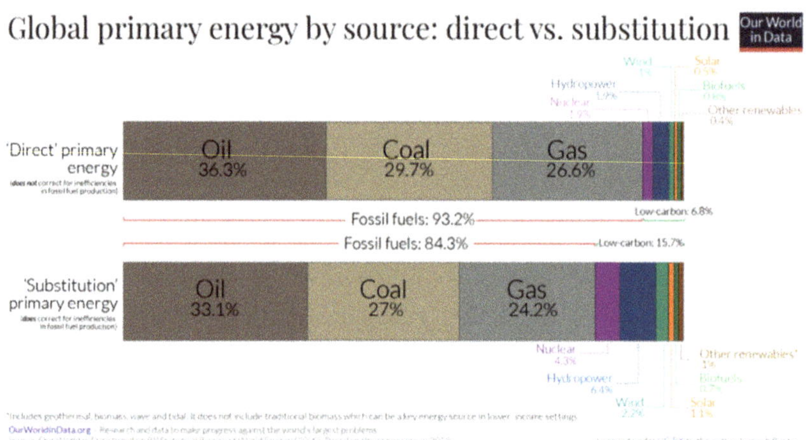


**Figure 6.7.** Comparison of direct primary energy use and primary energy use calculated by the 'substitution' method as described in the text. Reproduced from Ritchie (2021b). CC BY 4.0.

potential low-carbon energy sources and determine how much new infrastructure would need to be developed in order to meet this requirement. Note that this simple approach does not include the need to replace aging infrastructure at the end of its useable lifetime. In addition, if energy growth is required after 2070, additional low-carbon infrastructure will need to continue to be constructed, as needed. This is analogous to today's situation, in which additional energy requirements that largely result from industrial growth in Asia (see figure 1.3) are mainly accommodated by the construction of coal-fired power stations in China. Here, we look quantitively at the major low-carbon energy sources.

### 6.2.1 Solar energy

Based on the information concerning solar photovoltaics from chapter 1, we can consider a typical installation for which the annual average insolation is 200 W $m^{-2}$. For a typical photovoltaic efficiency of 20% for the conversion of solar radiation to electricity, the average electrical output is 40 W $m^{-2}$. To provide 34 TW by 2070, a total of about 850 000 $km^2$ of photovoltaic collectors would need to be constructed. To reach that goal, about 50 $km^2$ of collectors would need to be installed every day for the next 50 years. Recalling figure 1.20, which shows a 5 $km^2$ photovoltaic installation, gives an idea of the magnitude of this task.

### 6.2.2 Wind energy

A typical 2 MW wind turbine operating with a capacity factor of 30% would have a net output of 600 kW. About 57 million such turbines would be needed to provide 34 TW. Over the next 50 years, this would amount to the construction of about 3000 turbines per day. The current growth, as shown in figure 1.16, amounts to the equivalent of about 70 new 2 MW turbines per day.

### 6.2.3 Nuclear fission

A typical nuclear power station produces an output of around 1 GW of electricity. In order to meet the requirement of 34 TW, 34 000 reactors would be needed. This amounts to the commissioning of about two new reactors per day for the next 50 years. This requirement can be compared to the start of construction of about 35 new reactors per year during the maximum period of nuclear growth in the late 1960s. It is also relevant to note that current fission reactors typically require 5–10 years to construct.

### 6.2.4 Our future energy mix

Our future sources of energy are likely to be more diverse than those we have at present. Of the three energy sources noted above, solar, wind, and nuclear fission, only solar energy has the capability to provide all of our energy on a renewable basis (see table 6.1), and this source of energy will likely dominate our energy mix. The resources required for the development of wind energy are not sufficient to accommodate all of our energy requirements, as calculated above. In fact, over the next half century, wind will not likely be able to contribute more than 15% of our

**Table 6.1.** Approximate percentage of global power required by 2070 that can be provided by various renewable sources. Data adapted from Dunlap (2021).

| Source | % |
|---|---|
| Solar | >100 |
| Wind | 15 |
| Hydroelectric | 8 |
| Geothermal | 8 |

energy, as noted in table 6.1. Other renewable sources, particularly hydroelectric and geothermal, are currently important and will likely continue to be a significant component of our energy mix in the future. As discussed in chapter 1, nuclear resources are limited, and a variety of factors related to perceived safety, waste disposal, and infrastructure costs have been responsible for limited nuclear growth since around 1990. Thus, unless advanced reactors, including fast breeder reactors and thorium reactors, are commercially developed, it is unlikely that any significant nuclear power growth will occur in the foreseeable future.

The above discussion shows that electricity, however it is produced, will be the major form of energy that will supply our needs. In addition, liquid (or perhaps gaseous) fuels are likely to be appropriate for certain transportation applications. These fuels will either directly or indirectly rely on electricity. In the first case, electrofuels such as hydrogen and various derivatives of hydrogen (e.g. e-methanol) are merely electrical energy storage media, and are, in that sense, analogous to batteries. In the second case, biofuels are derived from biomass, either from dedicated sources such as corn used to produce ethanol or from waste material, such as agricultural waste, that is used to produce biosyngas. While the energy content of biofuels is derived from the solar energy that produces the biomass feedstock, electricity will be essential for much of the agricultural production and the fuel processing and distribution that currently depend on fossil fuels.

## 6.3 Vehicle efficiency analysis and grid capacity

Since about 95% of transportation energy currently comes from fossil fuels (table 1.4), it places relatively little demand on the electric grid. Electricity is used for some aspects of our transportation energy system, such as refining crude petroleum products and operating distribution stations, but the electrical energy used for these activities is small compared to the energy content of the fuels that are used for transportation. In the case of alternative carbon-free energy sources for transportation, this is not necessarily the case. In most scenarios described above for different transportation modes, electricity plays an important role. As the world transitions to more environmentally conscious methods of generating electricity by developing greater carbon-free capacity using sources such as wind power and solar power, it is necessary to realize that additional capacity will be required to accommodate the energy used by sectors that primarily use fossil fuel energy at present. Thus, although our overall primary energy use may remain more or less the same or may even decrease (as discussed further below), the relative electrical generating capacity will have to increase in order to keep up with the increased electricity needed for transportation. This is also true of other sectors besides transportation, for example, space heating and industrial processes that now rely heavily on fossil fuels. However, for the purpose of this discussion, we can confine our approach to the consideration of transportation energy only, while realizing that this is only part of the future situation. In order to understand the future requirements for electricity generation, it is necessary to carefully analyze the method by which transportation energy is obtained. While a complete analysis of this problem is far from simple or straightforward, we can get an idea of the effects of future modes of transportation on electrical generating capacity by looking at some simple examples.

We begin with the current total primary energy used by society, as given in chapter 1, of $5.8 \times 10^{20}$ J per year. Of this amount, about 25% (or about $1.45 \times 10^{20}$ J per year) (EIA 2017) is used for transportation in one form or another. Since transportation is almost entirely based on the combustion of some form of fossil fuel in a heat engine (see chapter 1), the net overall efficiency of converting this primary energy to mechanical energy for propulsion is about 20%. Thus, the net propulsive energy required to satisfy our transportation needs is about $0.2 \times 1.45 \times 10^{20} = 2.9 \times 10^{19}$ J per year.

Recent world electricity production and its sources are illustrated in figure 6.8. This figure shows a current annual production of about 26 000 TWh, which can be

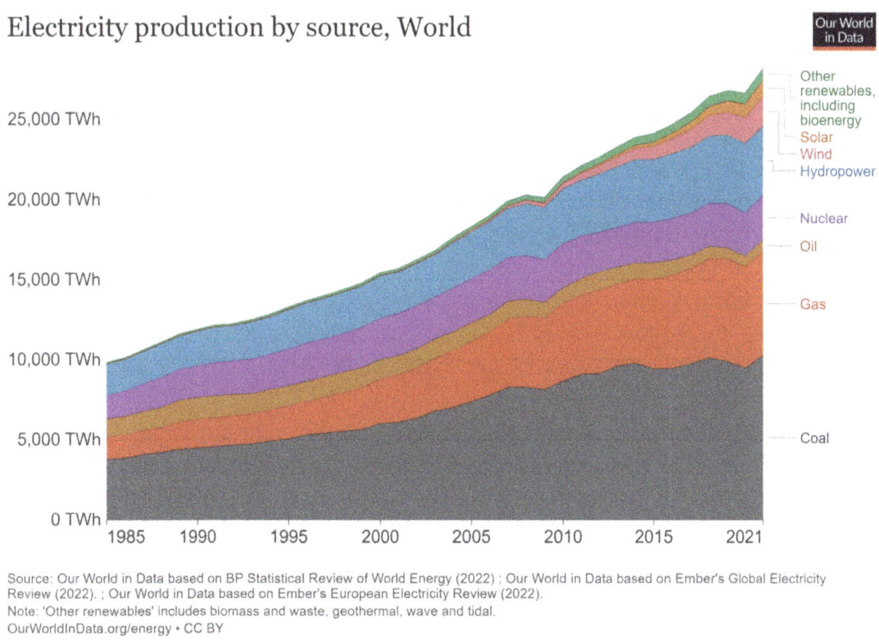


**Figure 6.8.** World electricity production and a breakdown of its sources from 1985 to 2020. Reproduced from Our World in Data (2022c). CC BY 4.0.

converted to $9.4 \times 10^{19}$ J. It is important to note that this value is the net electricity generated. By comparison, the primary energy used to generate this electricity is about a factor of two larger because of the efficiency of electricity production from primary energy sources. It is also important to note that this is the net electricity generated, which is less than the total that could be generated, because the capacity factor is less than 100%. Therefore, the total electricity generated could be increased now by increasing the capacity factor as discussed further below. While this could increase our ability to sustain alternative vehicle technologies from (primarily) fossil-fuel-generated electricity, it would not affect our future needs for net electricity generated, as discussed below.

From the above information, we can see that the net transportation energy requirements are about 30% of the current electricity generation. In order to provide meaningful information, we need to consider the details of the mode of transportation and the efficiency of converting electrical energy into mechanical propulsion energy. In general, this is not easy. However, as road transportation accounts for about 85% of transportation energy use (about $2.5 \times 10^{19}$ J per year) as seen in table 1.3 and in figure 6.2, an analysis of this mode of transportation will provide a very useful example.

Some of the more important approaches to possible future road transportation are summarized in figure 6.9. This summary includes all major methods of providing energy for road transportation as described in chapters 2 and 3, but does not deal with biofuels, as these are somewhat more difficult to analyze quantitatively. However, the analysis shown in figure 6.9 deals with battery electric vehicles, hydrogen-powered vehicles, and vehicles utilizing e-methanol.

Figure 6.9 shows a flowchart of the processes involved in each case. For each process in the figure, the typical efficiencies are given (as a percentage), as discussed in previous chapters. In order to determine the overall electrical energy required to yield one unit of transportation energy, the overall efficiency of the process is determined as the product of the individual efficiencies (expressed as a fraction). Table 6.2 shows the net efficiencies for each of these technologies. The values in the table can be put in context by comparing them with the typical efficiency of a gasoline internal combustion engine vehicle, which is around 20%. Viewing these technologies from a strict efficiency standpoint, only battery electric vehicles and, to a lesser extent, hydrogen fuel cells, offer an efficiency advantage over gasoline-powered vehicles, while the other technologies in the table offer similar or worse net efficiencies.

If we make the assumption that the $2.5 \times 10^{19}$ J per year for road transportation that is mostly provided by fossil fuel combustion is replaced using electricity through one of the approaches shown in table 6.2, then we can determine how much additional electricity needs to be generated. This assumption, on the one hand, ignores the small contribution made by electricity (for battery vehicles) and biofuels (for bioethanol or biodiesel blending) to today's road transportation, and is, therefore, an overestimate. On the other hand, it ignores the fact that additional electrical capacity needed to account for the electrification of other fossil fuel uses, such as other modes of transportation, as well as heating and industrial processes,

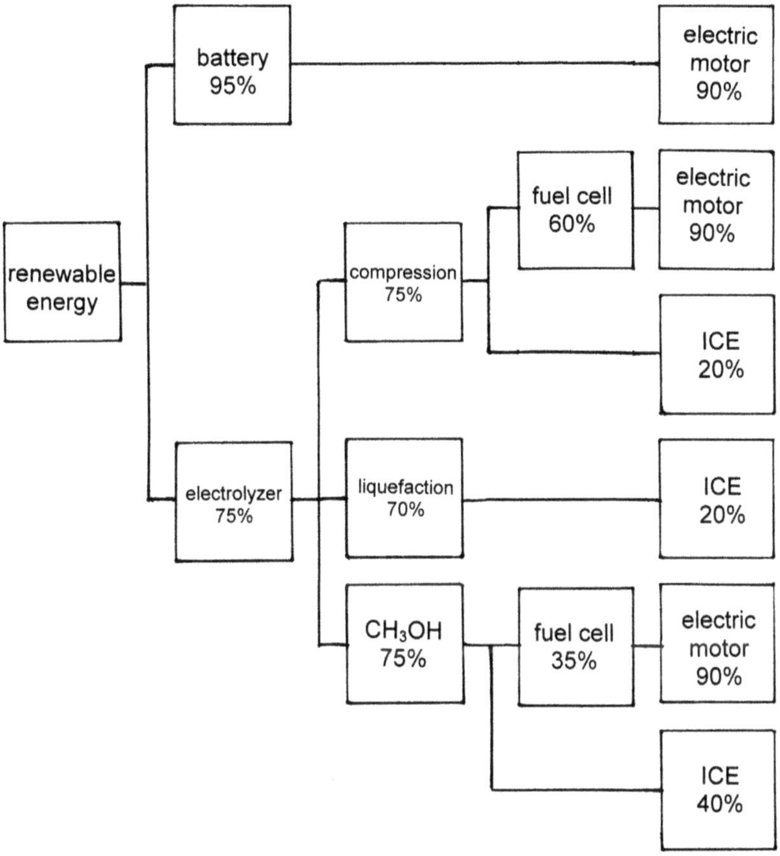


**Figure 6.9.** Approaches to the use of electricity produced by renewable sources for road transportation. The efficiency of each process is given (ICE = internal combustion engine). From the top, the technologies represented in the right-hand column in the figure are: (1) battery electric vehicle, (2) hydrogen fuel cell vehicle using compressed hydrogen, (3) hydrogen internal combustion engine vehicle using compressed hydrogen, (4) hydrogen internal combustion engine vehicle using liquefied hydrogen, (5) methanol fuel cell vehicle, and (6) methanol internal combustion vehicle.

and is, therefore, an underestimate. Since the latter factor is probably more significant than the former, the net result of our simple calculation is likely to be a bit of an underestimate.

The additional electricity is determined from the total energy needed for propulsion divided by the efficiency of converting electricity to mechanical energy (expressed as a fraction). The necessary additional new production is expressed as a percentage of the current production in the final column of table 6.2. The table shows that the necessary increase ranges from a fairly modest value (e.g. 31% for battery electric vehicles) to very substantial (e.g. internal combustion hydrogen vehicles), for which the total required production would be nearly four times the current production. In between these extremes, hydrogen fuel cell and

**Table 6.2.** Net efficiencies of different carbon-free road vehicle technologies as obtained from figure 6.9 and the new annual electricity production relative to the current production required to replace the use of fossil fuels for global road transportation. As noted above, the current annual production is $9.4 \times 10^{19}$ J.

| Technology | Net efficiency (%) | Additional annual production needed ($10^{19}$ MJ $y^{-1}$) | Additional production required as a percentage of current production |
|---|---|---|---|
| Battery electric vehicle | 85 | 2.94 | 31% |
| Hydrogen fuel cell vehicle using compressed hydrogen | 30 | 8.33 | 89% |
| Hydrogen internal combustion engine vehicle using compressed hydrogen | 11 | 22.7 | 241% |
| Hydrogen internal combustion engine vehicle using liquefied hydrogen | 10 | 25.0 | 266% |
| Methanol fuel cell vehicle | 18 | 13.9 | 148% |
| Methanol internal combustion vehicle | 22 | 11.4 | 121% |

methanol internal combustion engine technologies approximately double the necessary electricity production.

It is important to note that the above analysis is based on electricity production and not on generating capacity. Production is determined by the capacity and the capacity factor. It has been argued that our current grid infrastructure can accept a sufficient increase in capacity factor without the need to increase the total capacity to accommodate the conversion of a substantial fraction of our existing fossil fuel road transportation infrastructure to battery electric vehicles (Engel *et al* 2018). This approach is largely possible because facilities such as coal-fired thermal generating stations run at less than their maximum capacity for much of the time due to decreased demand during periods such as nights and weekends. Increasing the capacity factor during these low-demand periods can significantly increase overall production. While this may be a suitable approach to the short-term growth of battery electric vehicles, it is not sustainable in the long term. Ultimately, fossil fuel generating facilities will need to be replaced by sustainable low-carbon sources. The total amount of electricity produced will depend on the total capacity as well as the capacity factor. The efficient utilization of sustainable electricity sources along with adequate energy storage will need to accommodate the total electricity needs as described above.

In this simple approach to estimating the effect of transportation energy needs on the required electric grid capacity, it is obvious that the efficiency of the transportation technology is a crucial factor. From the table, we can better appreciate the importance of transportation efficiency and the need to make informed choices about transportation technology. This model omits a number of details related to transportation energy needs and it also underestimates the additional generating capacity needed in other energy utilization sectors, such as industry and space heating. However, it does emphasize the importance of energy conversion efficiencies in optimizing our ability to mitigate climate change. Since energy for road transportation is the largest component of transportation energy use, the above analysis makes a strong case for utilizing battery electric road vehicles to the greatest extent possible.

Although most sustainable vehicle technologies rely on electricity as an energy carrier and therefore require an increase in overall electrical generating capacity, they actually result in an overall decrease in total primary energy requirements because they offer greater efficiency than current transportation technologies. Following the same lines as the analysis of grid capacity given above, we can look at the effects of transportation technology on our overall primary energy needs. As an example, we consider the situation in which road transportation utilizes battery electric vehicles, and we consider two electrical generating extremes: (a) the case in which all electricity is produced by thermal generation using coal and (b) the case in which all electricity is produced using renewable energy, such as wind.

Figure 6.10 shows a flowchart of energy conversion efficiencies for the two cases mentioned above and a comparison with conventional gasoline internal combustion vehicles. The chart shows the energy at each step of the relevant conversion process, starting with primary energy, that will ultimately provide 1.0 MJ of energy to the vehicle's wheels. From tables 1.2 and 1.3, we use the typical values of 23.9% of primary energy for transportation and 84.6% of transportation energy for road

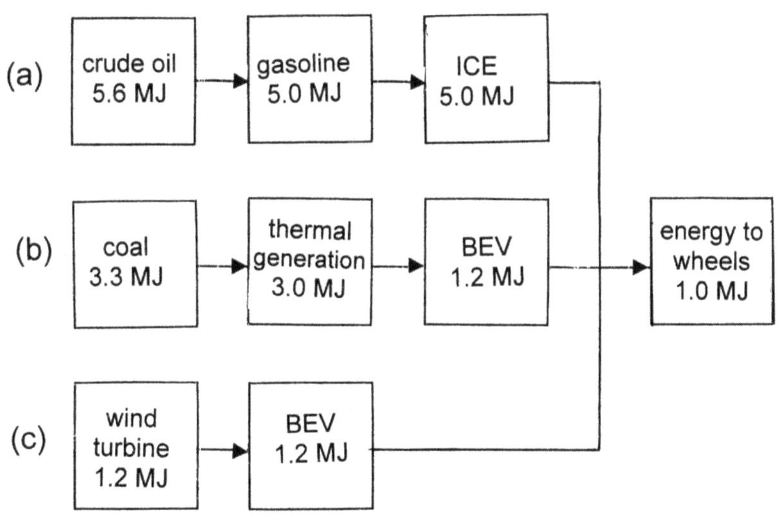


**Figure 6.10.** Energy flowchart for several approaches to vehicle propulsion: (a) traditional gasoline internal combustion engine (ICE), (b) battery electric vehicle (BEV) using electricity generated by a coal-fired plant, and (c) battery electric vehicle (BEV) using electricity generated from a wind turbine. The energy that is necessary to produce 1.0 MJ at the wheels is given at each stage.

transportation. Thus, (23.9%) × (0.846) = 20.2% of primary energy is used for road transportation. Figure 6.10 shows that the fractional reductions in primary energy that result from the replacement of gasoline vehicles with battery electric vehicles utilizing electricity generated from fossil fuels and from renewable sources are 100 × (5.6 MJ − 3.3 MJ)/(5.6 MJ) = 41% and 100 × (5.6 MJ − 1.2 MJ)/(5.6 MJ) = 79%, respectively. Since these reductions relate to 20.2% of primary energy use, the total reductions in primary energy for these two cases are (0.41) × (20.2%) = 8.3% for fossil-fuel-generated electricity and (0.79) × (20.2%) = 16.0% for renewably generated electricity.

Following the same lines as the analysis above, we can look at a less beneficial approach to transportation technology, namely the hydrogen internal combustion engine, as shown in table 6.2, and how it would affect our overall primary energy use. Using the information above, it can be shown that the net overall efficiency of the conversion of primary energy to energy at the wheels of a hydrogen internal combustion engine vehicle (using liquefied hydrogen) is 3.8% using electricity generated from coal and 10.5% using electricity generated from wind. This means, following the same lines as figure 6.10 for these two cases, that 26 MJ and 9.5 MJ of primary energy, respectively, are required to yield 1 MJ of energy at the wheels. Both of these values are greater than the amount of primary energy shown in figure 6.10 for the gasoline internal combustion engine vehicle. The net result of this analysis is that both of these approaches to vehicle propulsion increase the total primary energy requirement. In the case of fossil-fuel-generated electricity, the increase is 74%, and in the case of wind-generated electricity, the increase is 14%.

The above example again emphasizes the need to choose energy technologies that provide reasonable efficiencies not only for transportation but for all energy applications. This is necessary in order to minimize greenhouse gas emissions, as long as fossil fuels are used for electricity generation, and to facilitate our ability to develop the necessary energy infrastructure to eventually transition to renewable sources.

The discussion thus far has concentrated on the short-term consequences of different transportation technologies. As noted in section 6.2, the longer-term goal of mitigating climate change is an essential aspect of any future planning, and the future of electricity generation is at the forefront of such development. With regard to sustainable transportation, the additional generating capacity required to provide power for battery electric vehicles does not seem excessive; however, we need to look at the task ahead for electricity generation in general. Thus, as we go forward, there is a need to increase the electrical generating capacity of renewable sources for three reasons:

1. to increase overall generating capacity as a result of increased industrialization in underdeveloped countries;
2. to replace existing fossil fuel generation in order to mitigate climate change; and
3. to increase capacity in order to replace fossil fuel utilization in sectors such as transportation.

In the first case, the increases are exemplified by the trend shown in figures 6.6 and 6.8, which has been described in some detail in section 1.5. In the second case, renewable electricity has increased substantially in the past two or three decades, at least in a relative sense. This increase is illustrated for hydroelectricity in figure 1.15 (for which the development is almost entirely in China), in figure 1.16 for wind energy, and in figure 1.17 for solar photovoltaics. While these plots of increased renewable electricity in recent years are impressive, they need to be taken in context. In absolute terms, the increases in renewable energy are less than the overall increase in primary energy use in recent years. Figure 6.6 shows that most of the increased overall energy use came from increases in all three of the fossil fuels (coal, oil, and natural gas). It is the third point above that poses the biggest challenge as we go forward, and the magnitude of the infrastructure development required will depend critically on our technology choices, as summarized in the previous section.

## 6.4 A summary of road vehicle technologies

We can now look at the approaches to transportation that have been discussed in previous chapters and consider which are the most sustainable for the future. We begin in this section with a consideration of road transportation. The evaluation of specific transportation technologies requires the analysis of a number of factors, many of which have been discussed in this and previous chapters. Specifically, some points that should be considered are:

- vehicle efficiency;
- the cost of energy;
- the cost of vehicle production;
- infrastructure development and maintenance costs;
- environmental effects, vehicle performance, and convenience.

The development of any new technology inevitably involves compromises, and sustainable transportation technology is no exception. Some attractive road transportation options, for example electric roadways or ethanol internal combustion engines, suffer from severe drawbacks, namely excessive infrastructure costs and questionable environmental impact, respectively. The ideal solution may not be obvious, but in this section, we try to identify the most reasonable way forward for sustainable road transportation. Other modes of transportation are considered in the next section. We might also view approaches for the development of sustainable transportation technologies on the basis of two timescales: (a) what we can do as quickly as possible to mitigate the effects of climate change and (b) what the ideal longer-term approach would be to making our transportation system as practical and as environmentally sustainable as possible.

Road vehicles represent the largest single use of transportation energy. Road vehicles also represent one of the most significant challenges for the development of a sustainable transportation technology. This is partly because of the technical challenges of developing small, portable energy sources but also because of the magnitude of the infrastructure required. The most appropriate way forward in the development of technology for road transportation can be considered in the context of the criteria given above. This section deals largely with light-duty road vehicles, but much of discussion is also relevant to heavy-duty road vehicles, such as trucks and buses, and specific references to these modes of road transportation are noted as appropriate.

### 6.4.1 Vehicle efficiency

The efficiency of different transportation technologies, with particular reference to light-duty road vehicles, has been considered in the previous section. While this previous discussion has focussed on those technologies that utilize carbon-free (or at least low-carbon) electricity as a starting point, the same approach can be applied to any transportation technology (usually with more complexity and less accuracy). In general, the efficiency of a vehicle provides an inverse measure of the quantity of primary energy that is needed to yield one unit of energy at the vehicle's wheels, and this has direct consequences for future primary energy requirements.

### 6.4.2 Fuel cost

Fuel cost is certainly an important factor in consumer acceptance of a given vehicle technology. Fuel cost is closely related to efficiency, although it also depends on the cost per unit energy of the fuel. Table 6.3 provides an overview of some potential alternative vehicle technologies, as well as a comparison with gasoline. It should be noted that the information in the table is based on the current situation for the availability of fuels. This means that electricity is produced primarily from fossil fuels (as shown in figure 6.8), hydrogen is produced primarily by the steam

**Table 6.3.** Cost per km for different technologies suitable for light-duty road vehicles. The cost is the typical cost in the United States in U.S. dollars as of September 2021. For vehicle efficiency refer to figure 6.9. The energy needed at the wheels is assumed to be 0.5 MJ $km^{-1}$, as discussed above. The energy contents are 132 MJ/US gallon for gasoline, 142 MJ $kg^{-1}$ for $H_2$, 3.6 MJ $kWh^{-1}$ for electricity, 88.9 MJ/US gallon for ethanol and 67.8 MJ/US gallon for methanol. Hydrogen vehicles are assumed to utilize high-pressure gas.

| Technology | Fuel | Cost (per unit) | Cost (per MJ) | Vehicle efficiency (%) | Cost (per km) |
|---|---|---|---|---|---|
| Gasoline internal combustion engine | Gasoline | $3.18/US gallon | $0.024 | 20 | $0.060 |
| Battery electric vehicle | Electricity | $0.13/kWh | $0.036 | 85 | $0.021 |
| Hydrogen fuel cell | Hydrogen | $16.51/kg | $0.116 | 54 | $0.108 |
| Hydrogen internal combustion engine | Hydrogen | $16.51/kg | $0.116 | 11 | $0.527 |
| Methanol internal combustion engine | Methanol | $1.33/US gallon | $0.020 | 22 | $0.045 |
| Ethanol internal combustion engine | Ethanol | $2.22 US/gallon | $0.025 | 20 | $0.062 |

reforming of natural gas, methanol is produced primarily from syngas derived from natural gas, and ethanol is produced from biological feedstock. While these data do not precisely represent the situation for sustainably sourced fuels, they can serve as an approximate comparison of the economic aspects of different vehicle technologies. Alternative fuels that can serve as replacements for diesel fuel in compression ignition engines, such as biodiesel, renewable diesel, and dimethyl ether, are analogous in most ways to methanol and ethanol as fuels in spark ignition engines.

### 6.4.3 Cost of vehicle production

The technologies of some of the approaches shown in table 6.3 are very similar to those of traditional internal combustion engine vehicles. Specifically, methanol and ethanol internal combustion vehicles are basically gasoline-powered vehicles with minor modifications. The same is true for compression ignition vehicles designed to operate on biodiesel or renewable diesel. Hydrogen internal combustion engine vehicles are similar but require specialized tanks for compressed hydrogen storage, which add considerably to the vehicle cost and occupy additional space.

At present, battery electric vehicles tend to cost more than comparably sized and equipped gasoline vehicles, and this was discussed in some detail in chapter 2. Since a major component of the cost of a battery electric vehicle is the cost of the battery, it is interesting to examine how battery cost has changed over time since the earliest modern battery electric vehicles. Figure 6.11 shows that the cost per kWh for lithium-ion batteries has dropped substantially over the past three decades. The possibility of further price decreases in electric vehicle batteries has been discussed previously in chapter 2 and is considered further below.

Hydrogen (or other) fuel cell vehicles are the least cost-effective of the possibilities shown in table 6.3. While the cost of fuel cell systems has decreased somewhat since the mid-2000s, as shown in figure 6.12, the cost has leveled off in recent years. The cost of fuel cell systems, at least those that provide suitable power for vehicle use and have reasonable size and operating characteristics, is governed by the cost of materials, such as platinum, that are necessary for their operation. Without the further development of fuel cells that utilize less expensive components, it is unlikely, at least in the short term, that fuel cell vehicles will be able to complete financially with vehicles utilizing other available sustainable technologies.

### 6.4.4 Infrastructure development and maintenance costs

To consider the necessary fueling infrastructure for road vehicles, we can look at the following energy sources:

- electricity;
- ambient-temperature liquid fuels;
- low-temperature liquid fuels;
- low-pressure pressurized fuels;
- high-pressure pressurized fuels.

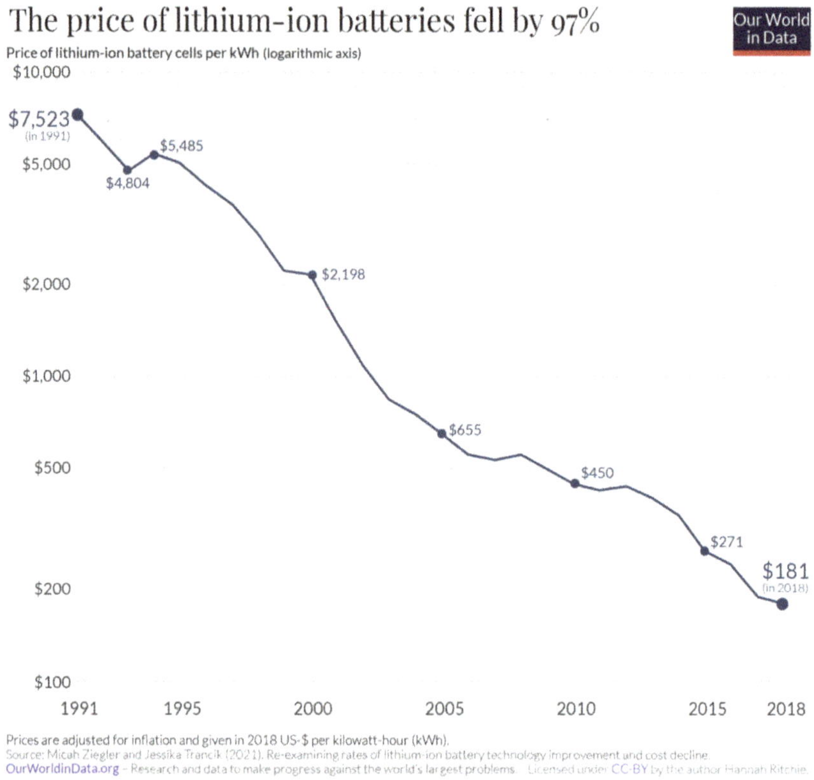


**Figure 6.11.** The price per kWh for lithium-ion batteries from 1991 to 2018. Reproduced from Ritchie (2021a). CC BY 4.0.

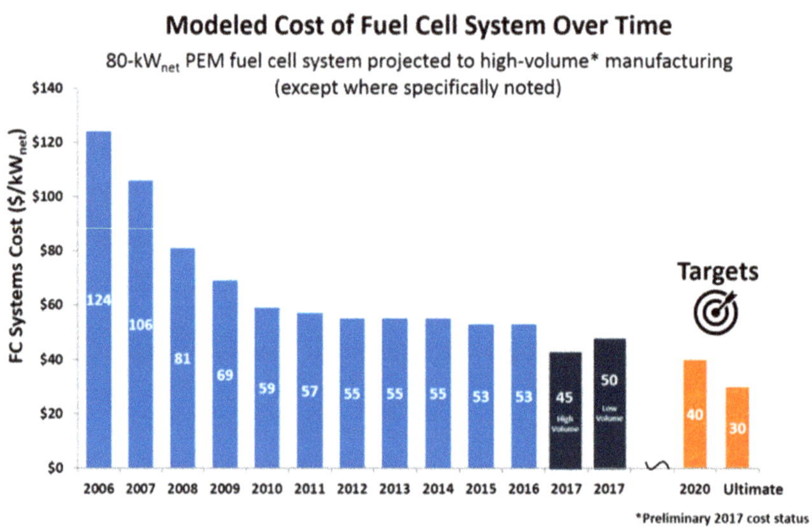


**Figure 6.12.** Cost per kW of an 80 kW polymer electrolyte membrane fuel cell as a function of time from 2006 to 2017. Reproduced from U.S. Department of Energy (2018). Image stated to be in the public domain.

### *6.4.4.1 Electricity*

The distribution of electricity for road vehicle use can take place either by charging onboard batteries or by supplying electricity to the vehicle from an external source. In the first case, the distribution network consists of either charging stations or an electric road that provides dynamic charging. In the second case, external electricity is typically supplied to the vehicle through an overhead catenary. While overhead catenary lines have been in use for many decades for powering buses in some cities, their application to light-duty vehicles or for road vehicles traveling outside urban areas is not practical. At present, electric road vehicles are almost exclusively battery electric vehicles, and these require a suitable charging infrastructure.

The growth of charging station infrastructure has followed the growth of battery electric vehicle sales. An example of the rapid growth in public charging infrastructure is illustrated in figure 6.13. As noted in table 6.4, there are now over 44 000 public charging stations in the United States alone, and growth has paralleled the growth in the number of battery electric vehicles on the roads—around 1.8 million as of 2020. As each public charging station typically has more than one charging outlet or port, the total number of publicly accessible ports in the United States is over 100 000. The relatively low cost of electric vehicle chargers has been a very positive factor in the growth of this infrastructure. Today, the cost of a 22kW AC charger is in the range of $5000–$10 000, and that of a 400kW DC fast charger is in the range of $100 000.

### *6.4.4.2 Ambient-temperature liquid fuels*

Ambient-temperature liquid alternative vehicle fuels include ethanol, methanol, and sustainable diesel fuels. At present, public filling stations in North America (particularly in Canada) that provide alternative liquid fuels are limited, as noted in table 6.4. This can be seen by a comparison with the number of public filling stations that sell gasoline—about 115 000 in the United States and about 12 000 in

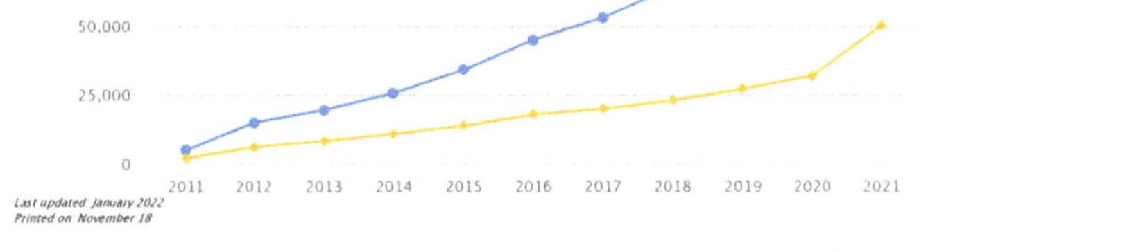


**Figure 6.13.** Number of public electric vehicle charging stations and available charging ports (EVSE) in the United States from 2011 to January 2021. Reproduced from U.S. Department of Energy (2022). Image stated to be in the public domain.

**Table 6.4.** Number of public stations in Canada and the United States offering alternative fuels and battery charging as of 2021. For electric vehicle charging stations, the number in parentheses is the total number of charging ports. Data adapted from https://afdc.energy.gov/stations/#/analyze.

| Fuel | Canada | United States |
|---|---|---|
| Electric charging stations | 6159 (13 893) | 44 172 (108 528) |
| Hydrogen (compressed $H_2$ gas) | 5 | 48 |
| Ethanol (⩾E85) | 0 | 3972 |
| Biodiesel (⩾B20) | 3 | 328 |

Canada. This number has decreased substantially over the past two decades, due to increased competition that has forced the closure of many smaller stations. Most of the stations shown in the table are gasoline stations in the United States that provide E85 for flex-fuel vehicles. Methanol has limited availability in some locations outside North America, e.g. Iceland and China, as discussed in section 3.10. Although extensive infrastructure has not yet been specifically developed for alternative liquid fuels, the existing infrastructure that has been established for traditional liquid fuels, such as gasoline and diesel fuel, can readily be adapted for use with alternative liquid fuels. This approach not only facilitates the development of an alternative fueling infrastructure but also utilizes existing facilities by repurposing equipment and allows for a gradual transition away from fossil fuels.

#### *6.4.4.3 Low-temperature liquid fuels*

Low-temperature liquid fuels consist essentially of liquid hydrogen, as discussed in section 3.5. This fuel has previously been used for the BMW Hydrogen 7, although more recent hydrogen-powered road vehicles, such as those utilizing hydrogen fuel cells, have concentrated on the use of high-pressure hydrogen gas (typically now at 700 atm) as a fuel. Overall, hydrogen-powered vehicles face a number of issues, such as the current lack of an appropriate fuel distribution network. In addition, hydrogen internal combustion vehicles suffer from low overall efficiencies, as discussed in section 6.3, while fuel cell vehicles suffer from the high cost of vehicle production, as noted earlier in this section. Although compressed hydrogen has the advantage of easier fuel storage and handling compared to liquid hydrogen, the liquid form of this fuel has a somewhat greater specific energy as a result of its greater density, 71 kg $m^{-3}$ compared with 40 kg $m^{-3}$ for gas at 700 atmospheres. Nevertheless, some initiatives to develop liquid hydrogen as a fuel are planned. For example, a collaborative initiative by vehicle manufacturer Daimler AG and hydrogen producer Linde plc is underway to build fuel cell trucks that operate on liquid hydrogen along with an appropriate fueling infrastructure (see, for example, Randall (2020)). For commercial trucks, the economics of fuel cell vehicles may be less of a concern than for private light-duty road vehicles, and the convenience of the potentially longer range and reduced fueling time associated with hydrogen as a fuel, compared with those of batteries, may outweigh at least some of their disadvantages.

*6.4.4.4 Low-pressure pressurized fuels*

As discussed in section 3.10, dimethyl ether is a potential substitute for diesel fuel in compression ignition internal combustion engines for road vehicles. The properties of dimethyl ether are very similar to those of propane, and existing propane or liquefied petroleum gas distribution infrastructure can readily be converted for use with dimethyl ether. This means that the infrastructure for the transport, storage, and dispensing of the fuel in liquid form at ambient temperature and a pressure of about 10 atmospheres is already in existence. At present, the propane distribution infrastructure is much more limited than the gasoline or diesel infrastructures. There are currently about 1300 public stations that sell propane in the United States and about 630 in Canada. While this infrastructure is not extensive, the potential market for dimethyl ether as a fuel for heavy trucking is also limited compared to fuel for light-duty road vehicles, and additional filling stations can be added as needed.

*6.4.4.5 High-pressure pressurized fuels*

Chapter 3 dealt with the use of hydrogen as a fuel, and much of the work in this area has focussed on high-pressure hydrogen gas. As seen in table 6.4, the number of filling stations offering high-pressure hydrogen gas is very small. The worldwide situation, as shown in table 3.13, is quite similar to the situation in North America. As the price of installing a hydrogen fuel distribution station is of the order of two million USD, there is relatively little incentive for fuel companies to build new distribution infrastructure unless there is a clear market. Analogously, the lack of a widespread fueling infrastructure is a disincentive for potential automobile buyers to choose a hydrogen fuel cell vehicle over a battery electric vehicle. In fact, aside from specific locations in Japan, Germany, California, and South Korea, the owners of hydrogen fuel cell vehicles have to reside near a convenient filling station and are likely not inclined to travel far from home. The potentially greater range of fuel cell vehicles, in comparison to those of battery electric vehicles, is likely to be an irrelevant benefit unless the fueling infrastructure is greatly expanded. Added to this is the high cost of manufacturing fuel cell vehicles. If, in the future, manufacturers were to price fuel cell vehicles in an economically sustainable manner, then they would not compete favorably with other alternative approaches and the future expansion of fueling infrastructure would not be likely.

### 6.4.5 Environmental effects, vehicle performance, and convenience

In order to assess these final criteria for road vehicle technologies, it is useful to begin with an overall summary of the advantages and disadvantages of each technology. Table 6.5 provides an overview of the pros and cons of the various transportation technologies that have previously been discussed in this book as possible sustainable options for road use.

Battery electric vehicles offer an excellent combination of efficient primary energy utilization, vehicle cost, and performance as well as an economically viable infrastructure. Unfortunately, range and recharge time are, at present, potential drawbacks for certain applications. In addition, extensive worldwide battery electric

**Table 6.5.** Advantages and disadvantages of different road vehicle technologies (ICE = internal combustion engine).

| Fuel | Advantages | Disadvantages |
|---|---|---|
| Battery electric | Substantial current infrastructure<br>Economically viable infrastructure<br>Good for grid capacity<br>Reasonable vehicle cost<br>Home charging<br>Excellent power<br>High-efficiency use of primary energy<br>Low cost per km<br>Reliable technology | Limited range<br>Limited lithium resources<br>Long charging time |
| Hydrogen fuel cell | Fast refueling<br>Moderate efficiency of primary energy<br>Moderate range<br>Acceptable power<br>Convenient $H_2$/battery hybrid | Expensive vehicles<br>Use of rare materials in cell<br>Expensive technology<br>Relatively new technology<br>Complex fuel distribution<br>Expensive refueling stations<br>Virtually nonexistent infrastructure |
| $H_2$ ICE | Fast refueling<br>Good power<br>Convenient $H_2$/gasoline hybrid | Very low overall efficiency<br>Excess load on grid<br>Minimal range<br>Expensive per km<br>Self-defeating $H_2$/gasoline hybrid<br>Complex fuel distribution<br>Expensive refueling stations<br>Currently nonexistent infrastructure |
| Bioethanol ICE | Convenient fueling options<br>Conversion of current fueling infrastructure<br>Modified current vehicle technology<br>Good power<br>Reasonable range | Questionable environmental sustainability<br>Conflict with food agriculture<br>Questionable viability in many regions |
| E-methanol ICE | Convenient fueling options<br>Conversion of current fueling infrastructure<br>Modified current vehicle technology<br>Mood power<br>Reasonable range | Lower specific energy than other liquid fuels |

*(Continued)*

**Table 6.5.** (*Continued*)

| Fuel | Advantages | Disadvantages |
|---|---|---|
| Renewable diesel | Convenient fueling options<br>Conversion of current fueling infrastructure<br>Drop-in fuel<br>Good power<br>Reasonable range | Early stages of development<br>Uncertain availability of feedstock |
| Dimethyl ether | Convenient fueling options<br>Conversion of current fueling infrastructure<br>Modified current vehicle technology<br>Good power<br>Reasonable range | Overall efficiency unclear<br>Uncertain availability of feedstock for bioDME |

vehicle development may stress the world's grid capacity and lithium resources. Recent developments in battery and vehicle technology have certainly gone far in addressing the limitations of battery electric vehicles, and future research aims to improve specific energy and cycle life while reducing recharge time and cost.

Hydrogen-fueled vehicles suffer from several serious drawbacks. Renewable hydrogen, i.e. hydrogen produced from water using electrolysis or a similar approach, has a very low overall efficiency when used in an internal combustion engine and therefore places unnecessarily large demands on grid capacity. Hydrogen fuel cell vehicles suffer from the need to utilize expensive materials. In either case, the development of a suitable fueling infrastructure is expensive and depends on the availability of a sufficient market for hydrogen fuel. The lack of fuel distribution infrastructure discourages the growth of vehicle production, and minimal vehicle production is a disincentive for infrastructure growth.

Vehicles using alternative liquid or gaseous fuels fall into two major categories: those that use biologically based fuels and those that use electrofuels. Biofuels may be derived from dedicated agricultural feedstock, as is the case for bioethanol, or from waste material, as is the case for renewable diesel. In the first case, competition with food crops is a serious concern and the overall environmental impact, including land-use change factors, is questionable in many locations. In the case of waste-derived biofuels, material that might otherwise contribute to methane emissions can be utilized for energy production. As suggested in tables 6.2 and 6.3, methanol is one of the most attractive liquid fuels. Although methanol is currently produced almost exclusively from fossil fuel feedstock, it can also be produced using current technology from waste biological feedstock or from sustainably produced hydrogen. Thus, both biomethanol and e-methanol are attractive replacements for gasoline in spark ignition engines. The high efficiency of methanol internal combustion engines contributes to the desirability of this sustainable fuel. The general comments that have been made for methanol are also applicable to dimethyl ether as a replacement for diesel fuel in compression ignition engines.

### 6.4.6 Road vehicle technology analysis

From an overall assessment of the factors considered above, it seems that in the short term at least, an increase in the development of battery electric road vehicles is the most cost-effective and environmentally sustainable approach to light-duty road transportation. This development must take place in parallel with the increased utilization of sustainable low-carbon approaches to the generation of electricity. Today's battery electric vehicle technology is ideally suited to urban and other short-range transportation. A significant charging infrastructure is already in place, and its development will, no doubt, continue in the future. Appropriate short-term technical advances would include batteries with increased specific energy, improved charging time, increased cycle life, and a reduction in the use of expensive electrode materials. Although charging infrastructure will likely grow in the near future, range and charging time will remain concerns for certain applications. In this respect, sustainable liquid fuels for use in internal combustion engines may be a suitable complementary technology. Current hybrid vehicle technology suggests an appropriate path for such development, and the Ford battery/E85 hybrid shown in figure 4.6 is an example of this approach. The use of alternate fuels, specifically methanol, in an internal combustion engine as part of a battery/internal combustion engine hybrid can provide a means of addressing the current drawbacks of pure battery electric vehicles for specific applications that require long-range transport. Depending on the specific application, series plug-in hybrid technology or extended range battery electric vehicle technology would make the best use of resources. The development of cost-effective methanol fuel cells could make hybrid methanol fuel cell/battery electric vehicles such as the Gumpert Nathalie (see figure 3.36) attractive. They would have improved efficiency over methanol internal combustion engine/battery hybrids, as discussed above, and simplified fueling technology compared with hydrogen fuel cell hybrids, as discussed in chapter 3.

Battery electric heavy-duty vehicles are now becoming available (see figure 2.25, which shows the Tesla electric truck and figure 2.26, which shows the Proterra electric bus) to replace current vehicles that utilize diesel fuel in compression ignition engines. Urban buses can use external power sources through overhead catenaries, although battery electric vehicles which top up batteries through charging at bus stops using conductive rails or inductive charging may be a better approach. In some situations, long-range trucking may, as noted above for light-duty vehicles, benefit from a battery/internal combustion engine hybrid technology involving renewable diesel or dimethyl ether as a fuel. Such a unified approach would limit the diversity of the infrastructure that would need to be developed in regions where current plans to expand battery electric vehicle charging infrastructure could be augmented with methanol/renewable diesel distribution based on existing gasoline or diesel infrastructure and/or dimethyl ether distribution based on existing propane infrastructure.

In the long term, the development of electric roads and battery-swapping technologies may facilitate road transportation, although at present these approaches are economically nonviable and have not met with consumer acceptance, respectively. Ultimately, the availability of lithium may impose limitations on the sustainability of today's battery electric vehicle technologies. Continuing research into new battery technologies, such as sodium and magnesium battery chemistries, may lead to a reduction in the demand for lithium and provide more cost-effective batteries.

## 6.5 A summary of other modes of transportation

While there are certainly similarities between the energy requirements of road vehicles and those of the other modes of transportation, as discussed in chapter 5, there are also notable differences. These differences relate to factors such as vehicle weight, the required power, and an appropriate power-to-weight ratio. Although rechargeable batteries will almost certainly play a major role (or the major role) in the future of road vehicles, they are likely to be less of a consideration for other modes of transportation. Following the discussion in chapter 5, we consider the most appropriate approaches to rail, marine, and air transportation.

### 6.5.1 Rail transportation

From a global perspective, rail transportation is the closest to utilizing sustainable technology. In many parts of the world, railways are almost exclusively electric and use grid power supplied by an overhead catenary or a ground-level rail. In such cases, it is a matter of converting the electricity generation infrastructure from traditional fossil fuel power stations to low-carbon energy sources and not a question of developing new sustainable rail technologies.

In locations such as North America, where rail transportation relies mainly on the use of traditional diesel locomotives, new approaches to rail technology must be developed in order to establish a sustainable low-carbon infrastructure. This is particularly the case for freight rail lines that traverse low-population-density areas. One approach to the reduction of greenhouse gas emissions is the conversion of existing diesel-electric locomotives to operate on sustainable liquid fuels. In the short term, drop-in fuels such as renewable diesel are a simple choice. In the long term, the sustainability of fuels such as dimethyl ether may prove to be superior to those of fuels derived from biological feedstock.

The use of batteries in conjunction with a diesel engine operating on sustainable fuel would serve to lower fuel consumption. In recent years, hybrid battery/diesel locomotives, as shown in figure 5.7, have been in use. Conventional diesel-electric and electro-diesel locomotives already incorporate both an internal combustion engine and an electric drive motor and are, in a sense, already series hybrids, just without the batteries. The use of this technology could, in the short term, provide substantial reductions in the greenhouse gas emissions produced by rail transportation. The incorporation of short segments of electrified rail line, as customarily used for electro-diesel locomotives, could provide a means of periodically adding charge to the battery from the grid and could reduce the fuel consumption of the internal combustion engine accordingly.

In the long term, the electrification of the majority of railway routes and the utilization of grid power derived from renewable sources will likely be the best approach and should be considered an ultimate goal.

### 6.5.2 Maritime transportation

In the previous chapter it was shown that battery electric marine ferries were a viable option for certain short routes. However, the majority of commercial marine transport would not fit into this category and would need to rely on other sustainable modes of propulsion. As with non-electrified rail transportation, the most reasonable approach to energy for large ships which currently utilize compression ignition internal combustion engines operating on a fossil fuel petroleum product would probably be the use of a drop-in sustainable diesel replacement fuel that requires minimal engine modifications. While some conversions of existing ships to operate on liquefied natural gas have been undertaken, and these produce less pollution and greenhouse gas than ships using traditional bunker fuels, they still rely on a fossil-fuel-based energy source. Limited trials have been conducted using ships that utilize a fuel that can, at least in principle, be produced in a sustainable and renewable manner. As shown in figure 5.18, the *Stena Germanica* has been adapted to operate on methanol and, all factors considered, this may be the most appropriate route for future commercial marine transportation. Recent research into ammonia-fueled engines for ships has also shown some promise.

Other possible low-carbon energy sources include nuclear power and wind power. While nuclear power has been quite successful for powering military vessels, it is generally considered to be impractical and financially unviable for large commercial vessels. In addition, nuclear energy for passenger ships may be an option that garners minimal public approval.

Wind has traditionally been a major source of energy for marine vessels. However, it provides insufficient power to make it currently viable. Nonetheless, modern technologies for utilizing wind energy, as described in section 5.3, may be suitable as an additional source of power that could make a positive reduction in fuel consumption.

### 6.5.3 Air transportation

Battery electric aircraft have been suggested in recent years as an alternative, low-carbon approach to personal short-distance travel. While this may be viable at some point in the future, it is a relatively insignificant component of travel infrastructure, and urban commuter aircraft are unlikely to replace any appreciable portion of road transportation in the foreseeable future. The low specific energy of batteries, compared to those of liquid fuels, suggests that larger battery-operated aircraft, particularly on longer routes, would not be feasible.

For air transportation, hydrogen does not necessarily have the disadvantages that it has for road transportation. Since the energy used for air transportation is only a small fraction of that used for road transportation (see figure 6.2), the adverse effects of the low efficiency of hydrogen as an energy carrier on overall energy use are not as serious. In addition, hydrogen fuel distribution at a relatively small number of designated airports where hydrogen aircraft could be fueled is not as much of an infrastructure development concern as a network of public filling stations that could service road vehicles. Other factors, however, such as the energy density per unit volume, put hydrogen at a disadvantage compared to hydrocarbon-based fuels.

While there has been some experimental work on hydrogen-fueled aircraft, which has used both hydrogen internal combustion engine technology and hydrogen fuel cell technology, work on sustainable aviation biofuels is much more advanced. As described in detail in the previous chapter, development is underway for the production of sustainable aviation fuels derived from biomass feedstock. Such fuels have been successfully used in commercial aircraft for the past decade or so, in the form of blends with traditional fossil-fuel-based aviation fuels. This approach parallels the use of biofuels in road vehicles, particularly as it relates to biodiesel blends for compression ignition internal combustion engines. For aircraft, the options for a sustainable energy source are much more limited than they are for other modes of transportation, and sustainable aviation biofuels may be the most suitable technology going forward. In the longer term, electrofuels, as discussed in the previous chapter, may be a viable option as well.

## 6.6 The future

Figure 6.5 shows a clear picture of our need to take action against the dangers of global climate change, and it also shows that this action must be drastic and immediate. By 2030, the paths for recovery and disaster are clearly delineated. Beyond that point, the route to recovery by the end of this century will be very difficult or impossible. While past studies have suggested that a 2.0 °C increase compared to preindustrial temperatures would serve to minimize the effects of global warming, more recent studies have suggested that minimizing temperature increases to 1.5 °C is necessary. As shown in figure 6.5, our present path will lead to an estimated 2.7 °C to 3.1 °C temperature increase over preindustrial global temperatures by the end of this century, while 'pledges and targets,' as defined by (e.g.) the Paris Agreement, will lead to a 2.4 °C increase.

It is also highly questionable whether the terms of the Paris Agreement can even be met.

The role of transportation in greenhouse gas emissions is illustrated in figures 6.1 and 6.4, where it can be seen that transportation is not the largest contributor to global climate change. While dealing with transportation emissions alone is certainly not sufficient to mitigate the effects of greenhouse gas emissions, it is a necessary component of any approach that can successfully keep global warming within acceptable limits.

This book has outlined the possibilities for dealing with greenhouse gas emissions from the transportation sector and has looked at the most viable options for the future. It has also made it clear that transportation issues cannot be adequately addressed without a clear analysis of the interdependency of the energy use in all sectors. We cannot fully benefit from the development of sustainable transportation technologies without the accompanying implementation of sustainable approaches to the production of electricity and appropriate liquid or gaseous transportation fuels. For example, at present, the proliferation of battery electric road vehicles and the reduction in the number of fossil fuel vehicles will reduce greenhouse gas emissions from this component of our energy use. However, this reduction will only be about half of what it should be, because the electricity that is used to charge the batteries comes primarily from fossil fuel sources. It is only through the greater net efficiency of battery electric vehicles compared to the internal combustion gasoline and diesel engines that some benefits are obtained. To reiterate the discussion presented in section 6.3, it is essential that sustainable infrastructure for low-carbon electricity generation is developed to cover the additional demand as well as to replace existing greenhouse-gas-emitting facilities.

Developing new energy infrastructure to minimize the effects of greenhouse gas emissions is a formidable task. Although some progress has been made and continues to be made, the need for action is imminent. Transportation technology is an essential part of this effort and its development must coordinated with the development of technologies for electricity generation. While the vast majority of current transportation energy is derived from fossil fuels, future transportation will need to be more diverse, as different modes of transportation have different requirements that will likely have to be fulfilled by different technological approaches.

## References and further reading

Dunlap R A 2021 *Energy from Nuclear Fusion* (Bristol: IOP Publishing)

Engel H, Hensley R, Knupfer S and Sahdev S 2018 *The Potential Impact of Electric Vehicles on Global Energy Systems* (McKinsey & Company) (8 August 2018) www.mckinsey.com/industries/automotive-and-assembly/our-insights/the-potential-impact-of-electric-vehicles-on-global-energy-systems

EIA 2016 U.S. Energy Information Administration: International Energy Outlook 2016, Chapter 8, Transportation www.eia.gov/outlooks/ieo/pdf/transportation.pdf

EIA 2017 *International Energy Outlook 2017* United States Energy Information Agency Report Number: DOE/EIA-0484(2017). Data extracted from Table F1. Total world delivered energy consumption by end-use sector and fuel, Reference case, 2015-50 www.eia.gov/outlooks/ieo/excel/appf_tables.xlsx

IPCC 2018 *Global Warming of 1.5 °C. An IPCC Special Report on the impacts of global warming of 1.5 °C above pre-industrial levels and related global greenhouse gas emission pathways, in the context of strengthening the global response to the threat of climate change, sustainable development, and efforts to eradicate poverty* ed V Masson-Delmotte *et al* (Cambridge: Cambridge University Press) p 616 https://doi.org/10.1017/9781009157940

Our World in Data 2022a Greenhouse gas emissions per sector, world https://ourworldindata.org/grapher/ghg-emissions-by-sector

Our World in Data 2022b Primary energy consumption by source, world https://ourworldindata.org/grapher/primary-sub-energy-source

Our World in Data 2022c Electricity production by source, world https://ourworldindata.org/grapher/electricity-prod-source-stacked

Randall C 2020 Daimler & Linde to develop liquid hydrogen fuel stations electrive.com (10 December 2020). Available at: www.electrive.com/2020/12/10/daimler-linke-to-develop-liquid-hydrogen-fuel-stations/

Ritchie H 2020 Sector by sector: where do global greenhouse gas emissions come from? Published online at OurWorldInData.org. Retrieved from https://ourworldindata.org/ghg-emissions-by-sector

Ritchie H 2021a The price of batteries has declined by 97% in the last three decades. Published online at OurWorldInData.org. Retrieved from https://ourworldindata.org/battery-price-decline

Ritchie H 2021b Primary energy production is not final energy use: what are the different ways of measuring energy? https://ourworldindata.org/energy-substitution-method

Ritchie H and Roser M 2020a Emissions by sector OurWorldInData.org. Available at: https://ourworldindata.org/emissions-by-sector#total-greenhouse-gas-emissions-by-sector

Ritchie H and Roser M 2020b Energy. Published online at OurWorldInData.org. Available at: https://ourworldindata.org/energy

Ritchie H, Roser M and Rosado P 2020 $CO_2$ and Greenhouse Gas Emissions https://ourworldindata.org/co2-and-other-greenhouse-gas-emissions

U.S. Department of Energy 2013 Fact #795: September 2, 2013 – Electric Vehicle Charging Stations by State www.energy.gov/eere/vehicles/fact-795-september-2-2013-electric-vehicle-charging-stations-state

U.S. Department of Energy 2018 Fact of the Month April 2018: Fuel Cell Cost Decreased by 60% since 2006 www.energy.gov/eere/fuelcells/fact-month-april-2018-fuel-cell-cost-decreased-60-2006

U.S. Department of Energy 2022 U.S. Public and Private Electric Vehicle Charging Infrastructure https://afdc.energy.gov/data/10964

**IOP** Publishing



# Appendix A

## Powers of ten

Very large and very small numbers are commonly expressed in scientific notation. Multiples of basic units of measurement are given prefixes, such as kilometers (or km) for 1000 (or $10^3$) meters or millimeters (or mm) for 0.001 (or $10^{-3}$ m). The following table gives some of the most commonly encountered prefixes associated with different powers of ten.

| Number | Prefix | Abbreviation |
|---|---|---|
| $10^{-9}$ | nano | n |
| $10^{-6}$ | micro | μ |
| $10^{-3}$ | milli | m |
| $10^{3}$ | kilo | k |
| $10^{6}$ | mega | M |
| $10^{9}$ | giga | G |
| $10^{12}$ | tera | T |
| $10^{15}$ | peta | P |
| $10^{18}$ | exa | E |

doi:10.1088/978-0-7503-5306-9ch7

**IOP** Publishing

Transportation Technologies for a Sustainable Future
Renewable energy options for road, rail, marine and air transportation

**Richard A Dunlap**

# Appendix B

## Power and energy

The terms ‘power’ and ‘energy’ are often used interchangeably in common speech—in fact, Google’s English dictionary lists them as synonyms. However, they have quite different meanings in the scientific literature. We can draw an analogy with two other terms which we do readily distinguish in common speech—speed and distance. An automobile traveling on the highway at a speed of 100 km $h^{-1}$ for two hours will travel a total distance of 200 km. In other words, the speed is the rate at which distance is traveled and distance is the total distance covered. Mathematically,

$$\text{distance} = \text{speed} \times \text{time},$$

where distance is measured in units of (say) kilometers, while speed is measured in units of distance per unit time, (say) km $h^{-1}$. Power is analogous to speed. It is the rate at which energy is used and is measured in units of energy per unit time. Energy is analogous to distance and is related to power by the following expression:

$$\text{energy} = \text{power} \times \text{time}.$$

In the International System of units or (in French) the Système International (SI), which is a specific manifestation of the metric system, energy is measured in joules. So, when time is measured in seconds, the unit of power is the joule per second (J $s^{-1}$), which is defined as a watt (W). The term power, therefore, is used to indicate the rate at which energy is utilized. It can denote power that is generally thought of as being used or power that is being produced. The former may be electric power that is used to operate a heater, while the latter may be power produced by an internal combustion engine that is burning gasoline. In either case, power is the rate at which one form of energy is converted into another form of energy.

A simple example that contrasts the difference between power and energy is a light bulb. A standard incandescent light bulb uses 60 W of power. If it is on for

doi:10.1088/978-0-7503-5306-9ch8

1 second, it will consume (60 W) × (1 s) = 60 joules of energy. If it is on for half an hour, it will consume (60 W) × (1800 s) = 108 000 joules of energy.

In the next appendix, we present a more extensive overview of the units used for power and energy and an insight into the sizes of the various units that are in common use.

# Appendix C

## Units of power and energy

As noted in the previous appendix, the watt and the joule (which is 1 watt·second, Ws) are the standard units of power and energy, respectively, in the Système International. There are, however, numerous other units for power and energy that are in common use for specific applications, and a few of these have been utilized in this book. In this appendix we review some of these units and provide some insight into their specific use and their relationship to the watt and the joule.

The watt is in common use worldwide as the unit used to measure power. Like energy, power can be manifested in various forms. In some countries, such as the United States, the watt is generally thought of as a measure of electrical power, while mechanical power is often thought of as being measured in horsepower, and this is most commonly encountered in the designation of the power produced by an engine (e.g. a gasoline automobile engine). In much of the world that uses the metric system, the watt is also used as a measure of mechanical power, often in the form of the kilowatt; this approach has been used in this book and is becoming more common, even in the United States. It is easy to convert from horsepower to watts. The conversion factor is 1 hp = 746 W, so a 100 hp engine produces 74.6 kW. While from the standpoint of units it would be correct to refer to electrical power in units of horsepower, so that a 100 W light bulb could be said to use 0.13 hp, this is not considered a suitable approach.

The kWh (or kilowatt-hour) is a common unit that is used to measure electrical energy. It is one of the most commonly encountered units of energy in our daily lives, as it is the unit of energy that is used by public utilities and which appears on our monthly electricity bills. Although the cost of electricity varies between states in the United States, provinces in Canada, and other political and geographical jurisdictions, it typically represents electricity that costs about 0.13 USD. It is easy to understand the relationship between a kWh and the electricity used by common electrical devices. For example, 1 kWh represents the electricity consumed by a 10 W LED bulb over a period of 100 h or the electricity consumed by a 4000 W clothes dryer over a period of 15 min.

doi:10.1088/978-0-7503-5306-9ch9  

Another energy unit that is commonly encountered is the calorie. The calorie that is used to measure the energy content of foods is actually the kilocalorie, which is usually written with an uppercase 'C' as Calorie (or abbreviated as Cal). The Calorie is defined as the amount of energy (in the form of heat) that is needed to raise the temperature of 1 kg of water by 1 °C. We have a feeling for the size of a Calorie, as a typical daily caloric intake for a human might be around 2000 Cal.

A final energy unit that is often encountered is the Btu (or BTU) which is the abbreviation for British Thermal Unit. A Btu is defined as the amount of thermal energy needed to raise the temperature of 1 pound of water by 1 degree Fahrenheit and is used as a measure of thermal energy. It is commonly encountered in the United States and Canada and to a lesser extent in a few other countries. The Btu is generally encountered as a measure of the output of a furnace or an air conditioner. A furnace with a rating of about 100 000 Btu is typically appropriate to heat an average-sized house in a temperate climate. This is very misleading terminology, as the rating is the furnace's power, but the Btu is a unit of energy. What the rating actually means is 100 000 Btu per hour (which is a proper unit of power).

The kWh, Calorie, and Btu are in common use as units of energy in various parts of the world, but other units, such as the electron volt are commonly used in the scientific literature when referring to the energy associated with individual atoms or subatomic particles. However, the joule is the official metric unit for measuring energy. We probably have a feeling for the size of different units such as the kWh, Calorie, and Btu in specific contexts such as the electricity used in our homes, the energy content of food, or the output of a furnace, respectively; their absolute magnitudes are perhaps best understood by comparing them to the joule. Table C.1 gives the conversion factors for different energy units to the joule and the kWh.

As can be seen from the quantities in table C.1, the joule is really a very small amount of energy. In fact, in our daily lives it is difficult to think of something that would represent 1 joule of energy. It is the energy consumed by a 100 W light bulb that is turned on for 1/100th of a second. However, a kilowatt hour is easy to imagine: it is the energy consumed by a 100 W lightbulb that is turned on for 10 h. But it is interesting to see how 1 kWh compares to energies that we are not used to

**Table C.1.** Conversion factors for different energy units to the joule and the kilowatt hour. The values in the table give the number of joules or kWh that equal an energy of one unit in the leftmost column. Quantities are listed in order of increasing size.

| Unit | Abbreviation | Joules | kWh |
|---|---|---|---|
| joule | J | 1.0 | $2.78 \times 10^{-7}$ |
| calorie | cal | 4.184 | $1.162 \times 10^{-6}$ |
| British thermal unit | Btu | 1055 | $2.931 \times 10^{-4}$ |
| Calorie | Cal | 4184 | $1.162 \times 10^{-3}$ |
| kWh | kWh | $3.6 \times 10^{6}$ | 1.0 |
| tonne oil equivalent | toe | $4.187 \times 10^{10}$ | $1.163 \times 10^{4}$ |
| quad | Q | $1.055 \times 10^{18}$ | $2.931 \times 10^{11}$ |

**Table C.2.** Objects or actions that have associated energies that are equal to approximately 1 kilowatt hour (kWh).

| Object or action |
|---|
| The energy needed to lift a human from base camp to the summit of Mount Everest |
| The energy required at the wheels to move a passenger vehicle 7 km |
| The energy needed to lift an elephant 75 m (vertically) |
| The kinetic energy of a Formula 1 car at 360 km $h^{-1}$ |
| The energy content of 100 mL of gasoline |
| The energy content of 120 g of coal |
| The caloric content of two slices of pepperoni pizza |
| The energy required to raise the temperature of 10 L of water from 25 °C to boiling |
| The electrical output of a coal-fired power station for 0.0036 s |
| The electrical output of a large wind turbine for 4 s |
| The average sunlight incident on 1 $m^2$ of the Earth's surface for 6 h |

thinking about in electrical terms. Table C.2 gives some quantities which are equivalent to 1 kWh. Keep in mind that at average North American electricity rates, this energy has a value of about $0.13.

**IOP** Publishing



# Appendix D

## Energy contents of various fuels

A variety of liquid and gaseous fuels have been discussed in this book. Table D.1 provides a summary of the energy contents per unit volume and per unit mass for the fuels that have been discussed previously.

**Table D.1.** Volumetric and specific energy capacities of liquid and gaseous fuels ordered by increasing volumetric energy density.

| Fuel | MJ $L^{-1}$ | MJ $kg^{-1}$ |
|---|---|---|
| Compressed hydrogen (700 atm) | 5.7 | 142 |
| Compressed natural gas (250 atm) | 9.0 | 53.6 |
| Liquid hydrogen | 10.0 | 142 |
| Methanol | 17.9 | 19.7 |
| Liquefied dimethyl ether | 21.2 | 31.7 |
| Liquefied natural gas | 22.2 | 53.6 |
| Ethanol | 23.5 | 30.0 |
| Biodiesel | 33.0 | 42.2 |
| Gasoline | 34.8 | 46.4 |
| Diesel fuel | 38.6 | 45.6 |

doi:10.1088/978-0-7503-5306-9ch10  

# Appendix E

## Battery characteristics

When we put a new battery into a portable electronic device, the first thing we usually look at is the battery size. We might also check the battery voltage. But we would need a lot more information about a battery if it were used to power an electric vehicle. Figure E.1 shows the label on a rechargeable nickel metal hydride AAA battery that provides some information about its properties, specifically its voltage (1.2 V) and a measure of its capacity (1000 mAh, that is 1 Ah). By looking at how the battery stores energy, we can understand what these numbers mean and how they are relevant to the performance of a battery for (say) an electric vehicle.

Since batteries operate by transporting charges, we begin with some general information about electric charge. Charge is a fundamental property associated with some elementary particles, such as the electron, which has one unit of negative charge, and the proton, which has one unit of positive charge. The charge associated with the proton in the nucleus of an atom is responsible for forming the positive charge on the ion when one (or more) electrons are removed. Charge, in the Système International (SI), is measured in coulombs, as indicated in table E.1. The elementary charge, which is the charge on one electron or one proton, is $1.6 \times 10^{-19}$ C. This means that a charge of 1 coulomb corresponds to the charge on $6.25 \times 10^{18}$ electrons or protons.

An electric current represents a flow of electric charges. The SI unit of current, the ampere (A, see table E.1), is defined as a current flow of 1 coulomb per second (C $s^{-1}$). Multiplying a current by time yields a quantity with the unit of charge. According to the definition of the ampere, multiplying one ampere by one second gives one coulomb. In other words:

$$(1\text{ ampere}) \times (1\text{ second}) = (1\text{ coulomb/second}) \times (1\text{ second}) = 1\text{ coulomb}.$$

If we multiply one ampere by one hour, we get 1 Ah, which is the unit of charge capacity used on the battery label. It is easy to see that

$$(1\text{ Ah}) = (1\text{ Ah}) \times (3600\text{ s/h}) = 3600\text{ As} = 3600\text{ C}.$$

doi:10.1088/978-0-7503-5306-9ch11  

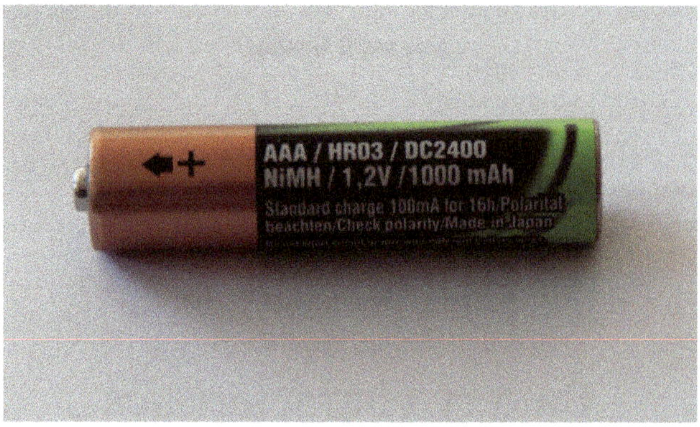


**Figure E.1.** Rechargeable nickel–metal hydride AAA battery. Image: Richard A Dunlap.

**Table E.1.** Quantities relevant to battery properties and their units in the SI.

| Quantity | Unit | Abbreviation |
|---|---|---|
| Charge | coulomb | C |
| Electric current | ampere | A |
| Electric potential | volt | V |
| Power | watt | W |

Voltage is a measure of electrical potential, as discussed in section 2.4, and the SI unit of this potential is the volt (V). According to Ohm's law, the current and voltage give the power as follows:

$$\text{power} = \text{current} \times \text{voltage}.$$

In the SI, we would write this as

$$1 \text{ watt} = (1 \text{ ampere}) \times (1 \text{ volt}).$$

If we multiply power by time, we get energy, as discussed in appendix B. This means that if we multiply both sides of the above equation by time in hours then

$$1 \text{ Wh} = (1 \text{ Ah}) \times (1 \text{ V})$$

which shows that

$$\text{energy} = \text{charge} \times \text{voltage}.$$

In table 2.1, the Wh has been used as the unit of energy stored in a battery. We can now see how the discussion above can be used to better understand the information on the battery label. The charge capacity of the battery is given as 1000 mAh (or 1 Ah), which is equivalent to 3600 C. The energy storage capacity of the battery is readily calculated from the charge storage capacity and the battery voltage (1.2 V), as follows:

$$(1\ \text{Ah}) \times (1.2\ \text{V}) = 1.2\ \text{Wh},$$

and this quantity can be used for analyses such as that in section 2.5 to determine the available energy for electric vehicle use.

www.ingramcontent.com/pod-product-compliance
Ingram Content Group UK Ltd.
Pitfield, Milton Keynes, MK11 3LW, UK
UKHW051903100926
8556IPUK00005B/749

9 780750 353045